Patterns of
Technological Innovation

Patterns of Technological Innovation

Devendra Sahal
New York University

Foreword by
Richard R. Nelson
Yale University

1981

ADDISON-WESLEY PUBLISHING COMPANY, INC.
Advanced Book Program/World Science Division
Reading, Massachusetts

London · Amsterdam · Don Mills, Ontario · Sydney · Tokyo

Library of Congress Cataloging in Publication Data

Sahal, Devendra.
   Patterns of technological innovation.

   Includes bibliographical references and index.
   1. Technological innovations.   I. Title.
T173.8.S24   608   81-7967
ISBN 0-201-06630-0   AACR2

ABCDEFGHIJ–HA–8987654321

To Maria

CONTENTS

FOREWORD

Sahal begins by remarking that, during the 1950s, economists came to recognize the dominant role played by technological advance in the economic growth process. The economists of the 1950s were rediscovering an old theme. Chapter 1 of Adam Smith's *Wealth of Nations* is mostly about what economists now would call technological advance and economic growth. Mill and Marx recognized the importance of technological advance to economic growth. But somehow, in the twentieth century, that knowledge, and that interest, faded from the economic textbooks and journals. The 1950s might be looked at, therefore, as marking a renaissance of interest in the topic.

Since that time there has been an outpouring of research, by economists and scholars in other disciplines, on the sources, influences, and nature of technological advance. Sahal's book both builds on, and is an important contribution to, that literature.

In this excellent book, many different topics are touched upon and a number of significant points developed. But there is one dominant theme, or collection of themes, which give power and focus. It is that in many fields technological advance reflects the gradual accretion over time of understanding on the part of technologists of the potential of a broadly defined technology, and the progressive exploitation of that potential. Put another way, a progressive flow of innovation results from a cumulative learning process. The flow takes particular directions channeled by the basic physical laws constraining what the technology can, and cannot, be nudged to do. Economic variables, like the demand for different kinds of products, and costs of different factors of production, play a role by stimulating or retarding certain kinds of innovative efforts, but only within the boundaries defined by the basic constraints on the technology and the internal logic of the learning process.

This may come as no news to historians of technology. But the analysis here represents a significant departure from, and improvement over, the view of technological advance built into many economic models. In much of economic modeling the kind of technological advance achieved is visualized as being determined largely by conditions of product demand and factor cost. Technological possibilities and learning processes are seen as constraining the

magnitude of advances, but not as influencing the kind of advances in an interesting way. The often tight constraints on what is technologically feasible
tend not to be considered. That technological advance is often a cumulative
process, with today's advances setting the stage for tomorrow's, usually is not
recognized at all.

Sahal's counterview, that technologies logically unfold over time, is
argued forcefully and documented well. Various implications are drawn out
skillfully; the discussion of implications for science and technology policy is
particularly well done. This is an interesting and important book.

RICHARD R. NELSON

PREFACE

Traditionally, the subject of technological innovation has existed in the shadow of other disciplines. It is built upon contributions from a wide variety of fields. Yet, or perhaps because of this, it remains a stepchild of modern science. The multidisciplinary parentage of the subject decidedly has been a mixed blessing. On the one hand, it has led to a field of inquiry that is uniquely rich in the diversity of viewpoints. On the other hand, it is singularly lacking in terms of a set of comprehensive, well-integrated themes. The subject is a confederation of many ideas with too few junctions to hold the structure together.

In recent years there has been a spate of investigations into technical change processes. While attesting to the importance of the subject matter, they have made notable advances in our understanding of it. Thanks to these works, we are left with a bounty of interesting and creative strands of thought in this area. At the same time, however, it must be admitted that they do not add up to a holistic body of ideas. Our knowledge of the subject is often opaque and fragmentary.

This book attempts to develop the basis for a science of technology in its own right. The litmus test of such an endeavor is, of course, whether it ultimately meets with the approval of the other already well-established disciplines. Indeed, sweet are the virtues of adversity! It is in this spirit that the book is addressed to scholars in three fields: economics, engineering, and public as well as private policy. Additionally, the general reader should be able to comprehend the essence of the volume from the concluding section of each chapter.

The book is a product of nearly seven years of my research in this field. It is, in part, based on my earlier articles published in various journals. The key papers are listed separately in the acknowledgment to the publishers; others can be found in the references to the individual chapters that follow. I am most grateful to the anonymous referees of these papers for their incisive comments on my previous investigations of the subject.

I should also like to thank a number of firms and individuals for their help in providing the data employed in the empirical analyses reported herein. The book's coverage of a wide variety of technical change processes owes much to their help, as acknowledged in the Appendix.

Earlier versions of the work were presented in the talks I gave at the Georgia Institute of Technology; the International Institute of Applied Systems Analysis, Laxenburg, Austria; Massachusetts Institute of Technology; National Defense Research Institute, Stockholm; and Resource Systems Institute of the East-West Center, Hawaii. I am grateful to the discussants of these papers for their helpful comments.

I am further indebted to Edgar Dunn of Resources for the Future, Inc., Richard Nelson of Yale University, Almarin Phillips of the University of Pennsylvania, and Josef Steindl of the University of Vienna for their helpful comments on earlier drafts of the book. I am especially grateful to Josef Steindl for both scholarly advice and encouragement during the culminating phases of the work. The final responsibility for the contents of the book of course rests with me.

My deepest gratitude is to my wife, Maria, for her enormous patience and understanding during this endeavor. My obsession with the subject merely initiated the work on this book; it was her personal sacrifice that brought it to completion.

DEVENDRA SAHAL

ACKNOWLEDGMENT TO PUBLISHERS

The book draws in part upon the following articles by the author, which are printed with the permission of the original copyright owners.

"Alternative Conceptions of Technology," *Research Policy*, **10**:2-24 (1981), [Chapter 2]. North-Holland, Amsterdam.

"The Distribution of Technological Innovations," *Proceedings of the International Conference on Cybernetics and Society*, IEEE, Tokyo: Kyoto, Japan, Nov. 1978, Part I, pp. 573-579, [Chapter 3].

"A Theory of Evolution of Technology," *International Journal of Systems Science* **10**:259-274 (1979), [Chapter 4]. Taylor and Francis Ltd., London.

"The Temporal and Spatial Aspects of Diffusion of Technology," *IEEE Transactions on Systems, Man and Cybernetics* **9**:829-839 (1979), [Chapter 5]. IEEE, New York.

"The Relevance of the Logic Theory Machine to Modeling of Evolutionary Systems," *Kybernetes, International Journal of Cybernetics and General Systems* **6**:49-53 (1977), [Chapter 6]. Thales Publications (W.O.) Ltd., U.K.

"Farm Tractor and the Nature of Technological Innovations," *Research Policy*, **10** (1981), [Chapter 6]. North-Holland, Amsterdam.

"Models of Technological Development and their Relevance to Advances in Transportation," *Journal of Technological Forecasting and Social Change* **16**:209-227 (1980), [Chapter 6]. American Elsevier, New York. ©D. Sahal, 1979

"Technological Progress and Policy" in *Research, Development and Technological Innovation, Recent Perspectives on Management* (D. Sahal, ed.), pp. 171-198, (1980), [Chapter 6]. D.C. Heath, Lexington, Mass.

"Technological Systems: An Exploration in the Dynamical Aspects of the Theory of Hierarchical Structures," *General Systems Yearbook*, **XX**, 1975, pp. 159-164, [Chapter 7].

"The Nature and Significance of Technological Cycles," *International Journal of Systems Science*, **11**:985-1000 (1980), [Chapter 7]. Taylor and Francis Ltd., London.

"The Determinants of Best-Practice Technology," *R & D Management,* **11**:25–32 (1981), [Chapter 8]. Basil Blackwell, Oxford.

"Structural Models of Technology Assessment," *IEEE Transactions on Systems, Man and Cybernetics,* **7**:582–589 (1977), [Chapter 9]. IEEE, New York.

"A Formulation of the Pareto Distribution," *Environment and Planning: International Journal of Urban and Regional Research,* **A10**:1363–1376 (1978), [Chapter 9]. Pion Publication, U.K.

"A Theory of Progress Functions," *American Institute of Industrial Engineering Transactions,* **11**:23–29 (1979), [Chapter 9]. AIIE Inc., Atlanta.

Patterns of
Technological Innovation

Chapter 1

INTRODUCTION: PROBLEMS AND APPROACHES

1. BACKGROUND: THE PRODUCTIVITY PUZZLE

During the late 1950s, a number of studies of long-term economic growth in the United States reached a remarkable conclusion: the lion's share of the observed growth of output in the American economy could not be accounted for by conventionally measured capital and labor inputs and must therefore be due to some form of technical progress. In a classical investigation in this area, Robert Solow attributed only 12.5% of the growth of output per man-hour to increased use of capital and the remaining 87.5% to technical change (Solow, 1957). A year earlier, Moses Abramovitz had come to essentially similar findings via a different approach to analysis of observed economic growth in the United States during the period from 1869 to 1953, stating (Abramovitz, 1956, p.11):

> This result is surprising in the lopsided importance which it appears to give to productivity increase, and it should be, in a sense, sobering, if not discouraging, to students of economic growth. Since we know little about the causes of productivity increase, the indicated importance of this element may be taken to be some sort of measure of our ignorance about the causes of economic growth in the United States and some sort of indication of where we need to concentrate our attention.

The challenge posed by these studies was quickly recognized. In particular, the causation of technical progress could no longer be ignored as in the past. To be sure, early classical economists—especially Adam Smith (1937)—regarded technical change as an important element in the process of economic evolution. But little further work was done in this area. To quote Bruton, from an essay entitled "Contemporary Theorizing on Economic Growth" (1956, p. 297),

Devendra Sahal, Patterns of Technological Innovation   ISBN 0-201-06630-0

1

Perhaps the most important deficiency—among a host of deficiencies—in the body of thought examined in this essay is the lamentable state of our understanding of the origin and process of technical change.

Yet, more than twenty years later, the deficiency in our knowledge about the process of technical change is perhaps even more glaring than it was before. Basically, there are two aspects of the problem posed by the unexplained residual in economic growth. First, there is the problem of accounting for the simple identity between the value of the output and the returns to the inputs. Thanks to some enthusiastic statistical work, this problem seems to have been more or less solved (Jorgenson and Griliches, 1967; Christensen and Jorgenson, 1970). Second, there is the much more fundamental problem of explaining the observed changes in the characteristics of both inputs and outputs. While some notable attempts have been made to come to grips with this aspect of the problem, it has remained essentially unresolved. As Star has put it (1974, p. 134),

Growth accounting measures *what* changes have taken place. The most interesting question—*why* the changes occurred—still remains to be answered.

Detailed analyses since the early studies of economic growth have revealed that a substantial portion of the observed growth of output is attributable to economies of scale and improvement in the quality of the inputs employed. Undoubtedly, these changes in scale and quality of inputs, in turn, have been made possible by advances in technology. However, little is known about how and why technical progress occurs. In the succinct statement of Nelson and Winter (1974, p. 889),

Research within the neoclassical theory now acknowledges the centrality of technical change in the growth process. The "indigestible" phenomena appear—the minute the neoclassical blinkers are removed—to be the basic characteristics of the technical change process.

There is another more recent problem of explaining the *decline* in the worldwide growth of productivity. For a period of nearly two decades, from the mid 1940s to the mid 1960s, the productivity of U.S. industry had been increasing by an average of slightly more than 3% per year. Thereafter it began to stagnate. On an average its growth has been barely 0.1% a year during 1973–1979. The productivity growth of other industrialized countries has likewise been slowing down, although the drop is not as steep as in the case of the United States. The output per worker in the seven largest economies of the world increased by an average of nearly 4.5% a year from 1963 to 1973, whereas it increased on an average of only about 1.5% a year during 1973–

1979. The observed slowdown in economic growth has been variously attributed to reduced research and development (R&D) expenditure, increased government regulation (e.g., in the form of environmental protection laws), lack of investment, etc. Further, there has been a great deal of discussion as to whether all this is a reflection of some lasting change in the pattern of innovative activity. However, the deeper reasons for the decreasing growth of productivity in recent years remain as mysterious as those for its increasing growth in earlier times. In a vital sense, what is missing is a thorough understanding of technological innovation processes.

In recent years a number of works have significantly contributed to our knowledge about the process of technological change. They have originated in many different disciplines and they go a long way toward testifying to the central role of technological decisions in all processes of long-term development, economic as well as social and political. However, they also point to the need for a more discerning analysis of the causes and consequences of technological change. A general theoretical explanation of the process of innovation is lacking. It is hardly an exaggeration to say that the problem of technological change has turned out to be one of the most vexing of all problems in the social sciences. In particular, there remain all too many missing links in our knowledge of the subject.

Hitherto, a great majority of studies of technological change have centered on the phenomenon of *diffusion*, that is, the process whereby a new technique is adopted over the course of time. For example, in an early work in this area, Griliches (1957) examined the pattern of the diffusion of hybrid corn in various parts of the United States. In an important follow-up of Griliches's work, Mansfield (1961) presented a number of findings on how rapidly the use of various innovations spread from one firm to the other in four industries: bituminous coal, iron and steel, brewing, and railroads. Since then, a host of works have shed further light on the determinants of the rate of adoption of new techniques. There are now more than 2000 published studies on various aspects of the diffusion of technology.[1]

Much less is known about the phenomenon of *innovation* itself—for example, what factors determine the very introduction of new techniques. Clearly, the topic is one of fundamental importance. However, even the limited knowledge available in this area is almost exclusively based on analysis of patent statistics and the chronologies of major innovations. The limitations of such data are well known (they are discussed in detail in Chapter 2). Suffice here to note that an explanation of change in the number of techniques is of necessarily limited value in an understanding of changes in the technology

---

[1] For a comprehensive bibliography see Rogers and Shoemaker (1971).

itself. In all, while a great deal has been written on the question of how new techniques are *adopted*, comparatively little systematic attention has been given to the question of how new techniques are *developed* in the first place. To put it somewhat differently, whereas the fruits of technological change have been widely examined, their roots have been accorded at best a step-motherly treatment. This neglect has a number of important implications.

Typically, widespread acceptance of a new technique is accompanied by a multitude of important changes in its design. The earlier versions of a new product or process often suffer from numerous flaws. The diffusion of most innovations therefore tends to be conditional upon extensive changes in their functional characteristics. In turn, it is not implausible that the identification and remedy of defects in the initial versions of many innovations become possible with the accumulation of feedback information received from their diffusion. A detailed discussion of this point is deferred until later in this book (see especially Chapters 5 and 6). For the present purpose, it will suffice to note that diffusion of technology is inherently a multidimensional phenomenon: while a new technology substitutes for the old, it usually undergoes numerous changes within itself (Sahal, 1977). Thus the diffusion of technology is inextricably interwoven with its development. The former cannot be adequately understood except in the context of the latter. It is apparent, for example, that the adoption of computers for a wide variety of purposes has been largely possible through changes in their various characteristics, such as multiplication time, memory, access time, size, and peripherals. Yet much of the contemporary theorizing on the diffusion of technology is based on the assumption that the characteristics of the "innovation" do not change during the course of its adoption. This is, of course, patently unrealistic. In reality, not only do changes in the characteristics of the innovation influence the *rate* of its adoption, they also make possible new uses for it, thereby significantly affecting the *extent* of its adoption as well. In turn, the diffusion of technology often exerts an important influence on its design. Our knowledge of the process of diffusion of innovations is necessarily incomplete without an understanding of the factors that govern the development of technology.

The issues raised above shed further light on the puzzle of long-term economic growth discussed here. As noted earlier, changes in the "quality" of inputs constitute one important factor in the observed growth of output. For example, it has been shown that growth in the productivity of U.S. agriculture during the postwar years was made possible in no small measure by advances in the fuel-consumption efficiency, horsepower, and numerous other characteristics of the farm equipment employed, such as tractors and combine harvesters (Griliches, 1963). This example is illustrative of a very general proposition: Economic growth largely originates from change in the *character* of techniques rather than from mere increase in the *number* of techniques employed. However, the present state of our knowledge in this area is woefully inadequate to explain how and why the traits of an innovation are what they are. The relevance of the issue to understanding the process of long-term economic growth is evident. Yet its treatment is conspicuously absent from the literature on the subject.

The problems outlined above also have a bearing on resource allocation in the management of research and development activity. It is a commonplace observation that investment in formal R&D has grown substantially in recent years. Already by 1970, the R&D sector of the U.S. economy employed over 1.5 million people, with a total expenditure of more than 17 billion dollars. However, there remain numerous unresolved issues regarding the management of R&D activity. In particular, one vital activity in this area is the determination of the feasibility of a proposed conception of a new technology. Is it worthwhile to allocate resources to the development of a small-scale model on the basis of a blueprint? Is the investment in the development of a prototype justified on the basis of small-scale demonstration? Is the final product likely to meet the desired level of performance? Is the innovation so much ahead of its time that it is unlikely to succeed? A framework for systematic analysis of these problems is generally lacking.

The formulation of a national technology strategy involves answering a series of analogous questions. A strategy for technological development ought to be based on careful consideration of the growth and stagnation characteristics of alternative systems. This is very rarely the case. One of the important obstacles in formulating such a policy is that very little is known about what constitutes the technological innovation potential (TIP) of a system.[2] The notion of technological innovation potential is one of central relevance to the planning of research and development activity. However, we have practically no idea how it can be conceptualized in an analytically meaningful way.

The foregoing discussion should not be regarded as a comprehensive inventory of the various unsolved problems in the study of technological change. Rather, the issues just raised are illustrative of the generally primitive state of our knowledge in this area. The inadequacy of the subject matter stems not merely from lack of theoretical work; it is due to lack of empirical analysis as well. It is remarkably difficult to obtain hard facts and figures in this field. Although some investigations have been made through the painstaking collection of real-world evidence, they are relatively few and far between. While there is abundance of folklore, there is little substantive information. To summarize, it is fair to say that the subject of technological innovation is yet to be baptized.

---

[2]The problem is of a very general nature. To quote Boulding (1977, p. 302), "Part of the system of evolution is the concept of evolutionary potential. It is extraordinarily tricky and hard to identify but it is fundamental."

2.   THE ANALYTIC FRAMEWORK

In the context of these important but unresolved issues, this book seeks to identify and explain certain regularities in the origin, diffusion, and development of new techniques. Two principal approaches are employed to this end. One is the standard mathematical-statistical approach to hypothesis testing as, for example, frequently used in econometrics. The approach is well known and needs little further explanation here. The other approach aims at the formulation of simple generalizations consisting of reasonably good abstracts of the data. In particular, it is concerned with the search for lawlike relationships (Ehrenberg, 1975; Kendall, 1961; Simon, 1968). Since application of this approach in the social sciences is of comparatively recent origin, it is described here in some detail.

It will be useful to have a concrete example as an aid to discussion of the search for lawlike relationships. Consider the well-known equation that the volume (V) of a given mass of gas varies inversely as the pressure (P) to which it is subjected at a constant temperature (T). In symbols,

$$PV/T = R \qquad\qquad (1.1)$$

where R is constant for any given amount of gas. The example is illustrative of a number of important features of relationships commonly regarded as laws. The first thing to be noted about a wide variety of laws is their simplicity; that is, they contain relatively few arbitrary parameters. Thus the example of the gas law involves only one parameter, namely, the constant R. Hooke's law that the extension of a spring is proportional to the load applied is another illustration, as is Poiseuille's law that the rate of flow of a given liquid through a tube is proportional to the pressure drop across the ends of the tube. Indeed, examples of such simple laws are ubiquitous (Davies, 1964). To be sure, the simplicity of these laws is often the result of much prior work in their formulation, since it involves a great deal of gathering, sifting, and scrutinizing of facts. Nevertheless, it is indeed remarkable that the behavior of an otherwise complex system can be adequately explained in terms of a very simple law. Thus the gas law holds despite the enormous complexity of the system, particularly in terms of the differences in motion and direction of its individual molecules.

Second, in the formulation of a law it is often just as important to determine the conditions under which it does not hold as it is to know when it holds. Thus the gas equation holds only at sufficiently low pressures and sufficiently high temperatures. Poiseuille's law of fluid flow assumes that the liquid consists of a solution of tiny, independent, spherelike molecules. In consequence, however, it holds for water and glycerol but not for solutions of certain polymers. There arise additional intramolecular forces of longer range that affect the internal friction in the liquid, and consequently the viscosity varies with the flow rate. In summary, the formulation of many important laws involves simplifying assumptions and finding just what are the limiting conditions under which they hold.

Third, it is of course important that a law provide a "reasonably good" approximation of the data in any given case. However, the utility of a law is determined by the variety of cases to which it applies rather than by the exact degree of its fit to any given case. Thus, the gas law is only approximately true and, what is even more, it can be decisively "rejected" at low enough temperatures or high enough pressures. The important point, however, is that it applies to some degree of approximation to the behavior of a wide variety of gases, ranging from noble gases to ammonia. One further implication of this is that the standard statistical methods are not wholly applicable to the testing of lawlike relationships. This is particularly true of simple generalizations that take the form of extreme hypotheses, in that they hold only under limited or idealized conditions.

Last but not least, a relationship cannot be said to describe the lawlike aspects of a system unless there exists a theoretical explanation of why the observed relationship is what it is. In principle, the explanation of an observed regularity need not be complicated. For example, the gas law may be derived a priori as a combination of Boyle's law that when temperature T is fixed

$$PV = K_1 \tag{1.2}$$

and the Charles law that when pressure P is fixed

$$V/T = K_2 \tag{1.3}$$

where K_1 and K_2 are constants. Ideally, of course, we would like to have a more general theory, such as the statistical-mechanical explanation of the gas law. This is also to imply that a lawlike relationship is not necessarily a causal relationship insofar as its origin lies in the meeting of chance and cause. In conclusion, unless we have a theoretical model of the process underlying a regularity, we have no way of determining its worth. Moreover, the consequences of such a model should be verifiable independently of the regularity itself. Finding a lawlike relationship without understanding why it should fit the data is like being left in the situation of Alice: "Somehow it seems to fill my head with ideas—only I don't know exactly what they are."

Put somewhat differently, knowledge of how well a law fits with other available information is just as important as the discovery of the law itself.

The additional information may well be of both a theoretical and an empirical nature. Frequently, its use leads to important new refinements in the form of a law. An example of such a refinement of an empirical nature for the case of the gas law is the van der Waals equation

$$(P + \alpha/V^2)(V - \beta) = RT \tag{1.4}$$

where the parameter α relates to intermolecular attractive forces and β relates to the size of the molecules. In this way, we can better account for the deviations from the gas law at lower temperatures and higher pressures. Moreover, evaluation of a law in the light of other available knowledge often paves the way for new discoveries while shedding new light on what is already known. As an example, the kinetic theory of gases relates the constant R in the gas law to the specific heats of a perfect monatomic gas at constant pressure C_p and constant volume C_v. Specifically,

$$C_p = \tfrac{5}{2} R, \qquad C_v = \tfrac{3}{2} R. \tag{1.5}$$

Maxwell's law of the equipartition of energy shows that any degree of freedom of a system possesses an energy of $\frac{1}{2} RT$ per mole, its contribution to specific heat being equal to $\frac{1}{2} R$. Thus, a perfect monatomic gas with three degrees of freedom and three independent directions of molecular motion has a specific heat C_v equal to $3R/2$. More generally, it may therefore be said that a formulation of a lawlike relationship requires a knowledge of both when and under what conditions it holds and why it should hold.

In the light of the above discussion, one plausible approach to the search for lawlike aspects of a system may be very briefly summarized as follows. To begin with, it is a matter of identifying simple relationships that fit the data to a reasonable degree. Next, the scope of these relationships—the *limiting conditions* under which they hold and the *variety of cases* to which they apply—must be determined. Finally, a theoretical explanation why the relationship fits the data is needed. That is, the significance of a pattern cannot be adequately determined until we have a model of the process giving rise to it. Clearly, the model must be based on plausible assumptions. The value of such a model depends on the extent to which it is capable of indicating any *new* facts and figures relevant to further discoveries.

It is apparent that there is a basic difference between the usual statistical approach to problem solving and the process of search for lawlike relationships as outlined above. The former has its focus on determining the *validity* of a hypothesis. Thus, very often it is concerned with the choice of one from a set of hypotheses on the basis of its goodness-of-fit to the data. In contrast, the latter has its emphasis on determining the *scope* of a hypothesis. More generally, it is concerned with the formulation of a simple relationship that holds for a wide variety of data.

The emphasis of this work will be on the formulation of simple lawlike relationships in the process of technical change. This approach will be supplemented by the use, almost throughout this book, of standard statistical devices. In part, the duality of approaches stems from the poverty of the subject matter under consideration. For the remainder, it reflects the viewpoint noted earlier that the informational context of a lawlike relationship is just as important as the relationship itself. Indeed, there is nothing permanent about the meaning of laws. What permits insightful new interpretations of existing generalizations is the background knowledge within which discoveries are made. One might characterize this process as a dialogue, or dialectic, between descriptive and explanatory models of the system producing the observed data. Thus we have often utilized the statistical approach to search for relevant information, which in turn is employed as an input to the search for simple laws of technical change. For example, the problem of development of technology is first studied by means of the statistical approach (Chapter 6), only to be taken up again in the search for lawlike relationships (Chapter 9). Regardless of the approach employed, the focus of the work is on the identification of simple patterns of technological innovation, and, what is more important, an explanation of *why* the patterns are what they are.

In closing this section, it may be legitimately asked whether there is a bona fide case for seeking simple lawlike aspects of technological innovation, or for that matter, of any phenomena that are outside the domain of natural sciences. Our answer is yes. It is true that there are comparatively few attempts to seek regularities in social phenomena. However, systematic efforts in this direction seem to have invariably yielded some very fruitful results. Examples of these are works in conflict and peace research (Richardson, 1960), economics (Ehrenberg, 1975; Ijiri and Simon, 1977; Steindl, 1965), linguistics (Zipf, 1949), and regional science (Israd, 1956), among others. The approach pursued in this book is very much in the spirit of these works. In our view, the existence of simple laws of complex systems can hardly be dismissed as a peculiar feature of physical phenomena. Indeed, it should be a sobering thought to the high church of systems analysis that simple laws of social phenomena can and do, in fact, exist. The Pareto law of income distribution is an example, as is Zipf's law of observed diversity in the linguistic text. What is even more, these simple laws are invaluable as instruments of planning and conscious change. To conclude, this study will be concerned with the identification and explanation of some such simple patterns in its quest toward understanding the process of technological innnovation.

3.   THE ORGANIZATION OF THE WORK

The starting point of this work is an attempt to clarify some of the difficulties involved in formulating an analytically meaningful conceptualization of technology. A detailed discussion of this problem will be found in Chapter 2, where it is concluded that an innovation is best conceived in terms of its functional properties. As an example, technical change in the turbogenerator over the course of time is best understood in terms of the change in its horsepower-to-weight ratio, fuel-consumption efficiency, etc. More generally, the innovation process is appropriately viewed in a technical systems context.

Next, a theory of the *origin* of new techniques is outlined, leading to a probabilistic model of the phenomenon in Chapter 3. This model is substantiated by an analysis of major inventions in the railroad, farm equipment, paper-making, and petroleum-refining industries during the years 1800–1957 and, to a lesser extent, by an analysis of major innovations in other fields. Variations in the rate of technical discovery are also examined, both spatially and temporally.

An attempt is then made to develop a general theory of the evolution of technology in Chapter 4. Reference to the biological world is inevitable here, and an examination of the detailed evidence shows that there are both important parallels and differences between the evolution of organisms and of artifacts. Evidence also strongly indicates that the existence of an equilibrium is an exception rather than a rule in the system of technological innovation. Expressed somewhat differently, any limit to the development of a technology often proves to be a stepping-stone to further development. In this respect, the long-term evolution of technology may be likened to the successive generations of the phoenix, each rising from the ashes of its predecessor. The *formal* specification of the process of technological development in the subsequent work is essentially based on the theory presented in this chapter.

The problem of *diffusion*, or how and why a new technique is adopted over the course of time, is studied in Chapter 5. The crux of the theoretical considerations advanced in this chapter may be summed up in the form of two simple relationships of a deterministic nature. One concerns the spatial, while the other concerns the temporal, aspects of the diffusion of technology. They are applied to ten cases of the adoption of new techniques in various fields, including electricity generation, farming, manufacturing, steel production, and the textile industry. The specific cases analyzed include the turbogenerator, combine with cornhead, oxygen steel process, shuttleless looms, use of tunnel kilns in brick making, new methods of steel-plate marking and cutting in shipbuilding, and automatic transfer lines for car engines. The spatial and temporal variations in the rates of diffusion of these techniques are also examined.

One recurring theme in Chapters 2–5 is that the processes of both origin and adoption of a new technique are intimately related to its development. Accordingly, a number of the following chapters in this book are devoted to studying the *development* of technology in one form or the other.

Two distinct but related perspectives on the process of technological development are presented. One, the macroperspective, is on how development of a technology is governed by the dynamics of its task environment. The other, the microperspective, is on how the development of a technology is governed by its internal dynamics. As an example, the theoretical treatment of technological change in the former case is at the level of a production plant, whereas in the latter case it is at the level of a single unit of physical equipment. Clearly, the difference between the two perspectives is one of how the system boundaries are drawn.

A detailed theoretical treatment of the process of technological development at a macrolevel is presented in Chapter 6. The theory is found to hold in various cases of technological innovation in farm mechanization, transportation, digital computers, and electricity generation over long periods of time. The results shed new light on a number of important policy issues, including technology transfer, tariffs, and other forms of subsidies aimed at infant industry protection and national technology strategy in general.

There are two other important facets of the technological development process besides the persistent trend of long-term growth analyzed in Chapter 6. One aspect of development concerns the phenomenon of (possibly recurring) fluctuations in the innovative activity. That is, while there exists a secular trend of technological development, it is often accompanied by oscillations. The other aspect of development has to do with the fact that there exist significant disparities in the different forms of the same technology. Typically, some designs tend to be more advanced than others, giving rise to a distribution of technological capabilities. This in turn raises the question of what determines the maxima (and minima) of the observed distribution of technological capabilities. A detailed theoretical treatment of each of these two aspects of technological development, together with a number of case studies of innovation, will be found in Chapters 7 and 8. Briefly, Chapter 7 examines the nature and significance of observed oscillations in the process of innovation and Chapter 8 presents an attempt to investigate the determinants of maximum capability of a technology. The results lend further support to the theory of technological development outlined at a macrolevel in Chapter 6. They also have implications for business cycles and, more generally, for economic stabilization policy.

Chapter 9 presents a theory of technological development at the microlevel. It is closely related to, but distinct from, the preceding treatment of the process at the macrolevel. The essence of the theory lies in its consideration of certain internal factors, such as scale and redundancy, in the development of a technology. Two simple lawlike relationships emerge in an attempt to specify the inner dynamics of innovative activity. One concerns learning in the process of design. The other concerns scaling of the object. Interestingly, although these relationships are of a deterministic nature, they are found to have an important probabilistic counterpart in the form of the Pareto distribution. Further, it becomes possible to formalize the hitherto elusive concept of the technological innovation potential of a system on the basis of these relationships. These theoretical constructs are substantiated by a variety of case studies of technological innovation in aircraft, computational devices, farm tractors, nuclear reactors, passenger ships, and tankers. One main implication of the results is for selection of an appropriate technology among a set of alternatives.

The foregoing description of the individual chapters of this book is deliberately confined to a brief discussion of some of the problems addressed, rather than the conclusions reached. The latter will be found in the final sections of the individual chapters themselves. There is necessarily considerable cross-referencing between the individual chapters. Nevertheless, every attempt is made so that each of them can be read independently of the other, sometimes at the expense of minor overlap between the text of different chapters. The data employed in the investigations are presented in the Appendix to the book except in a few cases where they are equally accessible elsewhere. A summary of the principal findings of this book together with their policy implications is given in Chapter 10.

4.   TOWARD AN EPISTEMOLOGY OF TECHNOLOGY

While a formal theoretical consideration of the innovation process is deferred until the later sections of the book, one main epistemological finding of the work may be briefly noted here. This finding is best stated by contrast with the two differing and often extreme views that have prevailed from time to time in the literature on the subject.

One eminent view on the process of technological innovation stems from the work of Josef Schumpeter in economics dating back to the early twentieth century (1934, 1939). Curiously, while the notion of innovative activity is central to Schumpeter's outline of economic development, it is largely an exogenous variable. That is, whereas technological change has important repercussions of an economic nature, it has essentially no economic causes. An equally eminent and somewhat related view of the process of technological innovation comes from the works of Ogburn (1922) and Gilfillan (1935, 1952) in sociology and related areas. In its essence, innovative activity follows an independent course that is predetermined by the forces of historical necessity, as, for example, demonstrated by the evidence that very often there are multiple sources of the same discovery. As will be discussed later in this book, there are important differences between these two viewpoints originating from economics and sociology. Nevertheless, both viewpoints regard the process of technological innovation as an essentially autonomous phenomenon.

The view of technical change as an uncontrollable process was first scrutinized in the late 1950s, particularly in the field of economics. As noted earlier in this chapter, this scrutiny was primarily a result of the sudden realization that it was impossible to determine how and why the process of long-term economic growth takes place without a causal explanation of technical progress. Against the background of this general awareness of the importance of innovative activity, a number of econometric studies of technological change were conducted. The broad general conclusion of these works, best articulated by Jacob Schmookler (1966), was that inventive activity not only responded to economic factors, but also could be wholly explicated in economic terms. With the general acceptance of this viewpoint, the theorizing on the subject of technology has come full circle. The earlier view used to be that technological change knew no master. The more recent view is that it is wholly a captive of the socioeconomic system.

If the findings reported in this book are any guide, neither of these two views noted is justifiable. According to the general epistemology that emerges from the considerations advanced in the following chapters, technical progress is neither wholly chaotic nor wholly controllable. Further, it is doubtful if the issue can ever be settled one way or the other in an unequivocal sense. This is because the process of technological innovation can be equally well specified in the form of either a probabilistic or a deterministic law. What is more, the probabilistic form of the law can be shown to arise from a deterministic mechanism, or vice versa. The question of whether technological innovation is an autonomous process is, in a fundamental sense, indeterminate.

The relevant problem therefore transcends consideration of what the nature of technological events is. It concerns as much the nature of the terminology employed in the description of these events. Here we are essentially confronted with the age-old issue of the relationship between matter and symbol, with profound epistemological implications. A detailed discussion of this issue would take us too far afield. Suffice here to note that our understanding of the former is governed by our choice of the latter.

The results of this study then indicate that relevant patterns of technological innovation are primarily physical and only secondarily of a socioeconomic nature. In particular, they remain unchanged over long periods of time despite changes in their environment. Clearly, the implication is that our grasp of the technological change process is necessarily incomplete without a comprehension of the inner dynamics of innovative activity. This is as much to indicate that the process under consideration cannot be *wholly* explicated in socioeconomic terms. None of this should be taken to imply that technological change occurs in isolation from changes in other spheres of human endeavor. It is rather to emphasize that in an abstract sense, the *logic* of technological innovation is not entirely reducible to variables of a socioeconomic nature. That is, while technological change takes place *pari passu* with other relevant forms of change, we must look for the generative process from inside-out rather than exclusively from outside-in.

My earlier works in this area have reflected a strong conviction that the process of technological change is endogenous to the economic system. However, during the course of this research it became apparent that the conclusions of these earlier works must be drastically revised (see, e.g., Sahal, 1975, 1977, and references therein). I should perhaps note that this revision of many years of earlier research has not been entirely unagonizing. The findings reported here nevertheless stand: The process of innovation cannot be understood in all its richness unless certain physical, technology-specific variables, as well as variables of a purely economic nature, are taken into account.

In summary, the socioeconomic studies of technological change are evidently important. However, knowledge of the inner mechanism of innovative activity is equally important. Moreover, although the two types of treatments of the phenomenon are complementary, neither is a substitute for the other. As a parallel example, molecular chemistry has made notable contributions to our understanding of living matter. However, knowledge of biology remains just as important. The irreducibility of statistical mechanics to thermodynamics is another example, as is the irreducibility of classical (Mendelian) genetics to quantum mechanics. In fact, the attempts to show one-to-one correspondence between them have all failed so far.

The new epistemology that emerges from the results of this study can thus be very simply expressed. The process of technological innovation is appropriately viewed from within rather than exclusively from without. It is then a partially controllable process with profound implications of a socioeconomic nature.

Chapter 2

THE CONCEPTION OF TECHNOLOGY

1.   THREE CONCEPTS OF TECHNOLOGY?

The term technology is often envisaged in terms of universals. It has numerous connotations, ranging from an object of material culture to the pool of applied scientific knowledge. The difficulty with some of these popular notions of technology is that they are so broad based as to defy any useful operationalization of the concept. Indeed, they are no more informative than is the truism that all uncles are males.

The formal concepts of technology, on the other hand, are relatively sparse. Broadly put, they follow two categorical viewpoints. First, there is the neoclassical conception of technology in the form of a production function. Second, there is what might be termed the Pythagorean concept of technology in terms of patent statistics, chronologies of major innovations, etc. Both viewpoints have been the subject of voluminous literature. Nevertheless, even the most enthusiastic proponents of these viewpoints concede that they suffer from a number of severe limitations. Their justification is that there exists little else by way of an adequate conceptualization of technology. It is a curious controversy in its rather narrow, almost caricatural view of the alternatives available, but it is indicative of the state of art. To sum it up, in spite of a number of attempts to come to grips with the problem, the concept of technology remains ambiguous and ill defined.

The main objective of this chapter is to clarify some of the problems involved in formulating an analytically meaningful definition of technology. This has led to the outline of a third concept of technology: systems viewpoint that a technology is best conceived in terms of its functional properties.[1] In what follows, all three concepts of technology are reviewed in detail. Their

Devendra Sahal, Patterns of Technological Innovation   ISBN 0-201-06630-0

Copyright ©1981 by Addison-Wesley Publishing Company, Inc., Advanced Book Program.

15

pros and cons are discussed. It is concluded that they are potentially complementary: each can possibly enrich the other.

2.   A GUIDED TOUR OF THE THREE CONCEPTS

2.1   The Production Function Concept of Technology

A production function expresses the relationship between various technically feasible combinations of inputs, or factors of production, and output. It is not, however, merely a specification of the possible combinations of inputs employed in a single production process. Rather, it is a specification of all conceivable modes of production in the light of the existing technical knowledge about input–output relationships. A classic example is the different methods employed in construction work: many men with shovels, fewer men with wheelbarrows, one man with a bulldozer (Cramer, 1969, p. 224). If such an activity were to be described in terms of a production function, the inputs would assume nonzero values for each of the three modes of construction.

A production function may be illustrated by means of a smooth, convex isoquant representing different methods that can be employed in the production of a *given* volume of output (Figure 2.1a). Any change in the proportions

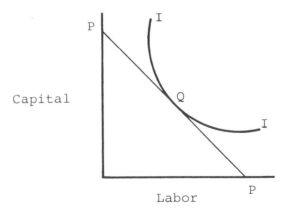

Fig. 2.1(a)   An illustration of the production function. II is an isoquant representing different techniques employed in production of the same output. PP represents the price of capital in relation to labor while Q is the equilibrium point.

---

[1]The term system is employed here in the classical sense originating in the works of Aristotle. Accordingly, a system is characterized by the unity of its parts. Thus the parts of a system, unlike those of an aggregate, constitute a *unitas multiplex*. The unity of a system lies in its functional properties, which cannot be adequately specified except with reference to the whole. This usage of the term system should be sharply distinguished from its casual use as exemplified by the so-called system-dynamics modeling. The proposed definition of a system is admittedly restrictive inasmuch as it does not make any of the tall claims of the modern-day priests of systems analysis.

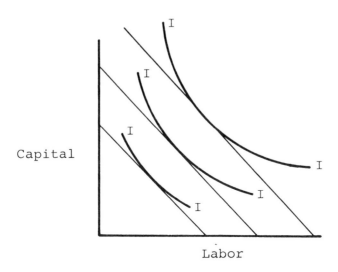

Fig. 2.1(b)   An illustration of technical change in the form of a shift in the production function toward the origin.

of the inputs employed due to changes in their relative prices—that is, factor substitution—corresponds to movements along the isoquant. In contrast, changes due to development of new techniques correspond to a shift toward the origin of the isoquant (Figure 2.1b). The implication is that advances in the technical knowledge make possible the same amount of production by a lesser amount of factors.

Further, it seems natural to classify a technological innovation according to its effect on the proportion of factors employed in the production of a given volume of output. (Figure 2.2). Considering only two factors of production, say, capital and labor, an innovation may be regarded as labor saving if it raises, and capital saving if it lowers, the capital–labor ratio in the production of a given volume of output. According to a more elaborate scheme for the classification of innovations, technical change is labor saving if it raises the marginal product of capital relative to that of labor at a *given* capital–labor ratio utilized in producing a given output, and conversely for capital-saving technical change (Hicks, 1932; Harrod, 1948). The definition of the neutral type of technical change naturally follows from the definition of labor-saving and capital-saving bias in technical change. A detailed discussion of the various classificatory schemes can be found elsewhere (Blaug, 1963). Suffice here to note that the essence of the production function concept lies in its separation of economic from purely technical factors in the choice of techniques of production. The reasons for this are, in part, historical.

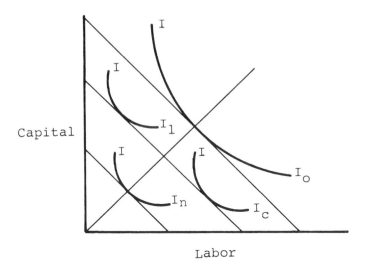

Fig. 2.2 Alternative types of technical change. II_0 is the original isoquant. II_1 represents the result of a labor-saving technical change, while II_c is a result of capital-saving technical change. II_n results from a "neutral" type of technical change. After Blaug (1963).

One of the earliest conceptualizations of technology originates in the work of Josef Schumpeter briefly referred to in Chapter 1. According to Schumpeter, the process of technological innovation consists of new ways in which factors of production can be combined. It is inherently a risky enterprise that is beyond the capability of all but a few pioneering individuals. Although only a few dare to take the initiative, there are many who can follow suit. The initial act of innovation by a few and its subsequent imitation by many are very different from each other. The distinguishing feature of innovation is its novelty. Accordingly, an innovation consists of change in the very *form* of a production function. As Schumpeter himself puts it (1939, p. 87):

> We will . . . define innovation . . . rigorously by means of production function
> . . . This function describes the way in which quantity of products varies if quantities of factors vary. If, instead of quantities of factors we vary the form of the function, we have an innovation.

Moreover, the process of innovation is of an autonomous nature. Hence a production function in its "logically pure" form is to be understood as "a planning function in a world of blueprints where every element that is technologically variable can be changed at will" (Schumpeter, 1961, p. 1031). That is, economic variables do not have any bearing on the origin of a production function.

All this, however, makes the concept of a production function much too obscure. Schumpeter is apparently aware of this difficulty. He therefore goes on to advance an alternative concept of the production function in the same breath (1961, pp. 679, 1026–1053): a "realistic" production function, which can be constructed from actual observations of the production process and which should be clearly distinguished from the "logically pure" production function described above.

The arguments of the Schumpeterian production function consisted only of land and labor. In contrast, the modern versions of the production function as employed in the neoclassical theory of economic growth admit capital and labor while generally excluding land and other inputs. However, notwithstanding a host of evidence that inventive activity may well be at least partially responsive to the forces of economic necessity (see, e.g., Schmookler, 1966), the essential aspect of the Schumpeterian scheme—namely, the separation of economic from technical factors—has lingered on. Moreover, evidence is typically lacking to support the conjecture that innovation and factor substitution are inherently different from each other in that the former entails far greater risk than the latter (Eckaus, 1963). Yet, the practice of distinguishing the one from the other has continued as a matter of sheer convenience if not as an act of faith. Economic theory regards it as customary to explain factor substitution due to a change in relative prices *solely* in terms of the demand and supply of inputs.

Nevertheless, curious difficulties arise in conceptualizing technological change in a production function framework. To begin with, consider the notion of a "logically pure," or "fundamental," production function exhibiting all conceivable techniques that *could* be designed by means of the existing theoretical knowledge. The presumption here is that advances in technology originate in pure sciences. Although this may be in keeping with the popular view of technology, it is open to a number of serious objections (see Section 2.3). Further, a specification of the fundamental production function in terms of known physical laws of a theoretical nature is seldom feasible in practice. As an example, it is theoretically possible to specify the process of generating electricity by steam by means of the gas equation. In reality, however, this is hardly feasible given that the behavior of real gases substantially deviates from the theoretical model at high temperatures and pressures, as in the case of a high-pressure steam turbine. Indeed, it is impossible to specify a production process by the use of pure theoretical knowledge in all but the simplest cases.

We must therefore resort to a "realistic" production function representing techniques actually in use rather than those that could conceivably be designed from existing theoretical knowledge. However, a production function based on actual empirical data entails numerous other difficulties (Salter,

1969). Basically, it purports to distinguish economic from purely technical factors in the choice of a production process. It is difficult to see how a production function can meet such an objective save in exceptional circumstances. For instance, it may reflect a labor-saving tendency arising from the dearness of labor relative to capital rather than from the nature of the technology itself. Moreover, following the so-called theory of induced innovations, suppose that changes in the relative price ratio govern *both* factor substitution and technical progress (Hicks, 1932; Fellner, 1961). In other words, as labor becomes dear relative to capital, firms are compelled to develop *new* techniques with the explicit objective of saving labor. In such a case, the whole scheme of distinguishing economic from technical factors virtually collapses. To complicate matters further, technical progress may be "capital embodied" (consisting of innovations that must be in the form of new machines) or "disembodied" (consisting of improvement in the efficiency of both old and new techniques due to a gradual increase in know-how acquired in their operation, changes in the organizational setup, etc.). In the former case, factor substitution cannot be distinguished from technical advances even if the latter are free from the influence of change in the relative prices.

The problem is compounded by the fact that any production process involves a number of variables besides factor substitution and technical progress. For example, even with constant factor prices, economies of scale may enable more to be produced from less of the factors. In that case, economies of scale cannot be distinguished from technical progress without further restrictions on the form of the production function. There is, of course, no shortage of theoretical assumptions for this purpose. The problem, however, is that these assumptions are often a matter of expedience rather than empirically justifiable hypotheses.

Measurement of the variables in the production function raises further difficulties because of the extreme heterogeneity of both inputs and outputs. It is common to employ concepts such as "capital stock" in the specification of the production function. That is, capital is assumed to be jellylike, homogeneous, and malleable. In reality, of course, capital goods exhibit a great deal of variation in productivity because of differences in cost, vintage, etc. How do we combine many different items in a single rubric and seek measurement? Clearly, aggregation of different goods is possible only if each item is weighted by a valuation coefficient, or its "price." This raises a number of serious problems: To begin with, there is the index number problem of deriving the most appropriate set of valuation coefficients. More important, capital, being a collection of machines rather than a fund of purchasing power, cannot be valued in its own technical units. This is because valuation of capital presupposes a certain rate of interest, while the interest rate in reality is determined by the value of capital. One way to circumvent this problem is to use *investment* instead of a capital *stock* variable, as in the analysis of disaggregated data. However, a number of other difficulties arise. For example, it is far from clear whether market prices should be used as such or whether they should first be corrected for market imperfections in estimating the investment costs of different techniques. In principle, similar difficulties arise in the measurement of labor. The present discussion of these difficulties is illustrative rather than exhaustive. Even so, it is apparent that the existing measures of variables in the production function leave much to be desired. Robinson (1954, 1956), for example, has undertaken an extensive review of the concept of capital. In her own words, By the time the problem has been specialized enough to yield a single composite measure of capital, the latter has already become 'a fifth wheel of the coach'.

A related problem concerns the concept of an aggregated production function at the level of an industry or the economy as a whole.[2] It can be shown that a wide class of well-behaved microeconomic production functions do not aggregate into a well-behaved macroeconomic production function. The incompatibility between the micro and the macro production functions by itself does not, of course, invalidate either of the two concepts. The two may well be complementary in the same way that psychology and neurophysiology are. The real difficulty is that a meaningful theoretical explanation of the observed production functions is generally lacking at both the micro- and macroeconomic levels.

Dissatisfaction with the neoclassical conception of the production function has led many economists to abandon it altogether. However, no satisfactory alternative has emerged as far as the theory of growth and capital goes. Kaldor, for example, has presented a "technical progress function" in which growth of capital per man is related to the growth of output per man at a given *rate* of change, rather than a given *level* of technical knowledge (1957, p. 596). The raison d'être for this new concept is that it is impossible to distinguish shift in the production function from movement along it (1961, p. 205). However, like the production function concept, the technical progress function fails to isolate the economic from the technical factors. As Blaug has put it (1963, p. 31, footnote added to the text):

> No one has yet managed to measure the state of technical knowledge, much less the rate of change of technological knowledge. The neoclassical idea of a given state of knowledge is admittedly an abstract one. But the concept of a given rate of change of knowledge is almost metaphysical.

---

[2]These problems are a subject of a 15-year-old controversy between two groups of economists: Joan Robinson, Nicholas Kaldor, and L. Pasinetti of Cambridge, U.K., representing the Keynesian tradition, and Paul Samuelson and Robert Solow of Cambridge, Mass., representing the Walrasian tradition. For a most readable account of this controversy, see Blaug (1975).

In summary, the above overview of the production function framework is admittedly a very brief one. While limitations of space preclude a detailed discussion, the production function concept has provided many useful insights into the causes of economic fluctuations, the contributions of variables such as growth of the labor force to economic growth over time, etc. In general, the concept has proven to be extremely useful in providing a number of broad general options for *macroeconomic* policy. Evidently, however, it has been much less useful in providing a framework for analysis of technological innovation processes. This problem will be further discussed in Section 2.3. Apparently, one main limitation of the production function concept is that it lacks a conceptualization of technology per se. Its relevance is primarily to an aggregated level of analysis. This is a source of both its weakness and its strength.

2.2   The Pythagorean Concept of Technology

What is termed here the Pythagorean concept of technology has a distinctly interdisciplinary origin. It is based on contributions from fields as diverse as economics, the history of science, sociology, and theoretical physics (Merton, 1935; Schmookler, 1966; May, 1966; Moravcsik, 1973). There are two characteristic features of this viewpoint. First, technological change is best conceived in terms of a count of relevant events, such as the number of inventions patented. Second, the uniqueness and novelty of an event is of crucial importance. Thus, for example, changes in an already existing technique are generally excluded. The viewpoint leads to the consideration of a potentially broad range of variables as appropriate measures of technological and scientific activity: the number of articles published in a given field as indicated by the various indexes of scientific and technical journals, the number of engineers in principal professional societies, the size of the professional grade staff in R&D, and the like (Freeman, 1969). For the present purpose, however, it will suffice to consider in detail two main indices of technical activity: patent statistics and the chronology of major innovations. Both have been widely employed and are generally preferred to other indices.

One great advantage of the patents statistics is that patents meet various explicitly designed criteria of originality, technical feasibility, and commercial worth (Kuznets, 1962). Moreover, they are readily available. However, they suffer from a number of well-known drawbacks (see, e.g., Sanders, 1962). An index of patented *inventions* is merely a list of blueprints available. It gives no information on *innovation*, that is, whether the new device is suitable for production and commercial use. By definition, it excludes the process of "development" of technology involving the translation of a blueprint into a working device amenable to mass production. The problem is compounded by the fact that, to a considerable extent, patented inventions do not result in innovations. Thus, it has been estimated that the proportion of inventions that are commercially utilized has risen from about one quarter in the nineteenth century to barely half in recent decades (Schmookler, 1966, p. 55). Patent statistics suffer from another defect: not all inventions are patentable, especially those made in otherwise essential routine engineering work. Moreover, many inventions are not patented for various reasons, such as inadequacy of patent protection, the delay and expenses involved in patenting, and legal problems stemming from the antitrust legislation. Yet other inventions are not patented even if they are of major importance (e.g., inventions that are the outcome of government-sponsored R&D). To complicate matters further, the propensity to patent an invention varies a great deal across both firms and industries. For example, there is considerable evidence that, contrary to the generally held opinion, large firms in the United States have a lower propensity to patent than their smaller counterparts (Schmookler, 1966, p. 33). Finally, patent statistics fail to reflect the varying technical and economic importance of different inventions.

These deficiencies of patent statistics have led to the use of chronologies of major innovations as an alternative. However, the procedure of assigning dates of occurrence of major innovations underlying these series is inherently unsatisfactory. An example may help make this clear. The first power takeoff equipment for vertical motion of the driven tool relative to farm machinery dates back to 1878, when a power-takeoff-driven reaper was exhibited at the Paris Universal Exposition (Zink, 1931). Yet a quarter of a century had to pass for the idea of power takeoff to take hold. Albert Gougis of France developed a power takeoff in 1906 that was capable of providing both lateral and vertical motion of the driven tool relative to the tractor. It was successfully used until 1918 in the driving of a grain binder. However, large-scale commercial use of power takeoff began only in 1922. Even so, hardly anyone seemed to realize the considerable scope for agricultural application of the power takeoff. In due course its use was extended to virtually every field operation, including the preparation of seedbeds, planting, cultivating, pest control, and harvesting. This was made possible by numerous changes in the various design characteristics of the power takeoff, such as capacity, drive efficiency, type of ending, safety features, and height above the ground. It is only fair to say that the technological development of the power takeoff is continuing even to this date. The moral of the example is evident: The making of an innovation is inherently a continuous process that does not easily lend itself to description in terms of discrete events. The example is further illustrative of a number of shortcomings of any chronology of major innovations.

First, even if a satisfactory date can be assigned to the advent of an innovation, an index of major innovations takes into account merely the initial development of the technology leading to its commercial introduction. It has no bearing on the *long-term* development of technology, consisting of various advances in the design and performance characteristics of the machine. By definition, it excludes a crucial phase in the development of technology after its commercial introduction, during which it is debugged in the light of information from its use, and its capability is enhanced with the experience acquired in the production process. Frequently, these changes involving the substitution of a great many new components for the old radically alter the character of technology, thereby increasing its capability by several orders of magnitude. However, a chronology of major innovations fails to take this into account.

Second, the index under consideration lacks a formal theoretical basis for distinguishing major from minor innovations. Rather, it rests on the assumption that a major innovation is a vehicle for a host of minor innovations (see Chapter 3). However, such an assumption raises a number of worrisome ambiguities. For example, on the basis of a very carefully done analysis of major inventions in the petroleum, paper, railroad, and farm equipment industries during the period 1800–1957, Schmookler (1966, p. 202) concludes that "at least as good a case could be made from the evidence for the proposition that long swings in all inventions in a field *induce* long swings in important inventions in the same field, as for the opposite proposition." Indeed, if the results from the detailed historical case studies of technology are any guide, virtually all innovations are made possible by a host of seemingly minor changes on a continuous basis until the desired degree of reliability and performance is secured. The previously cited example of the power takeoff is a case in point. There are, of course, many other examples of this phenomenon: what appear to be spectacular instances of advances in technology have been, in fact, made possible by changes of a prosaic nature. This is not to say that, in principle, it is not possible to distinguish major from minor innovations. It is rather that an explicitly specified theoretical criterion is a prerequisite to discrimination between major and minor innovations because the two are interlaced. No such basis exists for the chronologies of major innovations. Consequently, results from their use are often ambiguous.

Finally, while patent statistics have been criticized because they give no information on the possibility of the commercial use of technology, it must be noted that, to a lesser extent, this is also true of chronologies of major innovations. Thus, the latter fail to reflect the varying scope for adoption of different innovations.

In summary, patent statistics do provide a reasonable description of an otherwise continuous phenomenon in discrete terms. However, their coverage of technological change is somewhat limited. The chronologies of major innovations are relatively comprehensive in their scope. However, they are somewhat unsatisfactory in their approximation of the phenomenon in discrete terms. To a certain extent, indices of both patented inventions and major innovations are handicapped by their arithmomorphic nature. They describe only changes in the *number* of techniques. They fail to reflect changes in the techniques themselves. This is not to reject them as untenable. Rather, their usefulness in the analysis of technological change depends on a judicious choice of the other variables involved. Indeed, as indicated by a number of empirical studies, they can be of considerable value as relatively direct measures of the outcome from research and development activity (see, e.g., Schmookler, 1966; Reekie, 1973). The important point is that their relevance is primarily to the *origin* rather than the *development* of new techniques.

2.3   The Systems Concept of Technology

The origin of what is termed here a systems concept of technology has a curious history. It was simultaneously envisaged by two groups of scientists with little, if any, awareness of each other's work: economists (particularly those concerned with the problems of appropriate technology in underdeveloped countries) and systems analysts (particularly those concerned with the management of R&D in industrialized countries). At first sight these two groups would seem to have little in common. Nevertheless their viewpoints are broadly similar. Both hold that a technology is best conceived in terms of its performance characteristics. This leads to a consideration of variables such as the fuel-consumption efficiency of a device as one appropriate measure of the state of its technology.

The viewpoint of students of economic development apparently stems from their disenchantment with the neoclassical conception of technology (see, e.g., Stewart, 1977). As noted earlier, the relevant variables in the neoclassical framework are essentially factor substitution and the relative price ratio. All other variables are of minor importance which may well be ignored. In reality, however, many other variables enter into the choice of technology, such as material employed, availability of requisite skills, scale of output, and nature and type of product. This is, of course, particularly evident in the case of developing countries. Moreover, the variables that are generally excluded in the neoclassical framework may be far more important than those included in it. For example, some of the most important possibilities of factor substitution may not be at all between labor and capital. Rather, they may be between skilled and unskilled labor, between initial costs and maintenance costs, between new and secondhand equipment, between machine-paced and operator-paced activities, etc.

Second, techniques of different vintages may exist side by side. Suppose that the differences between different techniques are considerable, as in the case of developing countries, where it is commonplace to employ some of the most up-to-date equipment together with outmoded—even ancient—techniques. In the circumstances, it is unrealistic to assume the existence of *a* smooth isoquant. For example, the older techniques might require more of both labor and capital per unit of output rather than less of one at the expense of the other. The resulting situation does not correspond to either movement along, or a general shift from, a production function, as illustrated in Figure 2.3. It is scarcely surprising that the conventional form of production function does not apply to such cases. It is essentially an ahistorical formulation. The process of technological change, on the other hand, necessarily involves the dimension of time. In summary, the relevance of the conventional form of production function is somewhat limited in cases where the causes of technological change are numerous and of a historical nature. The conventional production function seems rather unsuited to problems of appropriate technology in developing countries. Hence the motivation to seek an alternative concept of technology.

One alternative concept of technology is provided by systems analysts, whose work in this area began on a clean slate. The expenditure on military R&D tremendously increased in the years following the end of the Second World War. While the development of advanced weapon systems proceeded rapidly, there existed little by way of a systematic body of knowledge for the allocation of resources. Thus, a number of practitioners in both industry and government felt motivated to seek a set of tools for the solution of their prob-

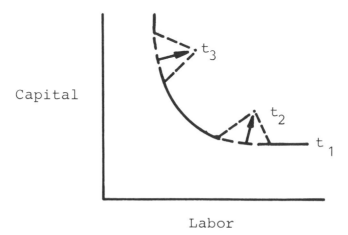

Fig. 2.3 Representation of techniques dating back to different time periods t_1, t_2, t_3.

lems. The early investigations were typically concerned with the capability of military technology (Martino, 1972; Lenz, 1962). For example, what might be an upper limit of the thrust-to-weight ratio of aircraft engines? What changes might be expected in the power-gain bandwidth and system noise of solid-state amplifiers? The range of investigations since then has considerably broadened to include topics ranging from the management of R&D at the firm level to sectoral and national technology strategy (Sahal, 1980). A review of these developments would take us too far afield. It will suffice to note that questions of change in the functional capability of technology occupy a central role in many of these investigations.

Although students of economic development are concerned with the choice of far less esoteric types of techniques, the focal point of their investigations is also the functional characteristics of technology. An example may help make this clear. In recent years a number of attempts have been made to develop biogas plants suitable for use in the rural areas of developing countries. However, the technology has failed to acquire a foothold due to certain obstacles (Garg, 1976). One important problem has been the relatively low generation efficiency of these plants. For example, the generation efficiency of plants built in India has typically ranged from 0.3 to 0.5 of a cubic foot per pound of cow dung in winter, and from 0.5 and 0.8 of a cubic foot per pound in summer. Second, the product characteristics are less than satisfactory. For example, the burning quality of gas is found not to be constant. Currently a number of attempts are being made to improve the technology by means of suitable alterations in the digester and ground temperatures; adjustment of the length of digestion cycles in various ways, including the development of a new species of bacteria; preheating of the slurry; and development of digesters of different scales for family and community use. The general implication of this specific example is that one principal problem of appropriate technology is that of improving product characteristics in relation to the production process. In particular, the crux of the appropriate technology is its capability to perform certain functions in an adequate manner.

The systems view of technology is appealing in many respects. First, the functional measures of technology have very clearly defined meaning and can be objectively measured. For example, the thermal efficiency of an electric power plant can be unambiguously defined as the ratio of the electrical to the total thermal output of the fuel. It can then be expressed either directly as a ratio or in terms of an inverse measure, such as the plant heat rate, that is, the amount of fuel in British thermal units required by a plant to produce one kilowatt-hour of electricity.

Second, functional measures of technology are of far more practical value for engineering and managerial purposes than are, for example, the estimates of the neoclassical production functions. This is because the most useful variables in industrial research activities are seldom of a purely *distributional* or economic nature. Rather, they generally tend to be *control* oriented. As an example of this, we find that the focus of R&D activity in the transport of crude petroleum by pipeline is typically on reducing the total cost of pipeline construction and of control of the oil flow by means of a change in "line diameter," "horsepower," etc., rather than on reducing the cost of capital or labor as such (Pearl and Enos, 1975). More generally, functional measures of technology are closely related to the actual objectives of innovative activity, a point so obvious that it is often overlooked.

Third, functional measures of technology make it possible to take into account both major and minor innovations and to assign appropriate weights to their importance according to a certain common denominator. An example may help make this clear. The development of farm tractor technology has been made possible by a host of innovations, such as the frameless or unitary form of construction, the power takeoff, pneumatic tires, the three-point hitch and control system, hydraulic lift, enclosed transmission, twin-disk clutch, removable cylinder lines, antifriction bearings, power steering, and torque amplifier. It is not easy to assign specific dates to any of these. The case of the power takeoff, cited earlier, is one example of this problem. As another example, the relatively perfected form of the three-point hitch developed in 1938 was a result of more than 17 years of cautious experimentation beginning with the design of an integral tractor plough. Nevertheless, its development has been still continuing. The essential point of these examples is that any attempt to assign a specific date to the origin of an innovation is bound to be unreliable. Further, numerous difficulties arise in distinguishing between an innovation and a minor variation in technology. Even if a list of innovations can be agreed upon, assigning weight to each of them according to some chosen measure of their importance remains a Herculean task. In short, as has already been noted, any conceptualization of technology in terms of the *number* of relevant innovations is fraught with numerous difficulties. However, when advances in technology are measured in terms of variables such as horsepower-hour per gallon of fuel used, ratio of drawbar horsepower to belt horsepower, or horsepower-to-weight ratio, not only does it become possible to take into account the various underlying innovations, but also the innovations themselves are evidently weighted according to their contribution to an objectively measurable characteristic of the phenomenon under consideration. For example, the chosen measure of fuel-consumption efficiency of farm tractors is an appropriately weighted measure of both major innovations, such as the successful use of pneumatic tires, improvement in the quality of fuels, and minor innovations, such as the use of more durable valve, piston, and ring alloys (leading to an increase in the compression ratio, thereby improving the utilization of fuel) and of diffusion effects such as the shift from gasoline to diesel tractors. Thus, the functional variables bypass many of the sins of omission and commission in the measurement of technology.

Fourth, it is evident that one focal point of the systems view of technology is changes in the product characteristics. This is in marked contrast with the neoclassical view, according to which the product characteristics do not change. However, it is in keeping with the objective most frequently stressed in actual R&D activity. Thus, it has been observed that the R&D effort of 90% of manufacturing firms is directed at the development of new products or the improvement of the old (Gustafson, 1962). This finding is confirmed by a number of other surveys indicating that three quarters or more of the results from R&D involved changes in the product characteristics (Bloom, 1951; Wagner, 1968). That is, the systems view appears to be more closely tied to the actual aims of modern-day R&D activity in comparison with other viewpoints.

Fifth, the systems view of technology has a number of important implications for a wide variety of problem areas. As discussed above, it is of particular significance to the problems of appropriate technology in developing countries and the management of R&D. It is further relevant to the multidimensional nature of the diffusion of innovations and long-term economic growth, as discussed in Chapter 1. In summary, the functional view of technology may well be a prerequisite to an adequate understanding of a variety of interrelated problem areas and policy issues.

All in all, the systems conception of technology has a number of advantages over other schemes. Nevertheless, it does suffer from three interrelated limitations. First, curious as it may seem, data on change in the functional characteristics of technologies *over the course of time* are typically lacking. This is not to say that they cannot be compiled; rather, they are not as readily available as, for example, patent statistics.

Second, it is evident that technical change in any given product or process tends to take place in several dimensions. As an obvious example, consider innovations in transportation technology, which have led to a *multiplicity* of advances such as increased speed, reduced fuel consumption, and increased thrust-to-weight ratio. This raises the very important question, how can the various functional properties of a system be appropriately weighted and combined into a single composite index? At one time, this problem seemed to defy solution. Results of the more recent research do make some headway in this area (Griliches, 1971; Sahal, 1977). However, it is evident that these methods require considerable amounts of data, which adds to the difficulty noted above. Thus a great deal of further effort is required in this direction.

Third, the systems approach is best suited to the microlevel of analysis. Its application at the macrolevel—for example, to a comparison of the performance of different industries, or sectors, or economies—is much less straightforward. This is not to say that it cannot be fruitfully pursued at an aggregated level of technological activity. Rather, more work is needed in this area. In particular, it seems clear that it may be useful to combine the systems approach with other concepts such as the production function. An example of this combination is the concept of an "engineering production function" pioneered by Chenery (1949). In principle, we can conceive of many other ways in which different approaches can be usefully pursued in conjunction with each other. However, while the value of an eclectic approach seems beyond dispute in this case, a great deal of further effort is required to find out just what all the possibilities of synthesis are.

3.   ON THE TWO-WAY STREET BETWEEN SCIENCE AND TECHNOLOGY

It is commonly held that scientific discoveries sow the seeds of technological innovation. According to the popular viewpoint, it is not only that there is a hand-to-glove relationship between advances in pure science and technology. What is more, science is the primum mobile of technological innovation. This is also presumably the justification for certain neoclassical and Pythagorean conceptualizations of technology, such as the fundamental production function or the number of articles published in the relevant scientific journals. Despite its intuitive appeal, however, it seems that the popular conception of technology as an outgrowth of advances in science is at best inadequate and at worst misleading.[3]

To begin with it should be noted that there are important similarities as well as differences in technological and scientific activity. In technological enterprise, synthesis of the available knowledge plays a far more important role than does its analysis. Witness the fact that design problems typically take the form of inequalities rather than equations (e.g., stresses must not exceed a certain specified limit which a given material can safely withstand; total costs must be kept down to a certain figure). In this activity the role of analysis is often confined to checking the reliability of the product and making the necessary modifications to it. Thus technology depends primarily on synthesis and only secondarily on analysis (Francis, 1961). In marked contrast to this, science relies mainly on analysis and only partly on synthesis. That is, although science and technology are not mutually exclusive, there is a basic difference in their respective orientations.

Moreover, success in the development of new techniques comes primarily from "getting one's hands dirty" rather than from conceptualizing alone. In

---

[3]For a somewhat similar viewpoint see Nelson (1974), Price (1965, 1980), and Schmookler (1966).

this respect, technical progress is fundamentally different from advances in pure science. This is not to minimize the importance of theoretical concepts in technical problem solving. Indeed, the perception of a problem is never a simple copy of reality. It invariably includes a process of assimilation of the observation to the preexisting structures (Piaget, 1974). Obviously, new techniques are conceived in the light of the available scientific knowledge of a theoretical nature. However, as is discussed at length in the following chapters, the course of their development depends to a large extent on accumulated experience of a practical nature.

We find therefore that a great many advances in technology have been made without the benefit of advances in science. The invention of the steam engine is a well-known example of this, as is the invention of the gas-filled electric bulb. The successful construction of the steam engine in 1816 necessarily embodied a wide range of theoretical principles from subjects such as thermodynamics, the kinetic theory of gases, and hydrodynamics. Yet very few of these principles were actually known at the time. Likewise, the successful design of the nitrogen-filled electric bulb in 1906 was made without any knowledge of thermionic emission into gases. A close examination of the evidence reveals that the list of such examples can be almost indefinitely multiplied. The development of modern aviation for the most part has proceeded in advance of progress in aerodynamics (Miller and Sawers, 1968, pp. 247–253). Indeed, among the major advances in aircraft technology, only the swept-back and delta wings appear to have directly resulted from scientific discoveries. Similarly, advances in the design of machine tools, the mainspring of some of the most important technological innovations, owe very little, if anything, to advances in pure science (Clausen, 1961). These examples are illustrative of a very general point. Contrary to popular opinion, there is very often no *direct* relationship between progress in science and technology.

There is an important exception to what has just been said: A number of innovations of far-reaching significance are demonstrably attributable to the intermingling of science and technology, as, for example, in the case of the chemical and electronics industries. Unfortunately, however, such instances tend to be comparatively few and confined to a handful of industries.

This brings us to the next point. Despite its meagerness, the traffic between science and technology is typically subject to the rules of a two-way street. The case of science-dependent technical progress is commonly recognized. However, while it is often overlooked, the case of technology-dependent scientific progress is equally important. It is indeed doubtful whether a great many scientific discoveries would ever have come to light were it not for the technological instruments made to discover them in the first place. A great deal of progress in biochemistry is attributable to the development of x-ray diffraction equipment and the electron microscope, as is progress in nuclear physics to the development of the electrostatic generator and synchrocyclotron. There are of course much more specific examples of the phenomenon under consideration. From the past we find that the bulk of advances in chemistry during the seventeenth and eighteenth centuries had their origin in the work of alchemists. In modern times, the chemical industry has substantially contributed to the science of polymerization in the development of synthetic fibers (Brozen, 1960). Likewise, the technological effort in the development of semiconductor devices has led to numerous advances in surface physics (Gazis, 1979). The conclusion to be drawn is that science and technology do get together in certain areas to the mutual and often enormous benefit of both. However, the cases of one-to-one correspondence between scientific and technical progress constitute an exception rather than a rule.

The relationship between science and technology will be further discussed elsewhere in this book (see especially Chapters 3 and 9). Suffice here to note that advances in technology are seldom an appendix to scientific discoveries. Rather, innovations depend on the gradual modification of existing techniques through a process of learning by experience. Technical progress is therefore primarily governed by the development of *empirical* rather than *theoretical* knowledge. In sum, the role of science in technological innovation is one of relative rather than absolute significance.

4.   A PRINCIPLE OF TECHNOLOGICAL GUIDEPOSTS

In the preceding sections we have outlined what seem to be three main conceptual schemes for the measurement of technology. The significance of these conceptualizations extends far beyond the quantification of innovative activity. Rather it seems that underlying each concept there is a certain general view of the nature of technical progress. As noted earlier, the neoclassical conception has its origin in the "heroic entrepreneur" theory of innovation. To a certain extent, this is also true of the Pythagorean view of technology that an innovation is essentially a novel product. In contrast, the systems view emphasizes the more or less continuous nature of technical progress while downplaying the role of the seemingly spectacular advances. Each of the three concepts also carries with it a differing methodological prescription. At the two extremes, while the systems viewpoint emphasizes the importance of examining the technology per se, the neoclassical framework is based on the view that technology is best conceived in terms of its *effect* on an aggregate of economic variables. In assessing these alternative concepts, we are therefore led to consider these questions: First, just how common are radical advances in technology? What accounts for them? Second, what is the extent to which the character of an individual technique in its own right plays an important role in technical progress? Clearly, these questions cannot be answered in an unequivocal sense. They are nevertheless worth examining in the light of available evidence.

One characteristic feature of the process of development is that it inevitably leads to the formation of a system. The system in turn sets the boundaries of further development. This interplay of development and system formation is perhaps the most essential key to understanding all processes of long-term change (Sahal, 1978). As regards the process of technological change, very often there emerges a pattern of machine design as an outcome of prolonged development effort. The pattern in turn continues to influence the character of subsequent technological advances long after its conception. Thus innovations generally depend on bit-by-bit modifications of a design that remains unchanged in its essential aspects over extended periods of time. This basic design is in the nature of a guidepost charting the course of innovative activity.

The notion of a technological guidepost is evidenced by the fact that very often one or two early models of a technique stand out above all others in the history of an industry. Their design becomes the foundation of a great many innovations via a process of gradual evolution. In consequence, they leave a distinct mark on a whole series of observed advances in technology. The following examples may help make this clear.

The first example is that of the farm tractor. It is often said that the design of the Fordson and the Farmall tractor models around the 1920s marked the beginning of a new era in the history of technology. There is indeed considerable justification to this statement. The Fordson was a product of straight-line assembly embodying the frameless type of design. Its low cost of production paved the way for widespread adoption of the tractor. The Farmall was the first general-purpose tractor adapted to a wide variety of farm operations. Prior to its introduction, a tractor was no more than a pulling machine unsuitable for most farm operations except plowing. The advent of the general-purpose tractor made it possible to utilize the machine for a host of activities including harvesting operations.

A detailed account of the history of farm tractor technology is given elsewhere in this book (see especially Chapter 6). For the present purpose, the main point of interest is that both the Fordson and the Farmall were an outgrowth of a series of development efforts spanning more than a decade. In turn, they have shaped the course of technological development to this date. The design of the Fordson, for instance, has left a lasting imprint on the form of the modern-day tractor. As one leading designer has put it (Reece, 1969–1970, p. 125):

Tractor production throughout the world has settled down into a small number of distinct tractor forms, skid-steered track layers, tool-frame tractors, and the conventional two-wheel (2 W.D.) machine and its four wheel drive (4 W.D.) variants . . . . Production is . . . totally dominated by the rigid frame, 2 W.D. tractor, with a small proportion of 4 W.D. adaptations. This form of tractor was first introduced by Ford in 1917, fitted with pneumatic tyres in the 1930's, and made into a thoroughly logical arrangement by Ferguson in the 1940's. Since then great progress has been made in detailed design and the machine has become much more complex, but no further really significant changes have occurred.

Similarly, the essential aspects of the Farmall have remained intact to this date. Indeed, there is a striking resemblance between the Farmall and the most up-to-date general-purpose tractor. The difference between the two lies in details rather than in fundamental aspects of machine design. In the words of a spokesman for the innovating company (International Harvester Spokesman, 1972, p. 9):

The Farmall has undergone many changes in power and utility since it was introduced. Though each year has seen important refinements, the essential features have remained the same.

There is, of course, no doubt that farm tractor technology has witnessed tremendous advances since the introduction of the Fordson and Farmall. The important point is that a great deal of this progress has been made possible by *gradual refinement of an essentially invariant pattern* of design. In the statement of a knowledgeable individual in this area (Baker, 1970, p. 15):

While power has increased, and will continue to do so, tractor configuration has changed little over the past 40 years . . . . The most probable course of development over the next 10–15 years is continuing evolution based on known technology rather than any revolutionary change.

The case of tunneling machines provides another admirable illustration of the thesis advanced here. According to one careful account of its history (Sandström, 1963, p. 293–295),

Fowle's rock drill [dating back to 1849] embodied principles of outstanding merit that permitted continuous development, and all the rock drills made in the world today derive from this machine.
   With Leyner's [airblown machine and hollow drills developed in 1897] had arrived the modern rock drill; nothing of significance has been added to it since then, except design refinements due to availability of better materials.

The examples above are illustrative of another point. It is that the emergence of a technological guidepost often depends on adaptation of the technology to the conditions of its use. The development of the steamboat for inland transportation in the western part of the United States during the period 1815–1860 clearly demonstrates this. In the statement of Hunter (1949, p. 64),

> At the time of its introduction in the West the steamboat was a poorly coordinated combination of seagoing vessel and stationary engine, neither of which was suited to the use to which it was put. Within twenty-five years this makeshift affair had been transformed into an instrument admirably adapted to the requirements of western river commerce. By 1840 the western steamboat had reached the form which in essentials it was to keep to the end of its days.

The subsequent advances in technology were made via little-by-little modifications of this basically unchanged form of machine design. As Mak and Walton put it (1972, pp. 635–636),

> Clearly, the story of improved steamboat productivity on inland waters was not simply one of technical advance, that is, advances in knowledge that lowered the input requirements per unit of output. The introduction of steam power had important initial effects, but most of the reduction of costs came from subsequent improvements after the introduction of steam . . . .
> These improvements came gradually, by a method of trial and error, and were the accumulated result of many minor changes and alterations, rather than a few major significant changes. More specifically, except for the initial introduction of steam, the alterations in structural design and mechanical equipment resulted primarily from learning the most suitable specifications of known principles of design and energy to shallow-water transportation. These changes resulted from many men's efforts to adapt seagoing vessel characteristics to shallow-water conditions and, for the most part, they were not the result of new knowledge about basic principles.

Innovations in so-called science-oriented or high-technology areas such as the aerospace and electrical machine industries follow the same general pattern outlined above. The history of aircraft technology is a good example of this. It is fair to say that a great many early advances in this area were centered around a single type of aircraft, the DC-3, which first entered service in 1936. It is interesting to note that from a strictly technical point of view, the DC-3 was not the most advanced aircraft of its time. Moreover, it was singularly lacking in novelty. Rather, it incorporated a number of previously successful techniques in such a way that the cost of its operation was lower than that of any other comparable airliner. It was essentially the culmination of a long series of development efforts (Phillips, 1971). Its design in turn provided an essential framework for the development of technology during the next two decades. The introduction of turbojet-powered aircraft in the 1950s clearly marked a point of departure in the course of innovative activity. Nevertheless, there has been a large element of continuity in technological advances throughout the history of aircraft. In the statement of Miller and Sawers (1968, p. 209):

> Looking back to the DC-3, the jet is technically much superior in its aerodynamic efficiency and structural efficiency. But half the drop in operating costs came from lower crew costs, which was a result of greater size and speed and so only indirectly affected the greater efficiency. Maintenance costs were the second largest source of savings. Here again the jet's productivity helped to lower the unit costs, though better design and the durability of the jet engine were also important factors. Yet for all these gains and all the changes in size and performance, the structure of a 707 or a DC-8 is startlingly similar to that of the DC-3; the line of descent is clear and direct, as is the debt that the present airliner owes to the innovators of the early 1930's.

The history of technological change in electric motors provides another illustration of the evolutionary nature of technical advances in science-oriented industries. Thus, the expert opinion is that (Laithwaite, 1977, p. 102),

> electric motors of popular sizes and speeds have not been designed *ab initio* for over half a century. When a better insulating varnish, or a new ferromagnetic material that can contain a high flux density is available, more elaborate calculations will be needed for the next generation of motors, but there is no questioning of basic shape or layout and it is by no means unreasonable to declare the modern electric motor to have evolved from the motors of 1910.

In summary, the examples of technological guideposts are ubiquitous. We find therefore that innovations in any given technology depend on the refinement of a basically one and the same pattern of machine design. Moreover, the characteristics of such a pattern are generalizable. First, the emergence of a technological guidepost often lies in the culmination of prior advances. It is seldom a matter of radical breakthrough. Rather, it is a result of synthesis as exemplified by the case of the general-purpose tractor and the DC-3 aircraft.

Second, it seems that the greater the variety of tasks to which a design is adaptable, the more likely it is to serve as a guide to the general direction of technical advances. The example of the general-purpose tractor is a case in point, as is the example of the steamboat cited above.

5. THE EVOLUTIONARY NATURE OF INNOVATIVE ACTIVITY AND THE VARIOUS CONCEPTS OF TECHNOLOGY

The preceding section has outlined the principle of technological guideposts. It is found to hold in a wide variety of advances in *product* technology. Additionally, it seems that advances in *process* technology are no exception to this general principle.

As an example, while the efficiency of the blast furnace technique of steel production has considerably improved over the last few decades, the basic layout of the technology has undergone little change. Thus, much of the progress has been made possible by gradual changes rather than by departing from the norm. According to one careful account of technological change in this area during the period 1950–1975 (Carlsson, 1978, p. 322),

> The main conclusions which may be drawn from the present study are that many improvements have occurred in blast furnace technology during the last 25 years, that these seem to have been of a step-by-step rather than revolutionary kind, that they have been adopted to varying degrees in various countries, and that these improvements seem to have had a much greater impact on input requirements and costs than increase in scale per se. Thus, given the periodic necessity of rebuilding existing blast furnaces, it appears to have been cheaper to install new technologies in old furnaces than to scrap them and build new ones incorporating both new technologies and scale economies.

Indeed, the bulk of technical progress is very often attributable to modifications of a *given* technique. To cite the conclusions from an in-depth study of technical advances in the petroleum-refining industry (Enos, 1957–1958, p. 180):

> In order to consider all aspects of technological progress, it is divided into two parts, that which occurs when a new process is introduced and that which occurs when the previously introduced process is improved. The evidence . . . indicates that improving a process contributes even more to technological progress than does its initial development.

The results from the investigation of a number of other areas, such as the aluminium production, electricity generation, and synthetic fiber industries, point to a substantially similar conclusion: Progress frequently takes the form of several minor innovations (Hughes, 1971; Hollander, 1965; Peck, 1962). Cumulatively, however, they are of major significance.

In summary, there is preponderance of evidence suggesting that the nature of technical progress constitutes an evolutionary system. This is not to say that radical advances in technology do not occur. Rather, the major innovations are made possible by numerous minor innovations. The cumulative impact of many seemingly minor changes in technology often tends to be quite substantial.

In the light of the above considerations, we are led to conceive of innovation as a process of learning by experience. This proposition is investigated in detail elsewhere in this book (see especially Chapters 5, 6, and 9). For the present, one main finding of the following analyses deserves attention: The process of learning tends to be technology specific. In general, there exist wide varia-

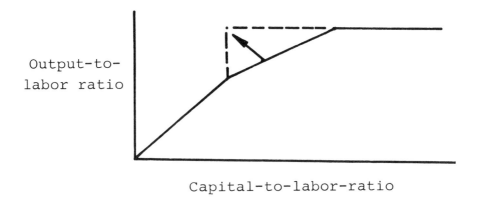

Fig. 2.4   Localized technical progress (cf. Atkinson & Stiglitz, 1969).

tions in the extent of advances due to learning in the case of different technologies.

The localized nature of learning has profound implications for the concept of the production function (Atkinson and Stiglitz, 1969). It is evident that the notion of a general shift in the production function as a consequence of technical progress is untenable, in that the fruits of learning are not equally shared by the entire spectrum of techniques. Basically, the situation corresponds to a selected modification of, rather than one of shift from or movements along, the production function. This is illustrated in Figure 2.4. That is, insofar as the process of learning tends to be localized, the very concept of a production function representing a near-infinite array of techniques sharing a common state of knowledge may have to be abandoned.

We arrive at roughly the same conclusion regardless of the nature of learning. Earlier it was pointed out that there are numerous methodological difficulties in distinguishing between movements along and shifts from the production function. If, however, advances in technology are of an evolutionary nature, as suggested in this chapter, it would seem to be pointless to even try to distinguish shifts from the movements along the production function. Rather the relevant notion of technical progress would seem to be of *evolution* of the production function.[4] Similarly, the Pythagorean view of invention of innovation as an antithesis of evolution would seem unwarranted in the light of actual evidence. Rather, the innovations would seem to depend on the development of technology.

---

[4]This confirms some of the conclusions reached by Nelson (1974) from a different premise.

6.   PRINCIPAL CONCLUSIONS

This chapter has outlined three principal concepts of technology. Two of these concepts are relatively well known: the neoclassical specification of technology in the form of a production function and the Pythagorean account of it in the form of, say, the chronologies of major innovations. A number of theoretical and methodological problems arise in pursuing the two concepts. Against the background of these problems, a third systems concept of technology is put forward. In its essence, technology is best understood in terms of certain measurable, functional characteristics of the phenomenon under consideration. A technology is as a technology does.

A detailed examination of the pros and cons of the various models shows that the systems concept of technology has a number of advantages over the other two concepts. Specifically, both the neoclassical and the Pythagorean viewpoints are founded upon an antithesis of evolution. However, if the considerations advanced in this chapter are any guide, much technological change tends to be of an evolutionary nature. In particular, there is considerable evidence to support the thesis underlying the systems concept that innovations depend on the development of existing technology. In summary, while the systems concept of technology is still in its infancy, it holds considerable promise for the future.

The above comparison of the various concepts of technology should not be taken to imply that there is some type of pecking order among them. The three concepts are not mutually exclusive. Rather, they are complementary not only in their points of emphasis but in their objectives and scope as well. The focal point of the neoclassical specification of technology is the process of production. However, it does not provide a direct means for the measurement of technological change. The Pythagorean viewpoint does provide a relatively direct measure of technology. However, it lacks a formulation of the production activity. The systems concept is, by and large, free from these limitations because it is closely tied to the actual technical basis of the innovation process. By the same token, it cannot be easily applied at an aggregated level like the other concepts. This suggests that each of the three concepts can be used to elucidate the other. Accordingly, while the emphasis throughout this book is on the systems concept of technology, the investigations reported in the following chapters employ a variety of technological measures.

Finally, this chapter has outlined a principle of technological guideposts. In its essence, the process of technological development invariably leads to a certain pattern of design. The pattern in turn governs the possibilities of further development. We find therefore that a great many innovations in any given area depend on exploiting the potential of an essentially invariant pattern of machine design by degrees. Some such basic design may be likened to a guidepost indicating the general direction of technical progress. What is more, it can be identified early in the history of a product or process. The implication for business policy is that it is possible to determine a priori the relative success of a technique in terms of the scope of its development.

Chapter 3

THE ORIGIN OF NEW TECHNIQUES

1. DESIGN AND CONTINGENCY IN TECHNOLOGICAL INNOVATION

It is a commonplace observation that some very important technical discoveries have been made in most fortuitous circumstances. As a classic example, it was a sheer accident that led to the celebrated eighteenth-century observation by Luigi Galvani that the legs of a frog, when connected to a set of different metals, contract due to the generation of electric spark. Although Galvani himself never fully understood the nature of the phenomenon he had discovered, his novel observation stimulated a number of systematic experiments. They eventually led to the successful invention of an electric battery. There are literally hundreds of similar cases—for example, the formulation of laws governing the polarization of light by reflection, the production of penicillin, and the discovery of radioactivity (Taton, 1957). These examples are illustrative of a central characteristic of innovative activity. On the one hand, it is a matter of quite some deliberate effort. On the other hand, it often involves an element of unpredictability.

The role of systematic experimentation in the process of technical discovery is generally well recognized. However, the fortuitous nature of the innovative process has received comparatively little attention. As admitted by Schmookler, author of one of the most extensive investigations in this area (1966, p. 197), "Chance received little attention in our enterprise." The investigation presented here attempts to fill this gap in the treatment of innovative activity.

It is customary to distinguish an invention from an innovation. An invention is essentially the creation of a new device. An innovation additionally entails commercial or practical application of the new device. An innovation, that is to say, is the first application of an invention. This distinction between

Devendra Sahal, Patterns of Technological Innovation
ISBN 0-201-06630-0

the two activities will be observed throughout this chapter. However, the two terms will be used interchangeably in the theoretical parts of the chapter that are applicable to both.

2.   A PROBABILISTIC MODEL OF INNOVATION PROCESSES

Two simple observations underlie the model of innovative activity presented here. First, the determinants of new techniques generally tend to be numerous and of an extremely varied kind. Frequently, there exist a great many different sources of technical discoveries in any given area. At the risk of laboring the obvious, we assert that some innovations are a result of systematic R&D activity whereas others originate in the course of day-to-day production or marketing activity. Very often, there is no single determinant of industry-wide innovative activity. Rather the origin of technical discoveries lies in complex interactions between a *multitude* of factors. If there is any aspect of innovative activity that stands out above all others, it is the extreme diversity of the context within which it takes place. In consequence, the individual innovation often seems to be a sport of chance.

Second, it is evident that the origin of new techniques depends to a large extent on the sheer passage of time, or more accurately, on the accumulation of relevant know-how. There are very few instances of technological change without a history of unsuccessful efforts. Although it is generally overlooked, the lessons learned from mistakes committed in the past constitute an important factor in the process of successful innovation. For example, Fulton's success in building a steamship was preceded by no less than 34 distinct attempts that resulted in failures (Gilfillan, 1935a, pp. 91ff). To give another example, as many as 62 aircraft types developed and available for order during the period 1927–1960 were never used commercially (Phillips, 1971, p. 73). It is equally true, however, that many of the technological failures proved to be valuable in other respects, notably as prototypes for many successful efforts subsequently. Very generally, success in technical problem solving depends on the acquisition of relevant experience over the course of time. The solutions of existing problems in turn add a new dimension to the previous experience. In this way, technical change takes place in a *cumulative* manner. Thus the process of technical change is not wholly chaotic in nature. There is a logic to it inasmuch as it is governed by a process of learning from past experience.

In symbols, let the probability of occurrence of a technological innovation during the differential dt be $\lambda\, dt$, and $P(n,t)$ the probability of n innovations in the time interval from 0 to t. As a very simple model incorporating the role of chance in the phenomenon under consideration, suppose that the process of innovation follows the Poisson distribution

$$P(n,t) = \frac{(\lambda t)^n}{n!} \, e^{-\lambda t}, \qquad n = 0, 1, 2, \ldots \qquad (3.1)$$

The central features of the Poisson process are well known. In the present context they imply that (1) the probability rate of the occurrence of innovations, λ, is invariant in time; (2) the chance of occurrence of an innovation is independent of whether previous innovations have occurred recently or not; (3) in a very small time interval no two innovations can occur simultaneously. This is not to say that the innovations in fact occur in a completely random manner, as implied by the Poisson representation of the process. Rather, it is hoped that the detection of a departure from randomness may shed light on the mechanism underlying the observed phenomenon.[1]

Thus, the above model may be regarded as a convenient starting point from which to consider the number of innovations in successive periods of a given duration. Upon expressing λ as a yearly rate of occurrence of innovations, we have $t = 1$ and (3.1) can be rewritten as

$$P(n) = \frac{\lambda^n}{n!} \, e^{-\lambda} \, . \qquad (3.2)$$

However, it may not be realistic to assume that the probability rate of occurrence of innovations remains constant insofar as the process of technological change is governed by causes of a *cumulative* nature. Instead, the parameter λ is best regarded as a random variable itself with probability density $f(\lambda)$; the relevant formulation of the process is then given by

$$P(n) = \int_0^\infty \frac{\lambda^n}{n!} \, e^{-\lambda} f(\lambda) \, d\lambda \, . \qquad (3.3)$$

The simplest frequency distribution that allows some variation in λ is the Eulerian distribution

$$f(\lambda) = \Gamma(\alpha)^{-1} \beta^{-(\alpha+1)} \lambda^\alpha e^{-\lambda/\beta}, \qquad \beta > 0 \, , \, \alpha > -1. \qquad (3.4)$$

The chief merit of this distribution is its flexibility, since it has two adjustable parameters. By combining Eq. (3.3) with (3.4), we obtain the negative binominal distribution (Cramer, 1969; Greenwood and Yule, 1920; Fisher, 1950)

---

[1] As Cox (1955, p. 130) has pointed out, such a "justification" for a completely random series is analogous to the central limit "justification" of the normal distribution.

$$P(n) = \frac{\Gamma(n+k-1)}{\Gamma(n)\Gamma(k-1)} \, p^k q^n \tag{3.5}$$

where $k = \alpha + 1$, $p = 1/(1+\beta)$ and $q = 1 - p$. The name comes from the fact that the above probabilities are derived from the expansion of $(1-q)^{-k}$. Further, it is evident that, unlike the Poisson distribution, the negative binomial has two parameters, k and p.

The first four cumulants of the distribution of n are

$$\varkappa_1 = \frac{kq}{p}, \varkappa_2 = \frac{kq}{p^2}, \varkappa_3 = kq\,\frac{1+q}{p^3},$$

$$\varkappa_4 = kq\,\frac{1+4q+q^2}{p^4}. \tag{3.6}$$

Hence the coefficients of skewness and kurtosis of the distribution are

$$\gamma_1 = \frac{1+q}{(kq)^{1/2}}, \qquad \gamma_2 = \frac{1+4q+q^2}{kq}. \tag{3.7}$$

These expressions can be easily derived from the characteristic function of the distribution.

The negative binomial distribution is always positively skewed and leptokurtic. The mean of the distribution is kq/p, which is less than the variance kq/p^2. It is often convenient to express the distribution of n in terms of the mean m and the exponent k; (3.5) can then be rewritten as

$$P(n) = \left(1 + \frac{m}{k}\right)^{-k} \frac{\Gamma(n+k-1)}{\Gamma(n)\,\Gamma(k-1)} \left(\frac{m}{m+k}\right)^n. \tag{3.8}$$

The most attractive feature of the above model is that it makes possible a probabilistic formulation of the *cumulative* mechanism underlying the process of innovation. This is also justified by the evidence from other fields. For example, two of the prominent empirical distributions to which the negative binomial model applies are (1) deaths in the case of a population subject to a cumulatively fatal disease or toxin (Yule, 1910), and (2) purchases of a product under the influence of repetitive advertisements (Ehrenberg, 1959). Thus, there is considerable further justification for regarding the negative binomial as an appropriate model of the phenomena under consideration.

In the present context, an alternative basis for the proposed model is as follows. It seems reasonable to suppose that every industrial activity is based on the use of a number of different but complementary technologies. Further, it is plausible that certain techniques are more amenable to innovation than others because of differences in their internal characteristics, such as scale and redundancy (see Chapter 9). Thus it may well be that the rate at which innovations occur *within* each homogeneous group of techniques is a constant, as implied by the Poisson distribution (3:2). However, the rate of innovations will vary *across* the different groups of techniques owing to the heterogeneity between them. Provided that the variations in the mean rate of innovation can be described in terms of the function (3.4), the observed industry-wide distribution of innovation will be the negative binomial distribution.

In summary, the proposed model of the phenomenon underlying the origin of new techniques is based on two simple probablistic schemes. The essence of the first scheme is that innovations grow out of a cumulative process of learning. The second scheme further postulates that advances in the various techniques do not materialize at the same rate. Rather, there exists an element of heterogeneity in the system of technological evolution. According to both schemes, a priori, the distribution of technological innovation is expected to be the negative binomial.

It should be noted, however, that the negative binomial distribution may also arise as a consequence of some very different set of assumptions. Among these are: (1) The case of "inverse binomial sampling," in which sampling from a binomial population continues until a specified number of successes are scored. The number of items sampled is then a negative binomial variable (Kendall and Stuart, 1947). (2) The case of events occurring in certain classes such that the number of classes are distributed according to the sample Poisson process and the size of the class is a logarithmic variable (Quenouille, 1949). As an example, suppose that the number of consumers of a product follows the Poisson distribution and the number of units bought by a consumer follows the logarithmic series distribution. The number of units sold is then a negative binomial variable. Such a model is consistent with the view that the case of a consumer with zero purchases is irrelevant.

The first of the above two models is of a purely statistical nature. It does not seem to offer any insights into real-world instances of innovative activity. The second model is somewhat more pragmatic. However, its justification for the case in hand is limited because it rules out instances of failure in the search for new techniques. It is not particularly appealing because it does not take into account the faux pas that, as discussed earlier, are inevitably present in innovation processes. Thus, neither of the two other hypotheses seems sufficiently rich to provide a genuine alternative to the theoretical schemes outlined earlier.

3.   CASE STUDIES: ORIGIN OF NEW TECHNIQUES IN THE FARM EQUIPMENT, RAILROAD, PAPER-MAKING, AND PETROLEUM-REFINING INDUSTRIES

This section presents the results of the following case studies (the time periods analyzed are indicated in parentheses):

1. Major innovations in farm tractor technology (1800–1971);
2. Major innovations in the farm equipment industry (1902–1949);
3. Major inventions in the farm equipment, railroad, paper-making, and petroleum-refining industries (1800–1957).

A description of the data employed in these case studies is presented in the Appendix to the book. For the present, it should be noted that the bulk of the data came from the work of Schmookler (1966). The choice of the specific cases themselves was dictated by empirical necessity. It is important to note that the following investigations are concerned with the origin of *major* new inventions or innovations, that is, techniques that have been economically important either in their own right or as the basis for subsequent techniques. This is, of course, as much to imply that the present analysis is based on what are highly selected rather than random samples. In the circumstances, it should not be surprising if there are large deviations from the theoretical model. If the distribution under consideration nevertheless yields a good fit to the data, the theory may be regarded to have passed an acid test.[2]

In the following analysis, our starting point is always a test of the pure chance hypothesis implied by the simple Poisson process (3.2). Richardson (1944) has developed a concise test for rejecting the Poisson law. Let $(x_1, x_2,..., x_N)$ denote the observed sample. The test statistic g is then defined by

$$g = \frac{1}{m(N-1)} \sum_{i=1}^{i=N} (x_i - m)^2, \qquad m = \frac{1}{N} \sum_{i=1}^{i=N} x_i . \tag{3.9}$$

The expectation of g in samples from a Poisson population is unity, for any N. Further, if the mean is large (e.g., $m>5$), the distribution of $(N-1)g$ in samples from a Poisson population is the same as that of χ^2 for $N-1$ degrees of freedom. Specifically, the probability that in samples of N from a Poisson population, g would be greater than its observed value is given by

$$L_{>g} = P(\chi^2, N-1). \tag{3.10}$$

---

[2]For a similar viewpoint see Kendall (1961, p. 6).

In this way, the statistic g can be employed to test the hypothesis that, apart from chance deviations, the N numbers $x_1, x_2,..., x_N$ would all be equal. Finally, it can be shown that for samples from a Poisson population, the variance of g is equal to $2/N$ for large N. The test statistic g is used throughout in this chapter to determine whether the observed data indicate an absence of any steady drift toward more or fewer innovations. When the results of such a test point to rejection of the law, we proceed to determine whether the observed distribution is consistent with the negative binomial hypothesis.[3] Finally, in the computations required for the usual form of the χ^2 test, we have pooled the tail group of each sample so as to avoid small frequencies. The extent of pooling is indicated by the degrees of freedom (ν) reported in each case.

Table 3.1 presents an analysis of innovations in tractor technology for the period of $N = 172$ years covered by the data. The observed frequencies of years with $n = 0,1,2,...$ innovations are compared with the theoretical frequencies derived from the Poisson law (3.2). The maximum likelihood estimate of λ is the sample mean $m = S/N$ where S is the total number of innovations. For the case in hand, it is equal to 0.134. As indicated by the χ^2 test, the agreement between theory and observations is excellent. Further, the computed value of g is 1.05, the expected value of g being in the range of 1 ± 0.107. The observed g is therefore credible for a sample of N from a Poisson population.

Table 3.2 presents the results of fitting the Poisson law to the empirical data concerning innovations in the farm equipment industry. The computed value of g is 0.95, the expected value of g being in the range 1 ± 0.204. Thus, once again, both the χ^2 and the g test indicate a good fit for the Poisson law.

One central implication of the Poisson distribution is that in a short interval of time δt it is "very rare" for an event to occur. However, this does not

Table: 3.1: Innovations in Tractor Technology (1800–1971)

Number of innovations per year	0	1	2	N
Observed frequencies	151	19	2	172
Theoretical frequencies	150.5	20.1	1.4	172

$$S = 23, \quad m = 0.134, \quad \chi^2 = 0.314, \quad \nu = 1, \quad P_{\chi^2} = 0.57.$$

---

[3]Jeffreys (1961) has provided a test for examining whether a distribution deviates significantly in the direction of a negative binomial. As Richardson (1944) points out, the test statistic g is related to Jeffreys' test. Our preference for the former is based on the fact that it is simpler to compute and its kith and kin can be traced in various historic treatises.

Table 3.2: Innovations in the Farm Equipment Industry (1902–1949)

Number of innovations per year	0	1	2	3	4	N
Observed frequencies	11	22	8	4	3	48
Theoretical frequencies	13.2	17	11	4.7	2	47.9

$$S = 62, \ m = 1.29, \ \chi^2 = 3.21, \ \nu = 3, \ P_{\chi^2} = 0.36.$$

mean that the distribution of every rare type of event follows the Poisson law. As an example, an industrial accident is a rare event. However, the Poisson law provides a very poor fit to an observed distribution of factory accidents (Kendall and Stuart, 1947, Vol. I, pp. 128–129). Thus the finding that certain empirical distributions of technological innovation follow the Poisson law is not merely a tautology.

Nevertheless, some caution is necessary in a generalization of the good fit of the Poisson law in the above two cases. In particular, the case of innovations in a tractor technology, while it does cover a long period of time, is restricted to a relatively homogeneous class of innovations. The case of innovations in the farm equipment industry, on the other hand, does consider a fairly heterogeneous class of innovations. However, its coverage is restricted to a rather short period of time. Indeed, according to the theoretical considerations of the preceding section, industry-wide distributions of inventions or innovations over long periods of time are not expected to conform to the Poisson law. This expectation is borne out by the results of the g test for data on inventions in the farm equipment, railroad, paper-making, and petroleum-refining industries during the period 1800–1957. The computed values of g in these cases are 1.42, 1.31, 2.42, and 3.52, respectively, the expected value of g being in the range of 1 ± 0.1125. Hence the pure chance hypothesis implied by the Poisson law must be rejected in these cases.

The next step is to test whether the negative binomial distribution holds in the above four cases. In fitting this model to the data, the maximum likelihood estimate of m is simply the sample mean. The procedure for estimating the parameter k is somewhat complicated. This study has employed two alternative methods for estimating the parameter k (Bliss and Fisher, 1953). According to the first method, k is determined from the mean m and variance v of the sample as

$$\hat{k} = \frac{m^2}{v - m}. \tag{3.11}$$

It can be shown that the solution above has an efficiency of 90% or more for small values of m where $k/m > 6$, for large values of m where $k > 13$, and for m in the intermediate range when $(k+m)(k+2)/m \geqslant 15$. When the requisite criteria for efficiency of 90% or more are not satisfied, we have employed an alternative estimation procedure based on the ratio of the total number of years in the sample (N) to the number of years without inventions (N_0). From Eq. (3.5), the expected probability for $n = 0$ is

$$P_0 = p^k .$$

Since $m = kq/p$ and $q = (1-p)$

$$p = \frac{k}{m+k} .$$

If P_0 is replaced by the proportion observed in the zero class, then

$$P_0 = \frac{N_0}{N} .$$

The required estimate of k is that which satisfies the equation

$$P_0 = \left(\frac{k}{m+k}\right)^k = \frac{N_0}{N} .$$

It is convenient to use the log-linear form of this expression, that is,

$$K \log_e\left(\frac{m+k}{k}\right) = \log_e\left(\frac{N}{N_0}\right) . \tag{3.12}$$

In the present study the value of k has been obtained as a solution of Eq. (3.12) by iteration.

Tables 3.3–3.6 present the results from fitting the negative binomial distribution to the data on inventions in the farm equipment, railroad, paper-making, and petroleum-refining industries. In each case the estimate of the variance (v) in the data is provided along with the values of m and k. As indicated by the χ^2 test, the agreement between theory and observations is excellent in all four cases. The empirical evidence therefore does not refute the theoretical model under consideration.

It should be noted, however, that the good fit of the negative binomial distribution to the observed frequency of inventions by itself does not "prove"

Table 3.3:   Inventions in Farm Equipment Industry (1800–1957)

Number of inventions per year	0	1	2	3	4	5+	N
Observed frequency	49	52	29	17	5	6	158
Theoretical frequency	50.63	48.67	30.54	15.76	7.26	5.03	157.89

$S = 215$, $m = 1.361$, $k = 3.255$, $v = 1.93$, $\chi^2 = 1.35$, $v = 3$, $P_{\chi^2} = 0.72$

Table 3.4:   Inventions in Railroad Industry (1800–1957)

Number of inventions per year	0	1	2	3	4	5+	N
Observed frequency	49	38	40	17	9	5	158
Theoretical frequency	43.91	49.32	33.50	17.79	8.13	5.27	157.92

$S = 232$, $m = 1.47$, $k = 4.802$, $v = 1.92$, $\chi^2 = 4.59$, $v = 3$, $P_{\chi^2} = 0.20$

Table 3.5:   Inventions in Paper-Making Industry (1800–1957)

Number of inventions per year	0	1	2	3	4+	N
Observed frequency	76	45	22	7	8	158
Theoretical frequency	76.02	41.69	20.92	10.17	8.99	157.79

$S = 159$, $m = 1.01$, $k = 1.190$, $v = 2.44$, $\chi^2 = 1.41$, $v = 2$, $P_{\chi^2} = 0.50$

the theory. Two alternative interpretations of the results are therefore worth considering. (1) The good fit of the negative binomial distribution to the data may be due to the existence of a "contagious" mechanism underlying the process of technological change. (2) The observed distribution of inventions may be consistent with some entirely different type of hypothesis.

According to the hypothesis of contagion, the occurrence of an innovation may *either* facilitate *or* retard the occurrence of some other innovations. This is plausible if, for example, the invention of a device such as the boring

Table 3.6: Inventions in Petroleum-Refining Industry (1800–1957)

Number of inventions per year	0	1	2	3	4	5	6	7	8+	N
Observed frequency	64	31	16	13	10	9	4	5	6	158
Theoretical frequency	64.02	31.22	19.34	12.83	8.80	6.15	4.35	3.10	6.48	156.29

$S = 311$, $m = 1.96$, $k = 0.650$, $v = 6.90$, $\chi^2 = 3.29$, $v = 6$, $P_{\chi^2} = 0.77$

mill increases the chance for development of another, such as the (conventional) steam engine. As Student observed (1919, p. 215): "If the presence of one individual in a division increases the chance of other individuals falling into that division, a negative binomial will fit best, but if it decreases the chance, a positive binomial." It should be noted that there is no easy way to test contagion in the data; neither evidence from a univariate distribution nor from correlation techniques can be wholly conclusive in this regard (Blum and Mintz, 1951). Perhaps the best one can do is to consider the Neyman's type A distribution, which is explicitly based on the notion of contagion (Beall and Rescia, 1953). However, the fit of this distribution to the data was found to be generally poor in comparison with the negative binomial. In the case of inventions in the farm equipment and railroad industries, the χ^2 values associated with the Neyman type A distribution were computed as 1.44 and 4.77, respectively, for three degrees of freedom (the values of P_{χ^2} being 0.70 and 0.25, respectively). However, it could not be fitted to the data with any success in the remaining two cases of inventions in the paper-making and petroleum-refining industries (the values of P_{χ^2} being less than 0.001 in both instances). In all, we can well agree with Feller that (1943, p. 398, italics in the original) *"contagion is not inherent in any phenomenon in nature, but simply in our method of sampling."* In the present context, the conclusion to be drawn is that the occurrence of an invention by itself does not guarantee that some other inventions will in fact occur. This point will be discussed further in Section 3.5.

As remarked earlier, the proposed model is based on the assumption that the distribution of λ's takes the form of a Pearson type III. In general, however, the variations in λ can take any form. For example, it may not be implausible that inventions occur by chance, as implied by the Poisson law (3.2); however, the mean rate of their occurrence is a function of certain economic variables such as profitability of invention. A number of such hypotheses incorporating the role of economic variables in inventive activity were tested by means of certain specialized statistical methods (Cox and Lewis, 1966). According to the results of these analyses, there is some evidence that economic forces have a bearing on the rate of inventive activity (Sahal, 1974). However, it is not possible to draw a clear-cut conclusion because of the paucity of relevant data.

In summary, the alternatives to the proposed formulation of inventive processes are not wholly satisfactory. Nevertheless, it would be improper to put forward the negative binomial as the only appropriate model of the phenomenon under consideration. Indeed, the choice of a specific model cannot rest merely on its good fit to the data. Thus, all that can be said is that pending further evidence, the negative binomial, on the basis of its simplicity and robustness, can be regarded as one suitable model of inventive activity.

4.   ON CERTAIN SPATIAL AND TEMPORAL ASPECTS OF INVENTIVE ACTIVITY

The first objective of this section is to investigate if there is any significant tendency for intervals between inventions to succeed one another in groups. The entire series of events can be broken up into groups of n successive intervals and can be tested for a significant difference in the k mean intervals (Maguire, Pearson, and Wynn, 1952). Let $\bar{t}_i$ be the mean interval in the ith group of n consecutive intervals ($i = 1,2,...,k$) so that $T_i = n\bar{t}_i$ is the sum of n intervals in the ith group. The test criterion to be calculated is

$$M = 2nk \left[\log_e \left(\frac{1}{k} \sum_{i=1}^{k} \bar{t}_i \right) - \frac{1}{k} \sum_{i=1}^{k} (\log_e \bar{t}_i) \right]$$

$$= 2nk \left[\log_e \left(\frac{1}{k} \sum_i T_i \right) - \frac{1}{k} \sum_i \log_e T_i \right] .$$

(3.13)

It can be shown that M/C is distributed approximately as χ^2 with $\nu = k - 1$ where

$$C = 1 + (k+1)/(6nk) .$$

(3.14)

The results are presented in Table 3.7. Regarded as χ^2, M/C is significant in all four cases at the 0.5% level. In the case of the petroleum industry, the last group contains more intervals than the others, but this is unlikely to have affected the results. It is quite clear, therefore, that inventions tend to occur in clusters.

Table 3.8 shows the results of estimating the mean rate of occurrence of inventions (λ) in the four industries for the time periods 1800–1899 and

Table 3.7:   Clustering of Technical Inventions

Industry	n	k	M/C
Farm equipment	10	20	44.07
Railroad	10	23	72.76
Paper making	10	15	110.15
Petroleum refining	20	15	279.54

Table 3.8: Variation in the Mean Rate of Inventions

	Inventions per year	
Industry	1800–1899	1900–1957
Farm equipment	1.76	0.69
Railroad	1.80	0.89
Paper making	1.39	0.34
Petroleum refining	0.67	4.18

1900–1957. The question arises whether the observed differences in the rates of inventions during the two periods are significant. This can be investigated by means of Kolmogoroff's test (Barnard, 1953). Let

$$nF_n(t) = \text{number of inventions occurring at or before time } t,$$

while for the theoretical cumulative

$$nF(t) = nt/T$$

where T is the total length of the interval considered and n the total number of events in time t.[4]

Then, let Kolmogoroff's statistic be denoted by D_n so that

$$nD_n = n \max_{0<t<T} \left| F_n(t) - F(t) \right|$$

$$= \max_{0<t<T} \left| (\text{number of inventions up to time } t) - \frac{nt}{T} \right|. \qquad (3.15)$$

The computed values of nD_n and the 1% point of nD_n (which is $1.6276\sqrt{n}$ for $n>80$) are provided in Table 3.9 (these computations are based on the values of $t=100$ and $T=158$). The results indicate that in each of the four industries the 1% point is exceeded, implying that the rate of invention has significantly

---

[4]As Maguire et al. (1952) point out, the Kolmogoroff test is powerful only in detecting certain kinds of variation in $\lambda(t)$; this is likewise true of other tests, such as the ω^2 test developed by von Mises and Smirnoff (Cramer, 1946).

Table 3.9:   Significance of Temporal Variation in the Mean Rate of Inventions

Industry	Computed nD_n	1% point
Farm equipment	28.92	21.56
Railroad	31.16	24.82
Paper making	38.37	20.55
Petroleum refining	129.20	28.69

changed in this century as compared to the last. Specifically, while the rate of invention has decreased in the farm equipment, paper, and railroad industries, it has increased in the case of the petroleum industry. Paper in its modern (wood-based) form is almost as new as petroleum. It is therefore somewhat surprising to find that the rate of invention in paper making has declined in contrast with the case of the petroleum-refining industry. However, as explained below, this is attributable to differences in the "technological base" of the two industries.

One final objective of this section is to determine whether there exist significant differences in the rate of invention in the various industries. This is investigated by means of the following test criterion (Cox and Lewis, 1966, pp. 235–236). Suppose that in fixed time periods $t_0^{(1)}, \ldots, t_0^{(k)}$, the number of events observed in k independent Poisson processes are $n_1, \ldots, n_k$ where n_i is the observed value of a Poisson variable, N_i, of mean μ_i. The test criterion to be calculated is

$$H = 2 \left[\sum_{i=1}^{k} n_i \log \left(\frac{n_i}{t_0^{(i)}} \right) - n \log \left(\frac{n}{t_0} \right) \right]. \qquad (3.16)$$

For the special case of equal $t_0^{(i)}$, the function H reduces to

$$H = 2 \left(\sum_{i=1}^{k} n_i \log n_i - n \log \bar{n} \right). \qquad (3.17)$$

where $n = \bar{n}/k$ is the mean number of events per series. Under the null hypothesis that $\mu_1 = \ldots = \mu_k = \mu$, H has a χ^2 distribution with $k-1$ degrees of freedom.

Table 3.10 presents the calculations in a summary form of the test statistic H for various cases. It can be seen that there exist significant differences in the

Table 3.10:   Significance of Spatial Variation in the Mean Rate of Inventions

Case	Industries	Period	*H*
1	Paper, petroleum, farm equipment, railroad	1800–1957	51.28
2	Paper, petroleum, farm equipment, railroad	1900–1957	306.79
3	Railroad, farm equipment	1800–1957	0.65
4	Paper, petroleum	1800–1957	50.05
5	Inventions and innovations in the farm equipment industry	1902–1948	6.98

mean rate of invention in various industries (case 1). Further, these differences have increased in this century in comparison with the last (case 2). It seems that the observed variations in the rate of invention across various industries are attributable to the corresponding differences in their technological and institutional characteristics. To begin with, whereas both the farm equipment and railroad industries are engineering oriented (group 1), the paper-making and petroleum-refining industries are science oriented (group 2). Moreover, while the industries in group 2 are highly vertically integrated, those in group 1 are not. Finally, the industries in group 1 share a common pool of *technical* know-how (in that both draw upon advances in mechanical engineering), whereas industries in group 2 share a common pool of *scientific* knowledge (in that both draw upon advances in chemistry) but lack a common technological underpinning. According to these considerations, a priori, the rate of invention in each of the two industries in group 1 is not expected to be significantly different from the other because the industries are alike in virtually every respect under consideration. The case of the two industries in group 2 is somewhat complicated because they are similar to each other in terms of their institutional background and their stock of scientific knowledge but not in terms of their technological characteristics. Accordingly, the rate of invention in each of these two industries is expected to be significantly different from the other if it is mainly dependent on technological factors rather than on purely scientific knowledge or institutional setup. The inference drawn from these a priori considerations is readily borne out by the results from the empirical analysis. Thus the mean rate of invention in the farm equipment industry is found not to be significantly different from that of invention in the railroad industry at the 1% level (case 3). However, significant differences exist in the rate of invention in the paper-making and petroleum-refining industries (case 4). In conclusion, it may be said that the inventive performance of an industry is determined mainly by the nature of its technology.

These findings are corroborated by the results of other unrelated investigations in this area. Thus, in a study of 448 of the largest firms in the United States, Scherer (1965) found that one of the most important causes of observed interindustry variations in patented inventions was differences in "technological opportunity," which he defined as being ultimately dependent on the broad advance of technical knowledge. In essence, the inventive output of the so-called high-technology industries, such as electrical equipment and chemicals, tends to be higher than that in the so-called traditional industries, such as food products and textiles, regardless of such other factors as the size and structure of the market.

Admittedly, however, while the foregoing conclusion is borne out by the results of other empirical analyses, its theoretical basis is somewhat unclear. One main difficulty seems to be that a precise formulation of the observed relationships between "potential for innovation" and the "nature of technology" (or the "technological opportunity") is generally lacking. This is an important problem in its own right and it is discussed at length in Chapter 9. Suffice here to note that the present conclusion is supported by the subsequent in-depth treatment of the basic issues involved.

Finally, the results indicate that the rate of invention is significantly different from the rate of innovation in the farm equipment (case 5). We have already presented the evidence that the observed distributions of inventions and innovations can be described independently of one another. In conclusion, there may be no *direct* relationship between the twin processes of invention and innovation.

5. TOWARD A PRINCIPLE OF TECHNOLOGICAL INSULARITY

Two noteworthy and somewhat related aspects of innovative activity emerge from the analysis reported in the preceding sections. First, there is comparatively little evidence of a technical relationship between the occurrence of successive innovations. On balance, the advent of one innovation does not seem to hasten or delay the advent of others. Second, innovations occur in clusters despite the lack of technical correlation in the origin of different techniques. That is, the occurrence of innovations is largely dictated by pure chance. However, the rate of occurrence varies both temporally and spatially.

There are two standard explanations of why innovations tend to cluster. The first one takes the form of the popular view that innovation is a process of chain reaction: technical triumph in one area paves the way for a quick succession of innovations in other closely related areas. It is therefore to be expected that the distribution of innovations across different fields is biased. There is the well-known saying that necessity is the mother of invention. According to the viewpoint under consideration, the converse is equally true: invention is the mother of necessity. Some such considerations point to the so-called technology push hypothesis. In its essence, technology creates its own needs. There is very little by way of either logic or evidence to support this hypothesis insofar as it implies that one innovation is a direct cause of the other. It is refuted by a number of detailed investigations of inventive activity (Schmookler, 1966, pp. 57 ff; Myers and Marquis, 1969). The results of the present investigation are consistent with the evidence found by these earlier studies based on a different premise. Whatever be the reasons for the clustering of innovations, the hypothesis of technological vitalism implied by the popular view of innovative activity is obviously untenable.

The second explanation of the clustering phenomenon is a somewhat sophisticated variant of the first explanation just noted. Basically, it states that a *major* innovation acts as an inducement for a series of *minor* innovations (Schumpeter, 1939). As Kuznets has put it (1972, p. 449),

> Technological change cannot be viewed as minor steps that are grouped randomly: they come in clusters, with a variety of complementary inventions and improvements grouped around a central major innovation.

Whatever be its merits, the above explanation must be ruled out in the present context. This is because the case studies reported here specifically exclude minor innovations. According to the results of the present investigation, major innovations themselves tend to come in clusters. We must therefore account for the observed phenomenon in some other way.

The alternative explanation of the clustering phenomenon turns out to be shockingly simple once the results are considered in their totality rather than on a piecemeal basis. There is, in fact, a complementarity between the two seemingly disparate features of technical change: Innovations tend to cluster because of—not in spite of—the lack of correlation between the origins of different techniques. If there were no barriers to the diffusion and intermingling of know-how acquired from the development of different techniques, we would expect the innovations eventually to be spread in a uniform fashion across different fields and time periods. Evidently, innovations tend to cluster when the transmission of technical know-how is stymied both spatially and temporally.

The conclusion that emerges from these considerations is that the know-how acquired in the development of one technique may not be wholly transferable to the development of another technique. An essentially similar inference is brought out by the results of further empirical investigations (see

Chapter 7). Together these findings suggest what may be called a principle of technological insularity. In its essence, a characteristic feature of technical know-how is the lack of its transmissibility. The reasons for this are discussed at length elsewhere in this book (see Chapters 6 and 9). Suffice here to note that unlike pure scientific knowledge, which is equally available to all, technical know-how is largely product and plant specific. The transfer of technology is therefore inherently a ticklish and toilsome process.

In economic theory it is customary to assume that technology is a public good. It is inappropriable and at the disposal of everyone at a nominal cost. One firm's gain of a technology does not imply its loss to the others. Rather, once a technical discovery is made, it becomes freely available to all producers in the industry. If the proposed principle is any guide, however, none of this holds water. Far from being exposed to view, much technical know-how tends to be concealed. Moreover, its efficacy is often circumscribed to a single environment. This is not to deny that an innovation may turn out to be useful in areas other than its initial application. Obviously, there are such "spinoffs" in the course of R&D activity. Moreover, the genesis of many important new techniques is demonstrably attributable to the dissemination of technical know-how across firms and industries (e.g., through the movement of trained personnel). The important point, however, is that the flow of technology across system boundaries is littered with obstacles. It is inevitably subject to various delays in time. The transfer of technology is therefore never a toll-free affair.

The proposed viewpoint is further supported by the results from a wide variety of other investigations in related areas. First, a number of empirical studies of economic development reveal that there are significant costs involved in the *search* for appropriate technology. As Cooper et al. put it on the basis of a comparative study of the choice of techniques for can making in Kenya, Tanzania, and Thailand (1975, p. 107), "The . . . concept of availability of different techniques needs to be more clearly formulated. What does 'availability' mean—in particular, what does it mean as far as private decision maker is concerned? In the first place, it seems clear that there are costs involved in establishing what is available. The curious notion that what is available is in some way immediately known to everyone in the industry, or can be discovered at no cost, is not valid." The same point is also brought out by studies of technological decisions in an industrialized setting (see, e.g., Magee, 1977). Very generally, there is a large element of latency in the knowledge acquired from the development of new techniques. The relevant form of technical know-how is therefore seldom available on tap. Rather, it must be systematically groped for.

Second, there are significant costs involved in the *adaptation* of the chosen technology to the conditions of its use. The importance of this issue may be gauged from the fact that there exist significant costs of technology transfer in relation to total project costs even within multinational firms where one would expect considerably fewer problems of appropriability and other obstacles to the flow of technical information. Teece estimates the average transfer costs to be some 19% of the total project cost of the recipient on the basis of one such study and thus concludes (1977, p. 247), "Clearly, the data do not support the notion that technology is a stock of blueprints usable at nominal cost to all."

Finally, in-depth studies of R&D activity reveal that technological spin-offs tend to be relatively few and are often subject to long time lags. The experience of NASA in the transfer of technology to nonspace use is a case in point (Welles, 1974). Moreover, contrary to the popular assumption, there is seldom any assurance of success in technology transfer. The nearly universal failure of the various attempts to develop a completely automated carding and drawing sequence for use in the textile industry is one among many examples of this (Catling, 1978). The essence of these specific illustrations is that transfer of technology generally occurs in a *partial* and *gradual* way.

This brings us to the conclusion. The conventional viewpoint in economic theory has it that technology is a piece of hardware or a method of organization which is freely and equally available to all. It is a common property and its transfer from one firm to the other a freewheeling process. In reality, however, the gist of technical know-how is often bottled up in the system of its origin. Moreover, it is deliberately held in a cryptic form so as to avoid its disclosure. Technology transfer is largely a game of hide and seek. Beyond that it is a matter of painstaking attunement of the chosen technique to the new environment of its use. Contrary to the standard assumption of economic theory, there is no free lunch in the course of technology transfer.

6.   PRINCIPAL CONCLUSIONS

This chapter has presented an attempt to examine the mechanism underlying the origin of new techniques. On the basis of some very simple observations concerning the development of new products and processes, the negative binomial distribution is proposed as an appropriate model of innovative activity. The model is applied to various cases of technical change in the farm equipment, railroad, paper-making, and petroleum-refining industries.

In interpreting the results of these case studies, it should be noted that the negative binomial model applies to a wide variety of empirical distributions. They include the distribution of (1) plants and species, (2) many types of industrial accidents, and (3) consumer purchases. The present investigation reveals that the (final form of the) negative binomial model also applies to the data on the origin of new techniques. The fact that one and the same model applies to a multitude of unrelated phenomena has deeper significance. It implies that the determinants of new products and processes tend to be very many and the effect of each rather small. In consequence, inventions often seem to be a matter of sheer chance.

These results are clearly at variance with the popular belief that an invention depends on the initiative taken by one or two progressive individuals or firms. They are nevertheless consistent with the evidence on "duplicate" inventions (Ogburn and Thomas, 1922). As an example, at least five independent attempts are known to have resulted in the invention of the same form of sail for ships. As another example, at least three independently successful attempts in quick succession underlie the invention of the screw propeller (Gilfillan, 1935a, p. 62, 134). Indeed, there are hundreds of such instances of duplicate inventions. According to the viewpoint advanced here, it is to be expected that an invention have several independent origins.

This is not to say that the development of new products and processes lacks the characteristics of a well-organized system. Rather, uncertainty is the hallmark of innovative activity. Its resolution is a matter of learning over time. The scene of technical progress includes elements of both variety and structure. Thus the results of our investigation indicate that there exists a well-behaved pattern in the origin of new techniques notwithstanding the apparently accidental occurrence of individual innovations. Here, as in so many other instances, there is a meeting ground between chance and law.

Against the background of shrinking economic growth in recent years, it is often asked whether the pace of technical advances is beginning to level off. Is there a technical stagnation in the offing? According to the results of the present investigation, the rate of invention in the farm equipment, railroad, and paper-making industries has significantly declined in the recent years. In contrast, the rate of invention in the petroleum-refining industry has significantly increased during the same period. Thus, it seems that the slowdown in the pace of technical progress is partly real. For the remainder, it is simply a reflection of change in the direction of inventive activity.[5]

The observed slack in technical progress may be attributed to the fact that the variety in the basic forms of techniques often tends to be limited. A detailed discussion of this point is deferred until Chapter 4. Suffice here to note that the fixity of form is an important barrier to the evolution of artifacts. The policy implication of these results is that the R&D expenditure required to develop major new techniques will have to be considerably more in the future than in the past in order to maintain the same pace of technical advances.

The results of the present investigation also indicate that the role of pure science in the process of innovation may be far less important than is common-

---

[5]For a similar viewpoint see Stafford (1952).

ly believed. The reasons for this are not difficult to explain (see also Chapters 2 and 9). The gist of science lies in indicating what is not possible. Thanks to thermodynamics, we are assured that a *perpetuum mobile* cannot conceivably exist. It has little bearing on the actual design of a steam engine. Scientific activity is largely concerned with *why* an object is what it is. Technical activity, on the other hand, is mainly occupied with discovering *how* an object can be built. Although the two are by no means antithetical, there is a fundamental difference in their character. It is therefore to be expected that scientific progress has comparatively little influence on advances in technology. The conclusion to be drawn is that the evolution of science and technology occurs within a single system. It is nevertheless a relatively decomposable system.

One fundamental issue remains: Is the very notion of a technology policy meaningful? Is there any reason to believe that a conscious formulation of technological decisions can be effective? Here a reference must be made to certain influential theories put foward by Schumpeter (1934, 1939) and Gilfillan (1935a, b). As noted in Chapter 2, the essence of Schumpeter's viewpoint lies in its distinction between the initial act of innovation by a few and its subsequent imitation by many. Only very few individuals are capable of undertaking new enterprises. Once someone has taken a successful initiative, however, many jump on the bandwagon. Thus, innovations seldom remain isolated events. They tend to cluster because "first some, and then most, firms follow in the wake of successful innovation" (Schumpeter, 1939, p. 100). Further, there is no direct relationship between invention and innovation. In the author's own words (Schumpeter, 1939, p. 84):

> Innovation is possible without anything we should identify as invention, and invention does not necessarily induce innovation, but produces of itself . . . no economically relevant effect at all.

In sum, each of the two processes of invention and innovation has a dynamics of its own.

While Schumpeter emphasizes the role of entrepreneurial genius in the process of technical change, Gilfillan is of the view that an individual act of invention is strictly subservient to the historical process. As the author himself puts it (Gilfillan, 1935b, p. 10),

> With the progress of the craft of invention, apparently a device can no longer remain unfound when the time for it is ripe . . . . Hence simultaneous *duplicate invention* often occurs . . . . There is no indication that any individual's genius has been necessary to any invention that has had any importance.

However, both Schumpeter and Gilfillan leave out a critical consideration: The apparently autonomous nature of technical change may well be due to the interplay of history *and* chance, rather than to the individual genius or historical necessity. While Schumpeter's conclusions are vindicated, the premise for his conclusions seems unfounded. On the other hand, while Gilfillan's evidence on duplicate invention is corroborated, his generalization of that evidence seems unwarranted. If the results of the present investigation are any guide, the question as to the effectiveness of a technology policy is one of degree rather than kind. The relevance of a technology policy seems beyond dispute.

Chapter 4

A GENERAL THEORY OF THE EVOLUTION OF ARTIFACTS

1. THE SELF-ORGANIZING NATURE OF EVOLUTIONARY PROCESSES

One identifying feature of evolutionary processes is that they involve qualitative changes of an irreducible nature. Typically, they impinge upon the complexity and other syntactical properties of the whole. That is, evolution is not just a matter of chop and change; rather, it pertains to the very structure and function of the object. As an obvious example, the growth of a biological organism eventually requires the development of respiratory and digestive functions. Likewise, the growth of a business firm beyond a small size requires new and specialized channels of communication both within itself and with the outside world. There are hundreds of similar examples. In essence, evolution is not merely aggregation. Evolutionary changes do not just pile up. They inevitably build up into a system.

Evolution determines the properties of a system. Conversely, the very existence of a system is a factor in governing the course of evolution. A striking example of this phenomenon is provided by the principle of technological guideposts outlined in Chapter 2. Thus it may be recalled that technical change often leads to a certain pattern of machine design that in turn sets the boundaries of further R&D activity. What is striking, however, is that the mutual causal relationship between evolution and system formation is a rule rather than an exception (Sahal, 1978). For instance, the growth of a plant is generally accompanied by hardening of its tissues so as to ensure adequate strength in relation to its increased size. In consequence, further growth takes place in only a few zones (the so-called meristems); the hardened tissues cannot themselves grow. Indeed, cell division in many plants takes place only in the

Devendra Sahal, Patterns of Technological Innovation ISBN 0-201-06630-0

apical meristems, that is, at the tip of the branches. Although no substantive analogy is implied, it seems apparent that there is a common principle at work here.

The question that remains to be answered is why this principle is what it is. It may be recalled that the principle of technological guideposts was found to hold in a variety of innovative activities. We will presently be concerned with determining what accounts for its ubiquity, and what if any significance there is to the fact that the same principle applies to many different types of evolutionary processes. In what circumstances might it not hold?

2.   SCALE OF TECHNOLOGY AND LIMITS TO GROWTH

The starting point of the theory developed here is the well-known observation that change in the size of a system generally necessitates change in the characteristics of its form and structure (Bonner, 1952; Thompson, 1917). If geometric similarity is preserved with an increase in the size of a system, its area increases as the square and its volume as the cube of its linear dimensions. Thus, if the characteristic length of a system is doubled, its area increases by four times and its volume by eight times. However, from a functional point of view, no system can afford to have such a preponderance of volume in relation to surface area. As an example, heat production in a system varies in relation to the system's volume, whereas heat dissipation varies in relation to its surface area. If the volume is greatly in excess of the surface area, the system acquires a greatly enhanced capacity for the generation of heat without comparable means for its dissipation. Thus, *ceteris paribus,* a system cannot remain unchanged both functionally and geometrically with change in its size. Rather, it must come to terms with the preponderance of its volume by selectively increasing the linear and areal dimensions of its subsystems. Put another way, the growth of a system generally involves a differential growth of its components (Huxley, 1932; Gould, 1966). That is, the parts and the whole of a system do not grow at the same rate. The rates of growth of different parts of the system in turn differ from each other. In all, it is to be expected that the growth of an object is generally characterized by a change in its form.

We therefore find that large objects often tend to be more elongated and convoluted than small objects so as to preserve constant area-to-volume ratios. Witness the fact that there is a basic difference between the roughly cubical form of a house and the relatively tapered form of a skyscraper. Thus it has been observed that if the Empire State Building were as slender as a wheat plant, it would be only about 1.8 meters wide at the base (Went, 1968, p. 411). Likewise, it is generally true that the larger the leaf of a plant, the longer it tends to be. Hence, we have the contrast between the large elongated leaf of the banana and the small rounded leaf of the berry. Similarly, the basins of large rivers tend to be more crenulated. In an essentially similar way, as a tree grows older, the ratio of its height to the trunk diameter generally declines. So, also, the geometric proportions of a little chick necessarily differ from those of a grown-up hen. Often, a system cannot adequately function beyond a certain size without a change in its shape.

Change in the scale of artifacts likewise necessitates change in their forms. As an example of this, there is a change in the shape of the large four-wheel-drive tractor such that the driver is positioned farther forward and the steering is done by articulation at the chassis. It has been proposed that the changed form of the machine should become a standard basis for the design of large tractors (Baker, 1970, p. 391). Similarly, increase in the size of passenger ships has been accompanied by a disproportionate increase in their beam length in relation to total length over a period of more than a century (see Chapter 9). Analogous departures from geometric similarity arise as the size of the system is reduced. Thus, in simulating the behavior of a dam on a small scale, it is mandatory to make the model much deeper than the actual river. This is because at small depths, the effects of viscosity and surface tension change the nature of the water flow to such an extent that functional similarity can be preserved only by means of systematic departures from geometric similarity.

Alternatively, change in the size of a system compels a change in the material employed in its construction. If two objects are built of the same material and their form is the same, then the one that is larger is obviously the weaker. Thus it is often necessary to use stronger materials in order to overcome the effect of increase in scale. For instance, in the previously cited example of simulating the behavior of a dam on a small scale, it is often necessary to use a material of much higher density, such as mercury, instead of water to obtain reliable results. The other way to circumvent the problem posed by increase in size is to incorporate inactive organic matter such as wood in case of higher plants. In the circumstances, while the bulk increases as the cube of linear dimensions, the increase in the "active" part of it remains well within tolerable limits. In general, a system cannot survive beyond a certain size without a change in its material characteristics. Thus large machines require materials of comparatively higher strength-to-weight ratios. Similarly, special alloys are almost always needed to ensure sufficient penetration of heat when manufacturing large steel parts. Such a need does not arise for smaller parts.

Finally, changes in the scale of a system often necessitate the complication of its structure. Thus increased differentiation of functions, such as the development of respiratory mechanisms in larger organisms, is to be expected because the quantity of the respiring tissues varies as the cube, whereas the surface of gas exchange varies as the square, of linear dimensions. Smaller organisms, however, do not require gills or lungs, for gas exchange can take place sufficiently rapidly for metabolism by means of diffusion alone. In an analogous fashion, large plants branch to a proportionately greater extent than small plants so as to keep their surface area at par with their considerably increased volume.

Similarly, large engines require duplicate ignition arrangements to prevent thermal cracking of cylinder heads and pistons due to increase in their size. Such provisions are unnecessary in small engines. Alternatively, large engines require more cylinders so as to avoid the combustion problems that become increasingly acute with increase in their size. In turn, the use of multiple cylinders frequently necessitates the use of multiple carburetors or of a multiple-barrel carburetor. The use of multiple cylinders also requires intricate layouts, such as the vee arrangement, in order to restrict the length of the crankshaft and overall length of the engine. These complications are an inevitable consequence of increase in size.

Frequently, scale-induced complications of technical structure feed on themselves through their implications for man–machine interactions. An example of this is the problem of intolerable noise created by the increase in the size of the farm tractor over the course of time. In turn, this drawback has led to numerous experiments aimed at increased differentiation of the machine, such that the engine may be placed on a flexible mounting suspended from the cab and chassis. There are, of course, many more examples of this phenomenon. The essential point is that change in the scale of technology is often accompanied by change in the division of labor among its components.

In summary, evolution of a system generally involves three processes: (1) disproportionate growth of its subsystems: (2) change in the material of its construction; and (3) increase in the complexity of its structure. However, these processes cannot continue ad infinitum without degenerating into functional inconsistencies. Thus, there exist limits to the growth of every system. As an example, the wings of a large bird must be proportionately bigger than those of a small bird so as to preserve a constant surface-area-to-volume ratio. However, such disproportionate increase of wing area cannot continue indefinitely without giving rise to absurdities. Indeed, the terminal form of large birds is known to be flightless. Similarly, the relatively tapered form of an apartment tower cannot continue to grow indefinitely because beyond a certain height the elevator space is bound to occupy a prohibitively large part of the total space available. The essential point of these examples is that the very processes that make the evolution of the system possible in initial stages eventually set the limits to its further evolution.

Some such considerations led Galileo to conclude that the conventional form of trees could not exceed a height considerably above 100 meters (Galileo, 1914). It is interesting to note that this height is very nearly the same as that of the *Sequoia* and *Eucalyptus,* the largest known trees. Similarly, there is a very definite limit to the size of man-made structures. Thus it has been observed that the tower of Babel could not have been as high as it is depicted (Nordbeck, 1965, p. 22). Likewise, the growth of a single power plant is limited by the augmented risk of breakdown as a consequence of increase in complexity. On the other side of the coin, size of a system cannot be indefinitely reduced. Thus there is a definite limit to which machines can be made lighter because of the disproportionately large (and often unpredictable) increase in vibrations. Similarly, miniaturization of electronic devices is ultimately limited by the problem of heat dissipation, which becomes increasingly acute as the size of the system is reduced.

Frequently, the evolution of artifacts is retarded much sooner than indicated by any ultimate limit to their growth. The reasons for this retardation are seemingly economic but their origin lies in the physical processes discussed above. As is well known, in certain circumstances the capital cost of an apparatus or a plant depends on its surface area, whereas production capacity is a function of the volume. Hence economies of scale are sometimes described in terms of the "six-tenths factor rule": Investment costs vary as the two-thirds power of capacity. The rule is found to hold in a wide variety of industries, particularly those employing continuous production process such as the chemical, petroleum, and cement industries (Bruni, 1964). A detailed discussion of this topic would take us too far afield. For the present purpose the salient point is that so long as the form and other characteristics of an object do not radically change with increase in its size, there must exist *internal* economies of scale as a matter of simple geometric necessity. Moreover, such economies are expected to hold with respect to both initial and operating costs. For example, the heat dispersion per unit of output of a big furnace is obviously less than that of a small furnace. In all, the trend toward the use of large-scale machinery in certain industries is to be expected.

By the same token, it is easy to see that the economies of scale under consideration cannot continue to exist forever. They are likely to be exhausted beyond a size at which the system cannot remain functional without *major* changes in its material, morphological, and structural characteristics. For example, as the size of a rotating machine is increased beyond a certain point, the stresses and strains increase disproportionately with its capacity. In consequence, it may be necessary to use more expensive material in the construction of blades and to employ special insulating devices or cooling methods. Often, benefits from increased size are outweighed by these increased costs of production. Further, a large increase in the scale of a system affects its operating costs as well. Thus the larger an electricity-generating unit at a given speed, the lower is its inherent stability during system disturbances. In consequence, it may be necessary to invest more in transmission facilities, or to employ such special provisions as dynamic braking and a high-response supplemental excitation scheme.

The economic considerations just outlined hold equally well as the size of an object is progressively reduced. For example, it is generally true that the lighter a reciprocating engine, the higher its crankshaft speed and its compression ratio are. The increased speed makes possible a disproportionate increase in the horsepower-to-weight ratio of the engine. However, the higher compression ratio generates more heat, thereby requiring more durable and expensive valve, piston, and ring alloys. Evidently, as the scale of a technology is continually changed, there inevitably arises a stage at which the increased manufacturing and/or operating costs limit further advances in its capability.

It has been observed that the development of a technology is eventually retarded because in the later stages of the product cycle concern about production costs begins to dominate concern about product characteristics (Hirsch, 1965; Vernon, 1966). According to the considerations advanced here, this is to be expected simply as a consequence of scale function relationships of a physical nature.

Several important propositions emerge from the considerations advanced above. First, the dynamics of a system is shaped by its history and size.[1] Second, evolutionary processes tend to be both self-generating and self-constraining. Seen in this light, the ubiquity of the proposed principle of technological guideposts is to be expected. Evidently, the basis of the proposed principle lies in the nature of space. Third, there exist upper and lower limits to change in the size of every system. Put another way, every system of a given form is characterized by limits to its growth. In all, evolution involves a process of equilibration governed by the internal dynamics of the object system.

3.   A PRINCIPLE OF CREATIVE SYMBIOSIS

The foregoing discussion of limits to growth must be qualified in one important respect. It is that evolution is a multidimensional process and its retardation in one dimension does not necessarily thwart it in other dimensions. Rather, the slowdown of evolution in one respect is often a spur to the building up of its potential in some other respect. It is a characteristic feature of long-term evolution that barriers to growth frequently prove to be temporary. In many instances, what would seem to be an insurmountable upper limit of

---

[1]Dobzhansky (1950) and Bonner (1952) have reached essentially similar conclusions in their studies of biological systems. As the former puts it (p. 160), "The structure of a gene is a distillate of its history. . . ." To add to this, we have the informed statement of Bonner (p. 15): "Size alone . . . is an aspect of adaptation, a vital cog in the wheel of evolution."

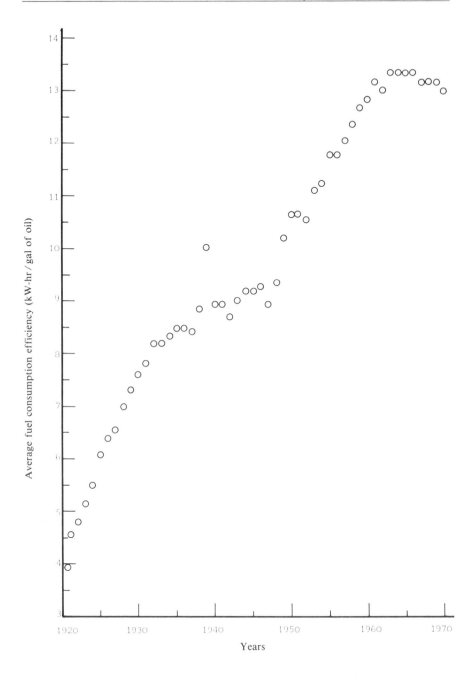

Fig. 4.1 Evolution of the fuel-consumption efficiency of fossil-fuel electric power plants, 1920–1970.

growth turns out in reality to be a stepping-stone to further development. Thus, very often, the pattern of evolution evinces a series of S-shaped curves. An example of this pattern is shown in Figure 4.1 for the case of growth in the fuel-consumption efficiency of fossil-fuel electric power plants over the course of time.

In general, it is evident that there tends to be more than one limit to the evolution of man-made devices. This is to be expected in view of the multiplicity of impediments involved. Thus certain limits to evolution originate in the limited possibilities of change in the material of construction. Others originate in the fixity of form. Yet other limits of evolution have to do with complexity of structure. There may well be a bottleneck in one area. Sooner or later, however, new possibilities present themselves in other areas. Thus, stagnation of a technology seldom lasts forever. But this is not all. As will become apparent in a moment, there is much more profundity to the disequilibration of technical progress.

In the biological world, once two lineages have been divergent for a certain time, they cannot normally rejoin. In the world of engineering, the opposite is generally true. It is commonplace to combine two technologies in the form of a single device. Examples of this are everywhere, ranging from nuclear-powered ships (e.g., combining the pressurized-water reactor technology and traditional marine engineering) to numerical control systems (e.g., combining solid-state electronic controls with modular circuit construction and a vertical milling machine). Indeed, it will be difficult to find cases to the contrary. Upon close examination of this mass of evidence, one class of examples stands out. It is illustrative of what may be called a principle of creative symbiosis: In certain instances, an integration of two technologies simplifies the outline of the overall structure, thereby circumventing the limits to its evolution. As discussed earlier, the growth of an individual system is often limited by its complexity. The point here is that symbiotic relationships between two or more systems streamline their overall outline such that the limits to their evolution are infallibly overcome. That is, the evolution of a system is subject to limits only insofar as it remains an isolated system. The potential for evolution of a coalition of systems is virtually unlimited. This would seem to be true of all human affairs. But speaking of artifacts, it may be noted that not all technological combinations imply a marriage of convenience. Certain cases of technological intermingling originate in common cause. In such cases, the system knows no limits to its evolution.[2] The following examples will help make this clear.

---

[2]This should not be taken to imply that the phenomenon of creative symbiosis lies outside the evolutionary process. Rather, it is a manifestation of evolution at a metalevel.

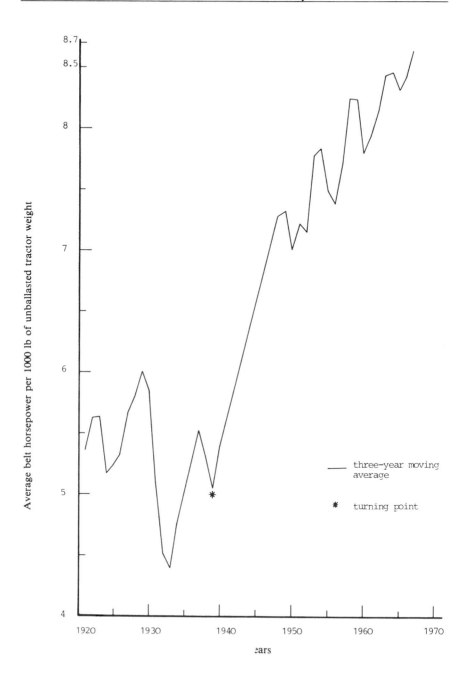

Fig. 4.2 Evolution of the horsepower-to-weight ratio of the farm tractor, 1921–1967.

The first example considered here is that of the farm tractor. The development of tractor technology had reached a dead end by 1940. The state of tractor design had matured, thereby making any further improvements unlikely. But then a sudden change took place, occasioned by the increased adoption of the three-point hitch for control of farm implements that had been developed a few years earlier. In consequence, the limit to the evolution of technology was effectively eliminated. This is illustrated in Figure 4.2 for the case of advances in the horsepower-to-weight ratio of the tractor over the course of time. The state of tractor design during the 1970s represents yet another instance of technological stagnation. It is generally agreed that considerable further progress in this area is nevertheless possible, depending on further understanding of the machine–soil relationship. Thus, in both instances, the limit to technological evolution appears to have been governed by the dynamics of successively larger systems: the tractor–implement system and the tractor–implement–soil system, respectively. In conclusion, while the short-term evolution is attributable to advances in tractor technology itself, the long-term evolution has been due primarily to the design of a more comprehensive system of tractor–implement technology.

As a second example, the earliest technique of harvesting crops essentially consisted of reaping them with a hand sickle. The next major step in the evolution of harvesting technology was undoubtedly the development of the cradle (which made it possible to integrate the operations involved in the cutting and collecting of crops) rather than improvement of the sickle itself. Subsequent advances came with the development of horse-drawn reapers in the 1830s, and these were widely regarded to have culminated in the development of wire and twine binders in the 1880s. Further development of harvesting technology was, nevertheless, made possible by the advent of the combine, which enabled the integration of harvesting with threshing techniques. In summary, long-term evolution of harvesting technology is attributable to the formation of increasingly comprehensive systems: the integrated system of operations within the process of harvesting and the integrated system of harvesting and threshing technology, respectively.

The evolution of weaving techniques in the textile industry provides another apt illustration of the principle proposed here. Initial advances such as development of the draw loom, flying shuttle, and mechanical loom were largely restricted in scope to the improvement of weaving per se. By the turn of the century, the technique had been considerably refined with apparently little scope for additional improvement. With the development of the ring spinning machine and automatic loom, however, it became possible to integrate the process of weaving with that of spinning. In this manner considerable technical progress was made possible. The inference to be drawn is that whereas the short-term evolution of technology was due to changes within the system of weaving, the long-term evolution was made possible by forming an integrated system of spinning and weaving.

A final illustration is from the field of architecture. It refers to the evolution of the Mousgoum hut as, for example, built by African tribesmen in the northern part of Cameroon (Alexander, 1964, pp. 30 ff; Denyer, 1978, pp. 92 ff). The hemispherical form of the hut provides the most efficient shape for heat transfer and keeps the inside fairly cool. The hut is supported by a series of vertical ribs that not only help sustain the main fabric, but also act as channels for rainwater while allowing the inhabitant access to the upper part of the outside during its construction and maintenance. Due to scarcity of wood, disposable scaffoldings are not used. Rather, the scaffoldings are built in as part of the structure. Thus it is apparent that the pattern of building design is governed by the salient variables of the object system: the peculiarities of the surrounding environment, the need for maintenance, the resources and skills available, etc. Indeed, together they form a coherent whole. But what is most striking is that the process of design has evolved through a cumulative mechanism of corrections, without the benefits of explicit principles of structural engineering and architecture. The long-term evolution of the design process can, however, be readily accounted for in terms of certain characteristics of the larger sociohistorical system. To an African tribesman, design of a hut is never an isolated activity. Specifically, a hut cannot be built *de novo*. Rather, its design and construction must be in keeping with the society's myths and traditions. Thus, without these constraints of the larger sociohistorical system, failures could have a ripple effect, limiting further development or even forcing the design process to begin from square one. Thanks to the various social customs, however, the evolution of the design process continues to take place until it reaches a homeodynamic state. The lesson of this example should be clear. Short-term evolution generally takes place in the context of the object system. Long-term evolution, on the other hand, depends on the existence of a more comprehensive or an integrative system.

In summary, as noted in the preceding section, short-term evolution is governed by the dynamics of the object system. According to the consideration advanced here, we have the complementary proposition that long-term evolution is governed by the dynamics of the whole system.

4. PRINCIPAL CONCLUSIONS

This chapter has presented a theoretical view of evolution in artifacts. According to the viewpoint presented here, the evolution of a system generally involves the disproportionate growth of its subsystems and an increase in the complexity of its structure. However, these processes cannot take place forever without giving rise to preposterous situations. Thus, the very consequences of a system's evolution eventually circumscribe the possibilities of its further evolution. In this way we can readily account for the principle of technological guideposts proposed in Chapter 2. Very simply, the essence of any design process lies in the creation of a system. In turn, the system governs the scope of further design.

Nevertheless, the principle of technological guideposts is not as infallible as it might seem to be. According to the considerations advanced here, there exists another equally important principle of creative symbiosis: In certain cases, two or more technologies combine in an integrative fashion such that the outline of the overall system is greatly simplified. This eliminates the previous obstacles to further evolution. In consequence, each of the members constituting the "coalition of technologies" can continue to evolve. This does not always happen, but when it does, totally new possibilities for further evolution present themselves.

Together, the two principles discussed above point to a very general thesis of evolutionary processes. It may be thus stated:

1.   Short-term evolution is a process of equilibrium governed by the dynamics of the object system.
2.   Long-term evolution is a process of disequilibration governed by the dynamics of the whole system.

The considerations advanced in this chapter further suggest that there are two key factors which govern the dynamics of a system. One is its history. The other is its scale. An attempt to formalize the role of these factors in technical progress is deferred until later sections of this book (see especially Chapters 6 and 9). Suffice here to note that the scale of a system is not merely an inert dimension of the present, and its history is not just a passive chronicle of the past. Rather, the two together provide a vital context within which evolutionary processes take place.

Following the Marxian world view, it is sometimes argued that technical progress occurs in the direction of increasing complexity. It sounds intuitively appealing that easier problems are tackled first, the more complex ones afterward. The implication, of course, is that the cost of innovation increases over time. In the light of our earlier discussion, there is clearly an element of truth in this viewpoint. Moreover, it is corroborated by the empirical evidence pointing to the necessity of increased R&D expenditure in the future (see Chapter 3). It would nevertheless be wrong to push this inference too far. We do not find any support for its farfetched versions, such as the frequently heard warnings in recent years that there is a technological doomsday in the wings. According to the considerations advanced here, trends toward increasing complexity reflect only one aspect of reality. There is a corollary to the proposed principle of creative symbiosis: While the evolution of a subsystem may well proceed in the direction of increasing complexity, the evolution of the whole system proceeds in the direction of increasing simplicity. This is true not only of the evolution of technology but also of language changes (Rapoport, 1974, pp. 71-72). This is exemplified by the loss of inflections in many languages, as in the case of English, over the course of time. Moreover, some of these changes take place much more rapidly, as for example the substitution of English nouns for verbs (to "Xerox" a document, to "research" a topic). Thus it is apparent that, in certain circumstances, the evolution of artifacts is characterized by a trend toward simplification.

Finally, there are two major epistemological implications of the theory presented in this chapter. First, hitherto socioeconomic systems have been conceived in terms of exogenously governed homeostatic systems that return to a constant state after a temporary perturbation. That is, the processes of socioeconomic change inevitably reach a state of equilibrium. According to a viewpoint advanced here, however, an appropriate formulation of socioeconomic systems is in terms of endogenously governed homeodynamic systems capable of seeking new paths of evolution through successive instabilities. That is, processes of socioeconomic change remain in a constant state of disequilibration. To put it differently, the traditional viewpoint is solely concerned with the *mechanism* of change. In contrast, the focus of the proposed viewpoint is on the *organism* involved in the change.[3] Second, a cardinal feature of the contemporary approach to the modeling of socioeconomic processes is that their course is completely determined by the present or the expected values of the relevant variables. If the considerations presented here are any guide, socioeconomic changes are best regarded in terms of cumulative changes of a systemic nature.

---

[3]For an example of the traditional equilibrium-centered viewpoint see Arrow and Hahn (1971). For an example of the alternative viewpoint which is complementary to the one proposed here, see David (1975), Dunn (1971), and Nelson and Winter (1973).

Chapter 5

THE DIFFUSION OF TECHNOLOGY

1. THE BACKDROP OF PRIOR RESEARCH

The process of diffusion of technological innovation has been a subject of voluminous literature. The great majority of investigations in this area are based upon a single premise: The first use of a technology and its subsequent imitation constitute two entirely different types of activities. As noted in Chapters 2 and 3, the innovation–imitation distinction originates from the work of Josef Schumpeter (1939). To recapitulate briefly, although only a few entrepreneurs have the capability to set the ball rolling, there tend to be very many who can mimic the pioneering attempts of the few. More recently, this viewpoint has been formalized by Griliches (1957, 1960) and Mansfield (1968a,b), whose work on the diffusion process has exerted considerable influence. In its essence, a central characteristic of the process of diffusion is the "bandwagon" effect. Accordingly, it is postulated that an individual firm's decision to adopt a new technique is dependent on the number of firms already using it. It can be shown that under some very simple assumptions this hypothesis implies an S-shaped curve of diffusion over the course of time.

For example, suppose that the rate of diffusion of an innovation at any given point in time is proportional to the existing level of diffusion and to the gap between its instantaneous and ultimate levels of use. In symbols,

$$\frac{dN}{dt} = rN(K - N) \tag{5.1}$$

where N denotes some measure of diffusion level, K the saturation (or equilibrium) level, t the time, and r the constant of proportionality. The solution of this differential equation is the well-known logistic function (see Section 5.3).

Devendra Sahal, Patterns of Technological Innovation   ISBN 0-201-06630-0

$$N = \frac{K}{1 + ae^{-bt}}, \qquad b = rK \qquad (5.2)$$

where a is a constant depending on the initial conditions and b is the "rate-of-growth" parameter. It can be easily shown that the logistic curve is a *symmetrical* S-shaped curve. According to the available empirical evidence (Griliches, 1957, 1960; Mansfield, 1968a,b), certain cases of diffusion processes can be adequately described in terms of the logistic type of growth.

Frequently, however, the data on the diffusion of innovations contain an element of skewness. In the circumstances, the use of the logistic function is clearly inappropriate. Rather, we must employ some other *asymmetrical* S-shaped curve such as the one generated by the Gompertz or the cumulative log-normal distribution function (Bain, 1963; Chow, 1967). In general, the empirically observed diffusion curves exhibit a wide *variety* of S-shaped patterns of growth. Accordingly, in the literature on the subject, widely different models have been employed to generate the S-shaped trends. Examples of these are the logistic function, the Gompertz function, the modified exponential function, the cumulative normal distribution function, and the cumulative log-normal distribution function. Each of these models is based on a different set of assumptions concerning the nature of the process giving rise to the observed pattern of diffusion. It is not possible to discriminate between alternative models on empirical grounds. Often, the fit of widely different models to the data turns out to be equally good. The choice of a specific model cannot therefore be properly made except on the basis of an *a priori* set of criteria. However, the present state of our knowledge in this area is woefully inadequate to provide a *theoretical* basis for specifying the functional form of the diffusion process.

The problem just noted has to do with the basic orientation of research work in this area. To date, the focal point of a great many investigations has been factors governing the *rate* of the diffusion process (e.g., the determinants of the parameter b in Eq. (5.2)). A number of attempts have been made to account for observed variations in the speed with which innovations are adopted between different sizes of firms within a given industry, across different industries, etc. (see, e.g., Davies, 1980; Mansfield, 1968a,b; Nabseth and Ray, 1974; Weiss, 1971). In contrast, there are relatively few studies of the mechanism underlying the observed form of the diffusion process (Bain, 1963; Chow, 1967; Sahal, 1977a,b). Most investigations have focused on the dynamics of the imitation process after the initial introduction of an innovation. Are larger firms quicker to imitate than groups of smaller firms in the same situation? Does an increase in industrial concentration enhance the rate of diffusion? Attempts to study these questions have shed considerable light on the question of why some innovations are adopted more quickly than others. However, much less is known about why some innovations are adopted on a much larger scale than the others. That is, while the *pace* of the diffusion process has been extensively examined, there is little by way of an adequate treatment of the *extent* to which an innovation is adopted. The question most often discussed is, what accounts for the time lag between the first adoption of an innovation and its widespread use? The more basic issue, which remains largely unresolved, is, why should diffusion of technology continue in the first place? To put it bluntly, after some two decades of research in this area, we must still ask why the observed pattern of diffusion is what it is.

The Schumpeterian view of diffusion in terms of the distinction between innovation and imitation is the point of departure for this chapter. The alternative view is advanced here that the process of diffusion is intimately linked with the process of technological development. Accordingly, the diffusion of an innovation is best conceived in terms of actual substitution of a new technique for the old. The implication is that diffusion is a process of disequilibration involving the transition from one equilibrium level, corresponding to the adoption of an *existing* technique, to another equilibrium level, corresponding to the adoption of a *new* technique.[1] In what follows, the viewpoint is further developed and illustrated.

2.   THE VIEWPOINT OF TECHNOLOGICAL SUBSTITUTION

It has been observed that an established technique improves radically when confronted with the prospect of being supplanted by a new technique. For example, it is commonly thought that the steamship replaced the sailing ship soon after its adoption in the 1850s. In reality, however, early use of the steamship was beset by a number of problems both in the development of its key components—the iron hull, the engine, and the paddle wheel or the screw propeller—and in their adaptation to each other. Meanwhile the sailing ship was continuously improving. As competition with the steam technology intensified, the development of the sailing technique gained momentum. The capability of the sailing ship was dramatically improved by a number of innovations, such as the selective use of iron in the construction of the hull, modifications of sails and rigging (thereby reducing the crew required), and increase in the cargo-space-to-tonnage ratio. This upgrading enabled sailing ships to compete effectively with the steamship for more than three decades after 1850 (Graham, 1956). The steamship went through a series of gradual advances of its own resulting from various innovations such as the substitution of the screw

---

[1] Starting from a very different premise, Nelson et al. (1967, pp. 97–98) arrive at an essentially similar disequilibrium-centered view of technological diffusion. For the alternative equilibrium-centered viewpoint, see Salter (1969, p. 60) and David (1969).

propeller for the paddle wheel, the development of the compound engine, and the use of steel in place of iron (Gilfillan, 1935). In particular, the adoption of the high-pressure triple expansion engine by 1885 substantially reduced the fuel consumption of steamships while considerably increasing their service speed. This paved the way for the dominance of the steamship over the sailing ship.[2]

The diffusion of the Solvay process in the production of soda provides another classical example of the phenomenon noted above. Although the advantage of the Solvay process was already apparent in the 1860s, nearly three decades elapsed before it replaced the older Leblanc process because the latter continued to improve in the face of competition from the former (Freeman, 1974). The essential point of these examples is that the advent of a new technique does not in itself spell an end to the use of the old technique. Rather, the diffusion of an innovation is a process of coevolution of old and new techniques involving numerous changes in their functional characteristics.

Conversely, the adoption of an innovation may be thwarted for want of necessary changes in its design and performance. Thus the failure to reduce the level of vibrations and noise associated with the combination of propeller and gas turbine appears to have been the root cause of the success of the turbojet over the turboprop airliner in the 1940s. Evidently, advances in the capability of an innovation play a pivotal role in its adoption. This point is so obvious that it is often overlooked.

As a final example of technological coevolution, the main outlet in the early adoption of the farm tractor was plowing, i.e., primarily the operation for which the older techniques involving animal power were being used. Thus, tractors were originally classified according to the number of plows they could pull rather than according to their horsepower. The focus on substitution of mechanical for animal power had important consequences for the adoption of the farm tractor. On the one hand, the efforts to make a horse out of the tractor greatly restricted its utility, thereby limiting the pace and scope of its adoption. On the other hand, the favorable performance of the tractor in plowing was a spur to its further development and adoption, since plowing was one of the most energy-consuming operations in agriculture (Walker, 1929). The subsequent development of the general-purpose tractor made possible its adoption for a wider range of farm operations. Nevertheless, as late as 1940 the tractor was essentially a draft machine. Its development into primarily a chassis for mounting various harvesting implements was to take place later. The current use of the tractor for a wide variety of operations not only in

---

[2]The observed phenomenon of improvement in the old technique against the background of its possible displacement by a new technique is thus termed the "sailing ship effect" by Ward (1967) and is further discussed by Rosenberg (1976), among others.

agriculture but also in industry owes a great deal to numerous changes in the characteristics of its design.

The specific examples cited above are illustrative of two fundamental features of diffusion processes. First, an innovation seldom remains unchanged during the course of its adoption. The growth in the use of a technique often hinges upon improvement in its functional aspects. In many instances, changes in the design and performance characteristics of an innovation are a *prerequisite* to its adoption. In yet other instances, changes in an innovation make possible new applications, thereby *facilitating* its adoption beyond the originally conceived scope of its application. Thus, diffusion of a technology is a multidimensional process (Sahal, 1977b). One aspect of the process has to do with the increased use of a technique. The other aspect concerns changes in the physical composition of the technique itself.[3] More generally, the diffusion of a technology is characterized by changes of both a quantitative and qualitative nature. These two types of changes are closely interconnected. One cannot be understood without reference to the other.

Second, it is evident that the adoption of a new technique is often related to the nature and significance of some comparable older technique in use. In principle, it is possible that an innovation has no closely related substitute. The point is that when comparable techniques do exist, each generally tends to affect the character of the other. The diffusion of an innovation does not take place in isolation. Rather, it is very much a matter of actual substitution of a new technique for the old.

In the light of these considerations, it is easy to see why the diffusion of an innovation is expected to follow an S-shaped pattern of growth. As is commonly recognized, there is a significant process of "learning" involved in the *use of a given technique* of production. This is exemplified by the well-known concept of the progress function, that the cost of production systematically declines as cumulated output goes up (see Chapter 6 for a detailed discussion). More generally, it may be said that the *installation of a new technique* is also a matter of learning insofar as adoption of an innovation involves actual substitution of the new technology for the old. Indeed, a new device cannot be incorporated into an existing production setup at one swoop. Rather, its installation requires new work methods, upgrading of production skills, modifications in product design and plant layout, etc. Frequently, change in any one process has a ripple effect throughout the system of production. Thus, a new technique can only be introduced gradually.[4] Initially, it may be used only for limited objectives. Its rate of adoption on an industry-wide basis is therefore

---

[3]For a belated recognition of this point in the literature on the subject see Rogers (1978).

[4]For a somewhat similar viewpoint see Nelson et al. (1967, Chapter 5).

likely to be slow at first. Subsequently, it may become possible to employ the innovation for a variety of purposes as the relevant know-how for its use is acquired. Thus, past an initial stage, the diffusion of technology is expected to take place at a rapid rate. The use of the technique may eventually reach the saturation point as the opportunities for its application are exhausted.

The process of learning in technological change has many facets. First, as discussed earlier, learning determines the scope for *utilization* of a technique. A second and potentially more important aspect of the phenomenon is that learning is a key factor underlying the *development* of the technique in the first place (cf. Chapters 6 and 9). Although this aspect is often overlooked, the role of learning in the *evolution* of a technique has profound implications for its *diffusion* as well. It is a truism that virtually every technology is characterized by numerous teething problems during the initial stages of its development. However, as experience in the design and production of a technology is gradually accumulated, it becomes possible to overcome various hiccups in its performance. This is essentially a process of "innovations within the innovation" involving not only the producers but frequently the users of the technique as well. A detailed discussion of this process will be found in the following chapters. For the present, it will suffice to note that a technology can rarely perform in a flawless manner from its very birth. It is therefore to be expected that the rate of adoption of an innovation will be slow in the beginning. In due course, the rate of adoption of the innovation rapidly picks up as its performance is shown to be superior to the existing techniques. Finally, the growth in the use of a technology may be retarded insofar as there is a limit to improvement in its performance.

The above considerations suggest what may be called a "diffusion via learning" hypothesis of technological change: The diffusion of a technique is conditional upon learning in its adoption and development. Further, the diffusion curves of both the new and old techniques are likely to follow *some* S-shaped patterns of growth. Last but not least, these patterns themselves are expected to be interrelated.

3.   TWO MODELS OF TECHNOLOGICAL SUBSTITUTION

This section presents an attempt to operationalize the proposed view of the diffusion process in the form of two interrelated models of technological substitution. The first model focuses on the *temporal* aspects of the phenomenon. It was originally presented by Fisher and Pry (1971), who also gave a number of its applications. The work reported herein provides a number of alternative derivations of this model. These new derivations further clarify the basis of the Fisher–Pry model while accounting for its empirical success. The second model of technological substitution focuses on the *spatial* aspects of the phenomenon. It was originally developed by biologists in their study of morphological changes in organisms (see Reeve and Huxley, 1945, and the references therein). To the author's knowledge, the relevance of this model to the process of diffusion of innovations has never been demonstrated before. Finally, it is shown that these two formerly disparate models are in fact complementary to each other.

In symbols, let $X(t)$ be the extent of adoption of a new technique at time t and $Y(t)$ the extent of use of an existing old technique at the same time. Clearly, the fractional adoption of the new technique at time t is given by

$$f(t) = \frac{X(t)}{X(t) + Y(t)} \tag{5.3}$$

and that of the old technique by

$$f'(t) = \frac{Y(t)}{X(t) + Y(t)} \tag{5.4}$$

so that

$$f(t) + f'(t) = 1. \tag{5.5}$$

Following the considerations advanced above, both X and Y increase according to *some* S-shaped pattern of growth. One way to represent such a pattern formally is in terms of the differential equation of the well-known logistic function

$$\frac{1}{Y} \frac{dY}{dt} = \frac{b_1}{K_1} (K_1 - Y). \tag{5.6}$$

This equation can be rewritten as

$$\frac{K_1}{Y} \frac{1}{(K_1 - Y)} dY = b_1 \, dt \quad \text{or} \quad \frac{dY}{Y} - \frac{-dY}{(K_1 - Y)} = b_1 \, dt.$$

Upon integrating we obtain

$$\log Y - \log (K_1 - Y) = A + b_1 t$$

or

$$\log \frac{K_1 - Y}{Y} = a_1 - b_1 t$$

or

$$Y = \frac{K_1}{1 + \exp(a_1 - b_1 t)} \tag{5.7}$$

where $a_1 = b_1 t_1$, and t_1 is the abscissa of the point of inflection. Thus the growth of Y and X can be described as

$$\log \frac{K_1 - Y}{Y} = a_1 - b_1 t \tag{5.8}$$

and

$$\log \frac{K_2 - X}{X} = a_2 - b_2 t, \tag{5.9}$$

respectively. The interpretation of the three parameters of the logistic function is straightforward. Briefly, as noted earlier, a_1 is a constant depending on the initial conditions, K_1 is the equilibrium level of growth, and b_1 is the rate-of-growth parameter. It can be readily verified that the logistic curve is a symmetrical S-shaped curve with a point of inflection at $0.5K$.

It is not necessary that the growth curves of the techniques be of the logistic form. For example, suppose that the pattern of growth is best described as a Gompertz function

$$\frac{1}{Y} \frac{dY}{dt} = J_1 \exp(I_1 - J_1 t). \tag{5.10}$$

The integral form of (5.10) is

$$\log Y = -\exp(I_1) \exp(-J_1 t) + \log K_1$$

or

$$Y = K_1 \exp[-\exp(I_1 - J_1 t)]. \tag{5.11}$$

Thus growth of Y and X can be described as

$$\log \log \frac{K_1}{Y} = I_1 - J_1 t \tag{5.12}$$

and

$$\log \log \frac{K_2}{X} = I_2 - J_2 t. \tag{5.13}$$

The three parameters of the Gompertz function can be intepreted in the same way as those of the logistic. Briefly, I_1 is a constant, K_1 is the equilibrium level of growth, and J_1 is the rate-of-growth parameter. Unlike the logistic, however, the Gompertz curve is an asymmetrical S-shaped curve with a point of inflection at $Y = K/e$ or at approximately 37% of the limit to growth.

There are a number of interesting implications of these simple systems of relationships. They are discussed in what follows.

3.1 Temporal Aspects of Technological Substitution

As discussed in Section 5.2, the essence of the technological substitution hypothesis lies in its emphasis on the disequilibrating characteristic of the diffusion process. One way to express the phenomenon under consideration is to assume that the system-wide disequilibrium caused by the gap in the use of two techniques is in some ways related to the disequilibrium within the use of each of them. This is merely to say that the dynamics of the whole system is a function of the dynamics of its subsystem. Following the earlier notations, use of one technique as a proportion of the use of the other is given by f/f', while the intraequilibrium gaps may be denoted by $(K_1 - Y)/Y$ and $(K_2 - X)/X$ since K_1 and K_2 are the equilibrium levels of the respective use of the two techniques. Specifically, suppose that the use of one technique as a percentage of the other is some fixed proportion g of the percentage of intraequilibrium gaps; that is,

$$\log f(t) - \log f'(t) = g\left[\log\left(\frac{K_2 - X}{X}\right) - \log\left(\frac{K_1 - Y}{Y}\right)\right]. \qquad (5.14)$$

Combining Eqs. (5.3), (5.4), (5.8), (5.9), and (5.14) immediately provides the following simple model of technological substitution over the course of time:

$$\log\left(\frac{f}{1-f}\right) = \alpha_1 + \beta_1 t \qquad (5.15)$$

where $\alpha_1 = g(a_2 - a_1)$ and $\beta_1 = g(b_1 - b_2)$.

Fisher and Pry (1971) have presented an identical model of technological substitution from a different premise. The starting point of their derivation is the assumption that the rate of adoption of a new product is proportional to the fraction of the old one still in use, that is,

$$\frac{1}{f}\frac{df}{dt} = b(1-f). \qquad (5.16)$$

Now let t_0 denote the time at which the fractional substitution reaches its midpoint, that is, $f = \frac{1}{2}$. Integration of Eq. (5.16) yields a special case of the logistic function

$$f = \{1 + \exp[b(t - t_0)]\}^{-1}, \tag{5.17}$$

which can be written in an alternative form as

$$\log\left(\frac{f}{1-f}\right) = \alpha_2 + \beta_2 t \tag{5.18}$$

where $\alpha_2 = -bt_0$ and $\beta_2 = b$. As can be seen from Eq. (5.17) one central assumption of this model is that the process of substitution eventually results in displacement of the old product by the new.

The simplicity of the model given by Eq. (5.18) is indeed striking, for it contains only two parameters. What is more, it has proven to be eminently successful in explaining a wide variety of data on the diffusion of technology. However, as can be seen from the comparison of (5.18) with (5.15), the two coefficients in the former equation are composed of five parameters. It is therefore easy to see why the Fisher–Pry model holds well in its explanation of actual data on diffusion. Moreover, according to the two alternative derivations of the model,

$$b = g(b_1 - b_2); \tag{5.19}$$

that is, the higher the coefficient g, the less the difference between the rates of adoption of the two techniques. This is an interesting result because it implies that the coefficient g is a measure of the speed with which movement from one equilibrium to another equilibrium takes place. Accordingly, Eq. (5.14) implies that the greater the disparity in the use of two techniques, the faster the speed of substitution is. This is analogous to the well-known principle of sequential sampling that *inter alia*, the expected length of an experiment is inversely related to the difference between the means of the populations under consideration (Griliches, 1960). In all, it is reassuring that the basic form of the proposed model given by Eq. (5.15) has a certain general, commonsensical interpretation.

It is interesting to note that the foregoing model can be obtained from yet another premise. From Eqs. (5.8) and (5.9)

$$\log\left(\frac{K_1 - Y}{Y}\right) - \log\left(\frac{K_2 - X}{X}\right) = (a_1 - a_2) + (b_2 - b_1)t. \tag{5.20}$$

When X and Y are small they become negligible in comparison with their final values K_2 and K_1. Equation (5.20) can be therefore rewritten as

$$\log\left(\frac{K_1}{Y}\right) - \log\left(\frac{K_2}{X}\right) = (a_1 - a_2) + (b_2 - b_1)t. \qquad (5.21)$$

Rearranging the terms in Eq. (5.21) immediately gives the final form of the model as

$$\log\left(\frac{f}{1-f}\right) = \alpha_3 + \beta_3 t \qquad (5.22)$$

where $\alpha_3 = a_1 - a_2 + \log(K_2/K_1)$ and $\beta_3 = b_2 - b_1$. Equation (5.22) is of course identical with Eq. (5.15) or Eq. (5.18). Thus three different premises lead to one and the same model of technological substitution over the course of time.

3.2 Spatial Aspects of Technological Substitution

The proposed model of spatial aspects of technological substitution can be obtained from many different starting points. One of its derivations is analogous to that of the preceding model of technological substitution over time as given by Eq. (5.22). Thus, consider the logistic type of diffusion process. Solving Eqs. (5.8) and (5.9) for t, we have

$$t = \frac{a_1}{b_1} - \frac{1}{b_1} \log \frac{K_1 - Y}{Y} = \frac{a_2}{b_2} - \frac{1}{b_2} \log \frac{K_2 - X}{X}, \qquad (5.23)$$

which immediately yields the expression

$$\frac{Y}{K_1 - Y} = C_1 \left(\frac{X}{K_2 - X}\right)^{b_1/b_2}. \qquad (5.24)$$

Clearly,

$$C_1 = \exp\left(\frac{a_2 b_1 - a_1 b_2}{b_2}\right),$$

which can be written in a simplified form as $C_1 = \exp[b_1(t_2 - t_1)]$ since, as noted earlier, $a_1 = b_1 t_1$ and $a_2 = b_2 t_2$ (cf. Eqs. (5.8) and (5.9)). When X and Y are small in comparison with their final values, Eq. (5.24) reduces to

$$\frac{Y}{K_1} = C_1 \left(\frac{X}{K_2}\right)^{b_1/b_2}.$$ (5.25)

Hence the following simple model of the spatial aspects of technological substitution is obtained:

$$X = A_1(Y)^{B_1}$$ (5.26)

where $A_1 = K_2/(K_1)^{b_2/b_1}C_1$ and $B_1 = b_2/b_1$.

Alternatively, consider the Gompertz type of diffusion process. Solving Eqs. (5.12) and (5.13) for t, we have

$$t = -\frac{1}{J_1}\left(\log\log\frac{K_1}{Y} - I_1\right) = -\frac{1}{J_2}\left(\log\log\frac{K_2}{X} - I_2\right)$$

or

$$\log\log\frac{K_1}{Y} = \log M + \log\left(\log\frac{K_2}{X}\right)^v$$ (5.27)

where $v = J_1/J_2$, $M = G_1/G_2^v$, $G_1 = e^{I_1}$, $G_2 = e^{I_2}$. Solving for Y, we have

$$Y = K_1 \exp\left[-M\left(\log\frac{K_2}{X}\right)^v\right].$$ (5.28)

If both variables under consideration grow at the same rate, that is, $v = 1$, then

$$Y = \frac{K_1}{K_2{}^M} X^M.$$ (5.29)

Further, $M = e^{I_1}/e^{I_2}$, since $v = 1$. Thus the value of M depends only on two constants; it does not involve the rates of growth of the variables under consideration. Equation (5.29) is, of course, identical with the earlier model of technological substitution given by Eq. (5.26) since it can be written as

$$X = A_2(Y)^{B_2}$$ (5.30)

where $A_2 = K_2/K_1^{1/M}$, $B_2 = 1/M$.

The above derivations might lead us to suspect that the proposed model holds only at initial stages of growth or as a special case when the rates of growth of the variables under consideration are equal (H. Lumar cited by

Reeve and Huxley, 1945). However, this is not so (A. J. Kavanagh and O. W. Richards cited by Reeve and Huxley, 1945). In fact, a variety of assumptions lead to the very same model. For example, suppose that X and Y increase according to different forms of S-shaped curves. Very generally, let

$$\frac{1}{Y}\frac{dY}{dt} = u_1(K_1 - Y), \qquad \frac{1}{X}\frac{dX}{dt} = u_2(K_2 - X^m) \qquad (5.31)$$

where $u_1 = b_1 / K_1$ and $u_2 = b_2 / K_2$. The solution of this system of equations is given by

$$\frac{1}{X}\frac{dX}{dt} = B_3\left(\frac{1}{Y}\frac{dY}{dt}\right) \qquad \text{or} \qquad X = A_3(Y)^{B_3} \qquad (5.32)$$

where $\log A_3$ is a constant of integration and

$$B_3 = \frac{1}{m}, \qquad m = \frac{K_1 u_1}{K_2 u_2}, \qquad A_3 = \left(\frac{K_2}{K_1}\right)^{1/m}. \qquad (5.33)$$

Clearly, the form of the Eq. (5.32) is identical with that of Eq. (5.26) or (5.30). This is a very reassuring result because it implies that the basic form of the proposed model holds regardless of the differences in the specific patterns of diffusion of the competing techniques.

3.3   An Overview of the Theoretical Relationships

To summarize the results, two principal models of technological substitution have been proposed. The simplicity of these models is remarkable in that they contain only two parameters. Moreover, both models seem to be quite robust: Their forms remain invariant under a variety of different starting points. They are also complementary to each other in their respective focus on the temporal and spatial dimensions of the process under consideration. The alternative derivations of these models have a number of important implications.

First, according to one of the derivations, both models seem to hold only during the initial stages of the process under consideration. As discussed in Chapter 1, a characteristic feature of many lawlike relationships is that they hold only over a limited range of data. Thus, Hooke's law of springs does not hold for too high values of applied load. Similarly, Ohm's law necessarily breaks down when the values of resistance are very high, as in the case of highly purified oils. These examples are illustrative of the fact that the formulation of many important laws necessarily involves certain simplifying assumptions. In consequence, they do not hold for too low or too high values of empirical observations. At the same time, however, the range of applicability of the proposed models may well be quite large for all practical purposes. This is because the two parameters of these models reflect the joint influence of as many as half a dozen different parameters connected with the individual patterns of growth of competing techniques. Moreover, the proposed models can also be derived in ways that do not require small-scale approximation of the variables involved. That is, under certain conditions, there are no theoretical restrictions on the range of data for which the models may be expected to hold.

Second, it is evident that a sufficient (although not a necessary) condition for the proposed models to hold is that individual forms of growth of competing techniques can be closely approximated in terms of *some* S-shaped curves. At the same time, the assumption of an S-shaped growth curve appears somewhat more predominantly in the model of the spatial rather than the temporal aspects of technological substitution. Empirically, it is very often the case that the data on stock (e.g., the total number of machines in use) correspond more closely to the S-shaped growth curve than do the data on annual sales. The implication is that each of the two models is peculiarly suited to different situations. The model of the temporal aspects of substitution is most appropriate to cases where the phenomenon under consideration is being investigated in terms of annual sales of the product. The model of the spatial process of substitution is most appropriate to cases where the relevant data concern the stock variable.

Third, one of the derivations of the model of the temporal aspects of substitution assumes that the old product is eventually displaced by the new. No such assumption is made in any of the derivations of the model of the spatial aspects of technological substitution. In fact, the implication of the final form of this model is that the new and the old product must coexist; if one vanishes, so does the other. The conclusion to be drawn is that the two models are best suited to two somewhat related cases of substitution involving essentially rival and complementary techniques. The difference between the two cases arises from the fact that some instances of substitution lead to the elimination of the older technology, whereas others do not. An example of the former is substitution of the catalytic cracking process for thermal cracking in petroleum refining. An example of the latter is the substitution of hydrocracking for catalytic cracking.

4.   CASE STUDIES: DIFFUSION OF NEW TECHNIQUES IN THE AGRICULTURE, ELECTRICITY GENERATION, MANUFACTURING, STEEL PRODUCTION, AND TEXTILE INDUSTRIES

This section presents results from an investigation into the following cases of diffusion of technology.[5] The choice of the specific variables, country, and time period will be defended on the grounds of empirical necessity.

1. The diffusion of the farm tractor in the United States, 1920–1960. Growth in the number of tractors on farms (X) is analyzed in relation to the number of horses (Y). Tractor power is measured in terms of both (a) the number of tractors and (b) the stock of tractor horsepower.

2. The diffusion of thermoelectric generating units in Canada, 1917–1972. The specific measure of technology employed is total installed capacity of thermal power (X) in megawatts, and it is studied in relation to the growth of hydroelectric power (Y). The growth of thermal power reflects the diffusion of both fossil fuel and nuclear power units.

3. The diffusion of the combine with cornhead in the United States, 1956–1971. The annual sales of the combine with cornhead (X) are analyzed in relation to the sales of mechanical cornpickers (Y). The former represents a technological advance over the latter in that the addition of cornheads to a grain combine makes it possible to harvest corn as shelled grain rather than as ear corn.

4. The diffusion of diesel-powered tractors in the United States, 1955–1971. The annual production of diesel-powered tractors (X) is analyzed in terms of the tractors' share in the total production (f) and in relation to the production of gasoline-powered tractors (Y).

5. The diffusion of the oxygen steel process in Austria, 1954–1967. The oxygen process uses pure oxygen (instead of air) in refining steel. It mainly replaces the acid and basic Bessemer processes and the open hearth process, all of which employ air as an oxidizing agent in refining steel. The main advantage of the oxygen steel process is that it makes possible the production of high-grade steel at a rapid rate. The main variable employed in the analysis of its diffusion is the share of oxygen process (f) in the total output of steel.

6. The diffusion of the continuous process of casting steel in Austria, 1952–1968. The continuous process of casting steel makes it possible to obtain the finished product, such as billets, bloom, or slabs, directly from molten steel by eliminating the soaking pit and blooming mill

---

[5]Further information on these case studies can be found elsewhere. For case studies 1, 3, and 4 see Sahal (1977a,b) and for case studies 5–10 see Ray (1969). Finally, see Chapter 8 for a discussion related to case study 2.

of conventional ingot casting. This in turn makes possible an increased yield of finished steel from a given charge while reducing the size of the rolling mills required. The main variable employed in this case study is the share of continuous casting steel (f) in the total national crude steel output.

7. The diffusion of shuttleless looms in the weaving of cotton and manmade fibers in Germany, 1958–1968. Shuttleless looms of all types essentially draw the weft from a stationary and continuous supply. They replace shuttle looms, in which the shuttle carries the weft and has to be constantly replenished. The principal variable employed in this case study is the ratio of shuttleless looms to all looms (f).

8. The diffusion of tunnel kilns in brickmaking in Sweden, 1950–1968. Tunnel kilns consist of stationary firing zones through which bricks move at a uniform speed on special wagonettes. They replace the traditional Hoffman kiln in which bricks are placed manually and are treated by firing in the successive chambers. Their main advantages over the conventional technique lie in the saving of fuel and the lesser skills required. The principal variable employed in this case study is the share of tunnel kilns in the total output of bricks (f).

9. The diffusion of new methods of steel-plate marking and cutting in shipbuilding in Sweden, 1950–1968. Three new techniques of marking are considered in this study: (1) the optical lofting technique, in which specially prepared drawings are projected onto the plate through an optical system, the marking being done by hand and the plate cutting with hand-operated burning machines; (2) the photoelectrically controlled cutting machine, which consists of fully automated cutting machines, thereby eliminating the marking process; (3) the numerical control machine, which consists of a computer program that controls the drawing and cutting machines and thus makes possible complicated designs for hulls and other three-dimensional surfaces. The principal variable employed in this study is the share (f) of steel cut by new methods in the total.

10. The diffusion of automatic transfer lines for car engines in Sweden, 1955–1968. Automatic transfer lines operate on a continous line of parts such as car cylinder blocks. This method replaces many individual machines and its main advantages lie in the saving of labor and standard time. The principal variable employed in this study is the share (f) of automatic transfer line engines in the total output.

In summary, the various cases of diffusion of technology represent a fairly broad cross section of new products and processes. The variables employed in these case studies fall into two broad categories: direct indicators of diffusion in terms of stock of, or annual investment in, the new techniques (cases 1–4, 7), and indirect indicators of diffusion in terms of the contribution of new techniques to the total output (cases 5,6,8–10). The studies of farm tractors and thermoelectric generators (cases 1,2) are based on an explicit recognition of the multidimensional nature of the diffusion of technology. Specifically, the chosen measures of technology in terms of tractor horsepower stock and total installed capacity in megawatts explicitly take into account changes in certain functional characteristics of the techniques during the course of their diffusion. In contrast, it has not been possible to take into account some such changes in combines, diesel-powered tractors, and shuttleless looms (cases 3,4,7) because of lack of data. Finally, the remaining studies (cases 5,6,8–10) circumvent the issue by choosing indirect indicators of diffusion as explained above.

The parametric estimates of the two models of technological substitution are presented in Tables 5.1 and 5.2. The first thing to be said about these results is that in a few instances the results from the Durbin–Watson test are inconclusive, while in some cases they indicate the presence of serial correlation in the residuals of the equations. In a majority of cases, however, the issue cannot be settled because the number of available observations falls short of the conventional requirement for the Durbin–Watson test. In conclusion, although the parametric estimates of the models are unbiased, the significance of the coefficients and the explanatory power of the equations have been overestimated in a few cases. The R^2 values and t ratio are nevertheless still very high. Thus in a majority of cases the models explain more than 90% variance in the data.

The performance of the models in various cases of technological diffusion is shown in a diagrammatic form in Figures 5.1–5.11. As regards the diffusion of thermoelectric generators, the model of the spatial process holds fairly well for almost the entire length of the time period considered, while the model of the temporal process holds only in the postwar years (Figures 5.1–5.2). This suggests that the complementarity of the two models can serve a valuable purpose in checking the results from either of them. The fit of the models to the data in various other cases is shown in Figures 5.3–5.11. In accordance with the theory, the values of $f/(1-f)$ over time result in approximately linear trends when plotted on a semilogarithmic graph. Likewise, when data on the absolute adoption of the new technique relative to the use of the old technique are plotted on double-logarithmic paper the resulting trend is approximately linear. The nature of the data in most cases favors the model of the temporal rather than the spatial process of technological substitution. Thus the case of

Table 5.1 Parametric Estimates of the Model of the Temporal Process
of Technological Substitution

Case	Estimated relationship[a]

1a $\log[f/(1-f)] = - \ 93.93 \ + \ 0.048T$
$(1.16) \quad (0.0006)$

$R^2 = 0.99, \ S = 0.04, \ F = 6495.12, \ \text{D–W} = 0.38, \ N = 41 \ (1920\text{–}1960)$

1b $\log[f/(1-f)] = - \ 103.27 \ + \ 0.05T$
$(1.17) \quad (0.0006)$

$R^2 = 0.99, \ S = 0.04, \ F = 7363.05, \ \text{D–W} = 0.37, \ N = 41 \ (1920\text{–}1960)$

2   $\log[f/(1-f)] = - \ 68.94 \ + \ 0.035T$
$(1.58) \quad (0.0008)$

$R^2 = 0.98, \ S = 0.04, \ F = 1860.24, \ \text{D–W} = 0.75, \ N = 30 \ (1943\text{–}1972)$

3   $\log[f/(1-f)] = - \ 192.16 \ + \ 0.098T$
$(13.71) \quad (0.007)$

$R^2 = 0.93, \ S = 0.13, \ F = 196.14, \ \text{D–W} = 1.03, \ N = 16 \ (1956\text{–}1971)$

4   $\log[f/(1-f)] = - \ 164.22 \ + \ 0.084T$
$(10.24) \quad (0.005)$

$R^2 = 0.94, \ S = 0.10, \ F = 256.94, \ \text{D–W} = 0.51, \ N = 17 \ (1955\text{–}1971)$

5   $\log[f/(1-f)] = - \ 212.33 \ + \ 0.108T$
$(75.13) \quad (0.038)$

$R^2 = 0.53, \ S = 0.57, \ F = 7.97, \ \text{D–W} = 1.38, \ N = 9 \ (1954\text{–}1967)$

6   $\log[f/(1-f)] = - \ 179.14 \ + \ 0.09T$
$(23.34) \quad (0.01)$

$R^2 = 0.89, \ S = 0.18, \ F = 57.44, \ \text{D–W} = 1.61, \ N = 9 \ (1952\text{–}1968)$

7   $\log[f/(1-f)] = - \ 159.99 \ + \ 0.08T$
$(11.79) \quad (0.006)$

$R^2 = 0.98, \ S = 0.05, \ F = 179.89, \ \text{D–W} = 1.01, \ N = 6 \ (1958\text{–}1968)$

8   $\log[f/(1-f)] = - \ 279.04 \ + \ 0.14T$
$(30.91) \quad (0.015)$

$R^2 = 0.93, \ S = 0.25, \ F = 81.51, \ \text{D–W} = 1.01, \ N = 8 \ (1950\text{–}1968)$

9   $\log[f/(1-f)] = - \ 252.97 \ + \ 0.13T$
$(24.97) \quad (0.013)$

$R^2 = 0.95, \ S = 0.19, \ F = 102.55, \ \text{D–W} = 1.09, \ N = 7 \ (1950\text{–}1968)$

Table 5.1 (Continued)   Parametric Estimates of the Model of the Temporal
Process of Technological Substitution

Case	Estimated relationship[a]
10	$\log[f/(1-f)] = -321.05 + 0.16T$
	$\hspace{3.2em}(38.21)\hspace{1.2em}(0.019)$
	$R^2 = 0.95,\ S = 0.20,\ F = 70.78,\ \text{D-W} = 1.52,\ N = 6\ (1955\text{--}1968)$

[a]Definitions: The variable f denotes the fractional adoption of the new technique as
defined in the text; R^2 is the coefficient of determination, S the standard error of the
estimate, F the ratio of the variance explained by the model to the unexplained variance,
D-W the Durbin–Watson test statistic, and N the number of observations from the
specified time period. The standard errors of the regression coefficients are given in
parentheses.

temporal substitution is more heavily illustrated here. In all, the fit of the two
models to the data is generally very good.

The coefficients of the explanatory variables in these models are of independent interest. Specifically, the coefficient of the time variable in the temporal model of technological substitution is a measure of the diffusion rate for
the innovation. It can be regarded as being inversely related to the length of
time required for the innovation to capture from, say 10% to 90% of the
market. Thus the results presented in Table 5.1 indicate that there are considerable differences in the rates of adoption of different techniques. In particular, while the rate of adoption of new techniques is highest in the manufacturing sector of the economy, it is lowest in the energy sector. Further, within
any sector of the economy, such as agriculture, there is considerable uniformity in the rates of adoption of different techniques. The question arises
whether the observed interindustry differences in the diffusion of technology
are significant. This can be easily determined by the analysis of variance. The
results are presented in Table 5.3. At the $\alpha = 0.05$ level, while the difference
between the rates of diffusion of new techniques in agriculture and the steel
production industry is insigificant, the corresponding difference between
manufacturing and the steel production industry is close to being significant.
Finally, as we would expect, there exist highly significant differences between
the rates of adoption of new techniques in agriculture and manufacturing.
Thus we have the important result that there generally tends to be quite some
heterogeneity in the diffusion of technology.

Next, consider the coefficient of the independent variable in the model of
the spatial process of technological substitution. It is essentially a measure of

Table 5.2   Parametric Estimates of the Model of the Spatial Process
of Technological Substitution

Case	Estimated relationship[a]

1a $\log X =$ 8.84 −   1.42 $\log Y$
                       (0.29)   (0.07)

$R^2 = 0.90$, $S = 0.11$, $F = 352.19$, D–W $= 0.14$, $N = 41$ (1920–1960)

1b $\log X =$ 8.04 −   1.62 $\log Y$
                       (0.33)   (0.08)

$R^2 = 0.91$, $S = 0.12$, $F = 382.08$, D–W $= 0.14$, $N = 41$ (1920–1960)

2    $\log X = -$ 2.62 +   1.46 $\log Y$
                       (0.28)   (0.07)

$R^2 = 0.88$, $S = 0.22$, $F = 389.77$, D–W $= 0.06$, $N = 55$ (1917–1972)

3    $\log X =$ 7.83 −   0.88 $\log Y$
                       (0.65)   (0.15)

$R^2 = 0.70$, $S = 0.14$, $F = 33.02$, D–W $= 0.74$, $N = 16$ (1956–1971)

4    $\log X =$ 8.88 −   0.79 $\log Y$
                       (0.91)   (0.18)

$R^2 = 0.55$, $S = 0.16$, $F = 18.72$, D–W $= 0.98$, $N = 17$ (1955–1971)

5    $\log f = -$ 1.65 −   3.72 $\log(1-f)$
                       (0.29)   (0.87)

$R^2 = 0.72$, $S = 0.37$, $F = 18.37$, D–W $= 1.35$, $N = 9$ (1954–1967)

6    $\log f = -$ 2.72 −   119.28 $\log(1-f)$
                       (0.15)   (28.53)

$R^2 = 0.71$, $S = 0.29$, $F = 17.48$, D–W $= 0.97$, $N = 9$ (1952–1968)

7    $\log f = -$ 2.34 −   66.56 $\log(1-f)$
                       (0.07)   (8.51)

$R^2 = 0.94$, $S = 0.08$, $F = 61.19$, D–W $= 1.06$, $N = 6$ (1958–1968)

8    $\log f = -$ 0.82 −   0.76 $\log(1-f)$
                       (0.12)   (0.19)

$R^2 = 0.73$, $S = 0.25$, $F = 16.14$, D–W $= 0.62$, $N = 8$ (1950–1968)

9    $\log f = -$ 1.19 −   1.81 $\log(1-f)$
                       (0.23)   (0.49)

$R^2 = 0.73$, $S = 0.34$, $F = 13.23$, D–W $= 1.28$, $N = 7$ (1950–1968)

Table 5.2 (Continued)   Parametric Estimates of the Model of the Spatial
Process of Technological Substitution

Case	Estimated relationship[a]
10	$\log f = -\ \ 0.53 - 0.43\ \log(1 - f)$
	$\qquad\quad\ (0.15)\ \ \ (0.18)$
	$R^2 = 0.48,\ S = 0.21,\ F = 5.54,\ \text{D–W} = 1.31,\ N = 6\ (1955\text{–}1968)$

[a]Definitions: The variables X and Y represent the extent of adoption of the new and the old technique, respectively. For the remaining definitions see footnote a, Table 5.1.

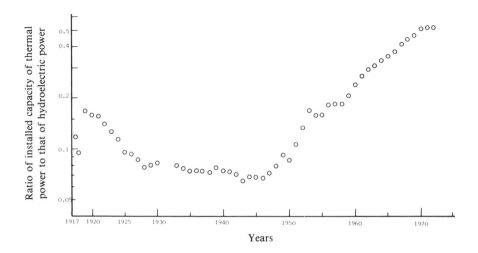

Fig. 5.1   Temporal representation of the relative growth of thermal and hydroelectric power, 1917–1972.

growth in the adoption of one technique relative to the other. That is, it is equal to unity if the rates of adoption of the two techniques are equal. The results in Table 5.2, however, indicate that in 6 out of 10 cases the value of the relative growth rate is significantly different from unity. In conclusion, the diffusion of technology is generally an allometric process of growth.[6] That is, substitution of one technology for the other generally involves disproportionate growth of one in relation to the other.

---

[6]The term "allometry" was introduced by Huxley to denote disproportionate change in the size of an organ as a consequence of change in the overall size of a biological system. The case of proportionate change, on the other hand, is referred to as "isometry" (Reeve and Huxley, 1945). For further discussion of these concepts, see Chapter 9.

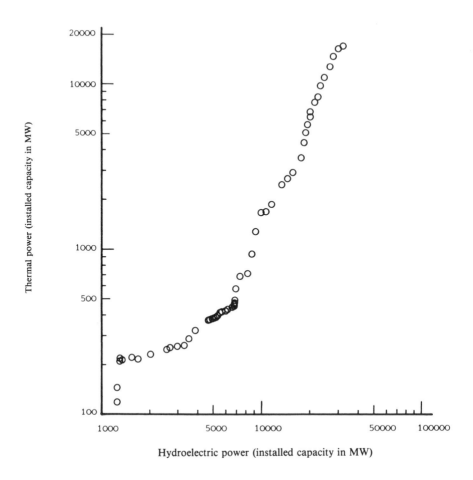

Fig. 5.2 Spatial representation of the relative growth of thermal and hydroelectric power, 1917–1972.

5. PRINCIPAL CONCLUSIONS

A great deal of contemporary theorizing on technological change is based on the assumption that the adoption of an innovation is fundamentally a process of imitation starting from the initiative taken by a few individuals. The alternative viewpoint advanced in this study is that the diffusion of a technique is intimately linked with its development. Accordingly, the phenomenon under consideration is best conceived in terms of the actual substitution of a new technique for the old.

There are a number of distinct advantages in regarding the diffusion of an innovation in terms of a technological substitution process. First, within the

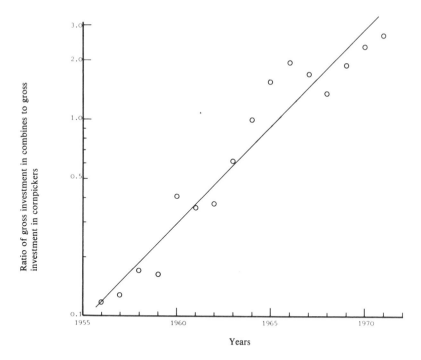

Fig. 5.3   Temporal representation of the substitution of combines for cornpickers, 1956–1971.

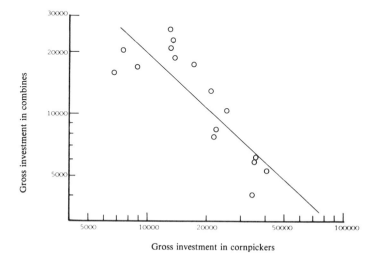

Fig. 5.4   Spatial representation of the substitution of combines for cornpickers, 1956–1971.

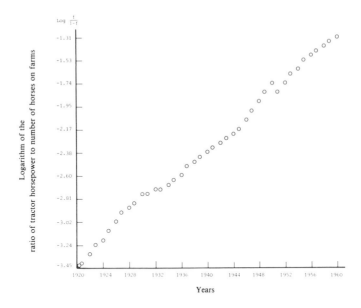

Fig. 5.5   Temporal representation of the substitution of tractor power for horses on farms, 1920–1960.

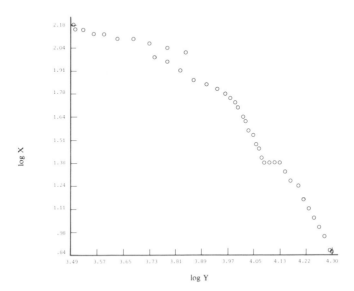

Fig. 5.6   Spatial representation of the relative growth of tractor horsepower (X) and number of horses (Y), 1920–1960.

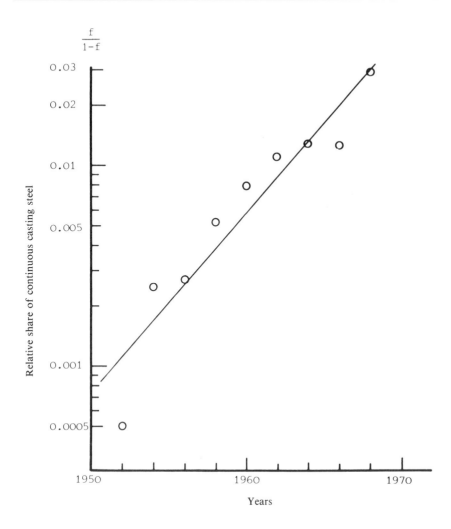

Fig. 5.7   Temporal representation of the substitution of the continuous casting of steel for conventional ingot casting in Austria, 1952–1968 (f is the share of continuous casting steel in the total national crude steel output).

proposed framework, the adoption of a new technique amounts to the obliteration of the equilibrium level associated with the use of the existing technique. This is in keeping with the fact that the limits to the growth of a system often prove to be temporary (see Chapter 4). The proposed viewpoint is especially suited to understanding the *long-term* characteristics of the diffusion process because of its emphasis on the disequilibrating aspects of the system. Second, a consideration of the phenomenon in terms of substitution yields a far more

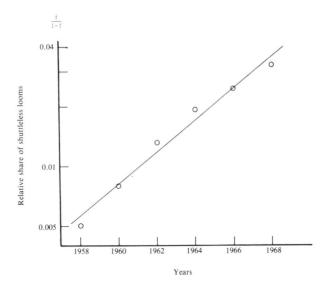

Fig. 5.8   Temporal representation of the substitution of shuttleless looms for other looms in Germany, 1958–1968 (f is the ratio of shuttleless looms to all looms).

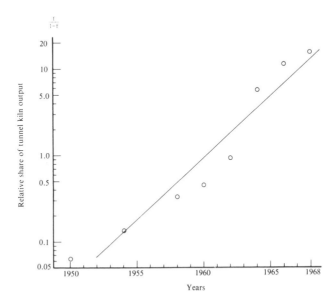

Fig. 5.9   Temporal representation of the substitution of the tunnel kiln for other methods of brickmaking in Sweden, 1950–1968 (f is the share of tunnel kiln output in total brick production).

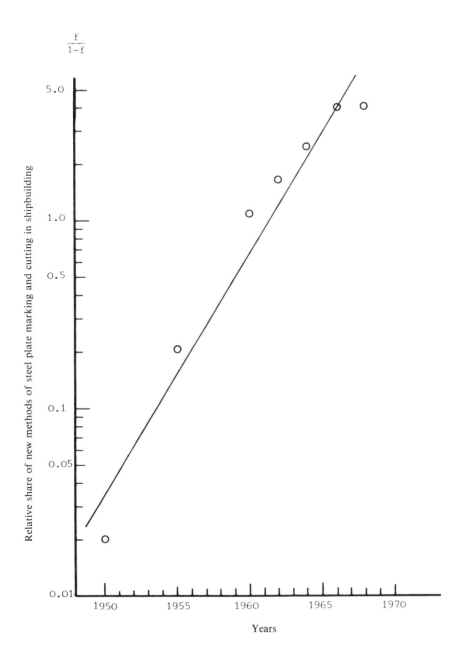

Fig. 5.10   Temporal representation of the substitution of new methods of steel plate marking and cutting for old in shipbuilding in Sweden, 1950–1968 (f is share of steel cut by new methods in total).

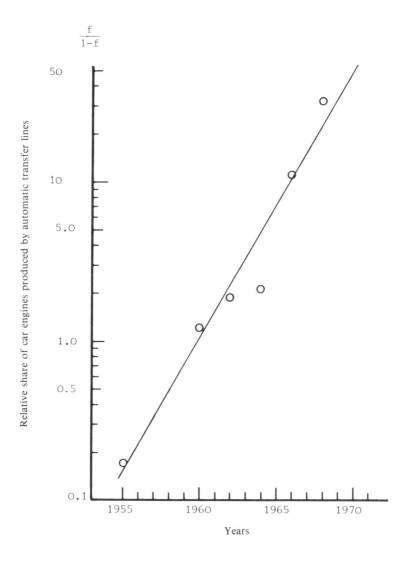

Fig. 5.11 Temporal representation of the substitution of automatic transfer lines for individual machines in Sweden, 1955–1968 (*f* is the share of output produced by automatic transfer lines).

stable relationship than does consideration in terms of diffusion per se. This is to be expected in that fluctuations in the use of interrelated techniques are smoothed out when they are considered together. Finally, the indicators of the substitution of a new technique for the old are more useful for *practical* purposes (e.g., marketing of a product) in comparison with other indicators, such as the rate of imitation in the use of a technology.

Table 5.3   Significance of the Differences in the Rates of Diffusion
of New Techniques in Various Industries

Industry	Mean rate of diffusion (β)	Source equations[a]
1. Agriculture	0.077	1b, 3, 4
2. Steel production	0.099	5, 6
3. Manufacturing	0.143	8, 9, 10

	F Matrix	
	β_1	β_2
β_2	1.52	—
β_3	17.68	6.38

Degrees of freedom: 1, 5.

[a]Numbers denote cases from Table 5.1.

The theory presented in this chapter is operationalized in the form of two models of technological substitution. One of the models has its focus on the temporal aspects of the phenomenon while the other has its focus on the spatial aspects. The two models are complementary in this respect. Moreover, one model is best suited to the case of rival techniques, the other to the case of complementary techniques. The former holds best when the data are in the form of annual sales of the two products, the latter when the data are in the form of stock of machines in use. Both models contain only two parameters. Further, their forms remain invariant against a number of changes in their basic premise.

The study has also presented a number of case studies of the diffusion of technology. The results from the empirical analysis indicate that the proposed models hold very well in their explanation of a wide variety of data on the diffusion of technology. Among other things, this has an important implication for the hitherto unresolved problem of what might be an appropriate specification of the observed pattern of diffusion. The past efforts in this area have generally cast this problem in terms of choosing a specific type of S-shaped curve from a large number of candidates. According to the results of this study, however, it seems doubtful that the diffusion process can be unambiguously specified in terms of *any* S-shaped curve.[7] Indeed, if the fit of the proposed models to the data is any guide, one and the same pattern of diffusion may be consistent with several alternative types of S-shaped curves. There are good reasons why the diffusion process evinces a *roughly* S-shaped pattern of growth. There seems no reason why it should rigidly adhere to any specific type thereof.

Two further generalizations about the nature of technological diffusion can be noted on the basis of the results presented in this chapter. First, there may exist significant interindustry differences in the rates of diffusion of new techniques. Second, the diffusion of an innovation involves a disproportionate relationship between the growth in the use of the new and the old techniques. That is, diffusion of technology is a process of allometric growth characterized by an element of heterogeneity.

---

[7]For a similar conclusion from a different premise see David (1969).

Chapter 6

THE LONG-TERM DEVELOPMENT
OF TECHNOLOGY

1. TOWARD A MACROVIEW OF INNOVATIVE ACTIVITY

It is commonly held that technological innovation is the linchpin in productivity performance. As we saw in Chapter 1, detailed empirical analyses in
this area confirm that innovative activity has been one of the main factors
underlying the growth in productivity over the course of time. The snag,
however, is that an adequate explanation of technical change per se has been
lacking. This has important implications for not only the temporal but also the
spatial variations in productivity. Since the temporal aspects of productivity
growth have already been discussed at length in Chapter 1, the following discussion focuses on its spatial aspects.

There exist tremendous differences in the productivity performance of
different industries. For example, the yearly growth in output per worker during the period 1948–1966 was 5.8% in communications and utilities but only
1.2% in services. The experience of the manufacturing and mining industries
provides another contrast. The growth of labor productivity in these two cases
over the same period was 2.9 and 4.6%, respectively (Kendrick, 1973). The intersectoral differences in productivity growth are sometimes explained in terms
of the corresponding differences in expenditures on research and development
(R&D) activity (see Terleckyj's study contained in Kendrick, 1961, and Mansfield, 1968). Further work in this area suggests that the basic issue concerns the
differences in not only the magnitude but also the type of R&D expenditure
across various industries. Thus there is some indication that self-financed
R&D has played a far greater role than government-financed R&D in promoting the growth of productivity in the manufacturing sector (Leonard,
1971). Apparently, this is also one of the reasons why the rate of growth of

Devendra Sahal, Patterns of Technological Innovation

ISBN 0-201-06630-0

total factor productivity in the nonmanufacturing sector has been nearly the same as in the manufacturing sector despite the fact that R&D expenditure in the former case has been much lower in comparison with the latter (Terleckyj, 1974). Although these works have provided many valuable insights into the role of R&D in the growth of productivity, they also raise a host of unresolved issues. As Nelson and Winter put it (1975, pp. 338–339):

> The studies [of the relationship between productivity growth and R&D expenditure] have been useful and provocative, but have not cut very deep.
>
> Clearly there are severe specification problems . . . . There is a tangle of causations, from R&D to productivity growth, from productivity growth and lowered prices to growth of output, from growth of output in the presence of scale economies to productivity growth, from expansion of industry to greater incentives for research and development, and so on.
>
> Even if it were granted that causation runs, at least in part, from research and development, what explains the great differences in the (direct and indirect) R&D spending across different industries?

More generally, one central issue that remains unresolved is: What accounts for differences in the technical progressivity of different industries?

This chapter presents a macroview of the technological innovation process in an attempt to provide a deeper understanding of spatial and temporal variations in productivity.

2. LEARNING, SCALING, AND TECHNOLOGICAL INNOVATION

There are two central propositions of the theory advanced here. First, the long-term evolution of a technology is governed by the accumulated experience of a practical nature. Second, the technology of a system is a function of the scale of its utilization. In what follows each of these two propositions will be developed in detail.

The starting point of the first proposition is the well-known concept of the progress function (also sometimes called the learning curve)—that the productivity of a plant gradually picks up as it becomes possible to remedy various bottlenecks in its operation through the accumulation of relevant experience. The phenomenon was initially observed in the aircraft industry, where the direct labor input per airframe was found to decrease at a uniform rate with increase in cumulative output (Wright, 1936). Since then, a host of analyses have confirmed the systematic nature of this relationship (Alchian, 1963; Andress, 1954; Hirschman, 1964; Yelle, 1979). An illustrative example of this is shown in Figure 6.1. It has been found that the concept of the progress function holds in such diverse industries as shipbuilding and petroleum refining. It is relevant to a wide variety of activities ranging from manufacturing to clerical opera-

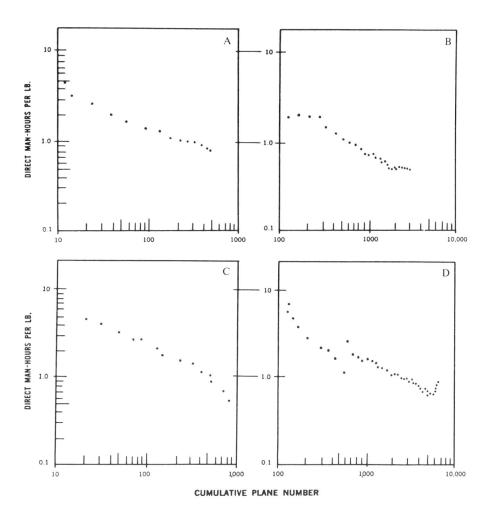

Fig. 6.1 An illustration of the progress function in the aircraft industry. Figures A–D refer to the production of B-29 at Martin–Marietta Corporation, Omaha; B-17 at Lockheed Aircraft Co., Burbank; B-24 at Douglas Aircraft Co., Inc., Tulsa; and B-17 at Boeing Co., Seattle, respectively. They are illustrative of a "normal trend," "leveling-off," "toe-down," and "scallop" and "toe-up" in the evolution of progress function. Source: After Reguero (1957) and Hirschman (1964).

tions. The observed phenomenon reflects the combined effect of a multiplicity of factors, for example, growth in the skill of workers, refinement of blueprints, improvement in plant layout, and improvement in methods of management (see Chapter 9). In its essence, experience plays a vital role in the evolution of productivity.

The progress function is appropriately viewed as a model of improvement in the productivity of any system embodying a *given* technology. A classic example is provided by the case of the Swedish steelworks at Horndal. For a period of 15 years after its construction, no further investments were made in the facility. Thus, there was no change in the production technology employed. Yet, its output per man-hour continued to grow at an average annual rate of about 2% over the same period of time (Lundberg, 1961). The observed growth in productivity can, however, be readily accounted for in terms of increased experience in the actual running of the facility (Arrow, 1962). The case of the Lawrence Cotton Textile Company of Lowell, Massachusetts, provides another illustration of this phenomenon (David, 1975). For a period of 22 years after its construction, one of its production facilities remained virtually unchanged. Except for maintenance and repair work, there was neither new investment for expansion of its production capacity nor significant replacement of the machinery employed. The productivity of the plant (cloth yardage per man-hour) nevertheless continued to increase at an average annual rate of about 2.25% over the course of this period. These examples illustrate that there is a significant process of learning in the *use of an existing technology*.

The thesis is advanced here that yet other learning processes underlie the *development of new techniques* in the first place. If there is any feature of innovative activity that stands out above all others, it is that the fruition of technical advances invariably depends on the accumulation of production experience (Arrow, 1962, 1969; Fusfeld, 1970; Rapping, 1965; Sahal, 1975; Shehinski, 1967). Witness the fact that many new techniques disappear, only to be conceived again at some later point in time when the necessary know-how has become available. As an example, in 1833, Babbage had already invented an "analytical engine" that had all the essential characteristics of a computing device. Nevertheless, real progress in this area became possible only in recent decades. Likewise, the design of radar was conceived much before it was pressed into service. The essential point of these examples—and there are literally hundreds of them—is that miscarriages tend to be commonplace during the process of innovation. Contrary to the popular history of technology, R&D activity does not always bring home the bacon. The outcome of initial efforts to develop a new technique is often left hanging for want of the relevant know-how.

To put it somewhat differently, innovative activity involves as many instances of deadlock as of progress. The modern form of the turbojet had already been patented in 1921 (Gohlke, 1942). Yet the first turbojet-powered flight was made possible only after nearly two decades of further experimental work, in 1939. The diesel locomotive was first used on American railroads in 1924, yet it became a viable means of transportation only in 1938, after some 15 years of further research and development effort (for further details see Section 5). Such examples can be almost indefinitely multiplied. They indicate that many seemingly instantaneous advances in technology are in fact based on prolonged development effort.

Specifically, as discussed in Chapter 2, a great deal of technical progress is made through the gradual refinement of certain basic patterns of design. Even major innovations must undergo extensive modification before their potential can be fully exploited. A new technology does not emerge like Minerva from Jove's forehead. Typically, it is the outcome of countless improvements in the capability of some earlier, less specialized device through the gradual acquisition of practical know-how. Success in technical problem solving is never just a matter of armchair theorizing. It depends primarily on some very down to earth considerations. Even the most carefully conceived blueprints seldom prove to be practicable at once. Rather, they must be tried out several times before they can be made operational. Typically, new designs tend to be unreliable, inefficient, and cumbersome. Moreover, their execution into new products generally requires special tools, jigs, and fixtures. The installation of these new devices is seldom possible without giving rise to unanticipated bottlenecks in the production process. As experience is gained, however, it becomes possible to identify and extirpate the bugs. The use of various production apparatus is harmonized and the reparative operations are gradually reduced. The work force becomes better acquainted with the task at hand and management finds new ways to improve plant layout and the scheduling of materials, labor, and equipment. All these factors contribute to the successful development of new techniques.

The thesis that technological innovation depends on the acquisition of relevant experience is further substantiated by the fact that design practices typically take the form of rules of thumb. For example, a common belief among naval architects, known as the "inch rule," is that service speed increases as the square root of the length of a ship. Needless to say, some such rules are an outcome of the learning process. There is, of course, nothing permanent about them. As new experience is acquired (especially as the scale of technology is changed beyond certain limits), the old rules of thumb give way to new ones. For example, until 1930 it was believed that the maximum safe speed in miles per hour of the steam locomotive engine was approximately the same as the number of inches in the diameter of its driving wheels. Subsequently, this estimate has been revised to the effect that the rule of thumb figure may be exceeded by about 25% (Bruce, 1952, p. 223). Further, the role of experience as reflected in these rules of thumb is often instrumental in radical advances in technology. For example, some of the most important advances in supersonic aerodynamic design were made possible by Whitcomb's "area rule" (Archer, 1967; Hartman, 1970, pp. 206–208) that supersonic drag is minimized when the longitudinal cross-sectional area distribution of airframe components including the wings can be represented by a smooth equivalent body. This example further illustrates that learning plays as important a role in the development of so-called science-oriented, high-technology items as it does in the case of traditional techniques.

These considerations suggest what may be called a "learning by doing" hypothesis of technological innovation.[1] In essence, technical progress is never a matter of touch and go. It depends on acquisition of practical experience over the course of time. The evolution of technology is not merely an outcome of a set of *replicative* events at work. Rather, it is governed by a process of *cumulative* change.

The operationalization of the proposed hypothesis requires a suitable measure of experience. Briefly, there are two main dimensions of the process whereby relevant experience is acquired. First, it involves a certain *activity*. Second, it takes place over *time*. Corresponding to these two dimensions, experience can be measured in terms of either cumulated production quantities or cumulated years of production. These two variables are, of course, interrelated; both are endogenous, system-specific measures of time. Nevertheless, the choice between the two has been a subject of considerable controversy (David, 1975; Fellner, 1969; Ishikawa, 1973). The case for the use of the cumulated output variable as a measure of experience is that the design of new techniques is integrally linked to the process of production. The main limitation of this variable is one of redundancy. For example, suppose that once a technical discovery is made, it becomes readily available to all the firms in the industry through imitation, movement of trained personnel, etc. In the circumstances, the relevant variable is the time required to assimilate the available know-how rather than cumulated production as such. Despite this possible limitation, however, the latter seems preferable to the former. As discussed earlier, success in technical problem solving comes largely from actual participation in the production activity. Indeed, one very general conclusion that is repeatedly brought out by various case studies of technological

---

[1]The expression "learning by doing" was coined by Arrow (1962). The term learning in this expression should not be understood in its traditional sense of improvement in the efficiency with which a given task is carried out. In the present context, it is properly understood as a long-term process involving accumulation of relevant know-how in an ever changing task environment. The expression is nevertheless employed here because of its widespread use in the literature on the subject.

development is that changes in the production process lie at the very heart of product innovation. Examples of this are everywhere. The biggest obstacle to the development of a satisfactory solar panel is the difficulty in manufacturing it beyond the production of silicon pellets. The development of the new computer microcircuit, the 64K random-access memory, has been retarded for want of a reliable and efficient process for its production. Numerous other examples of a historical nature will be provided in Section 5.

For the present, the main point is that product innovations often depend on successful changes in the production techniques employed. Further, there is little conclusive evidence that innovative activity tends to be unevenly distributed across the different firms in the industry. Insofar as the occurrence of innovation is a probabilistic process (Chapter 3), the production activity of every firm in the industry is of equal importance. Thus, as regards the *origin* of technical know-how, cumulative output seems to be an appropriate measure of experience. Once the relevant know-how is generated, its successful *adoption* is again a function of learning in the production activity. This is true even in the most unlikely case, where there are no obstacles to the industry-wide transmission of new discoveries. The reason, once again, is simply that product and process technology constitute an integrated system, so that one cannot be changed independently of the other (see also Chapter 5). Moreover, the mutual dependence between the two generally grows stronger over the course of time. Thus, successful assimilation of technical know-how is also a matter of experience acquired in the production process. In all, it seems that the learning by doing hypothesis is best operationalized in terms of a cumulated production quantity variable as an appropriate measure of experience.

Another point deserves attention. While experience plays a central role in the process of technological innovation, the benefits from increased know-how can seldom be obtained as soon as the know-how is acquired. There are many examples of this. With regard to the early history of farm tractor technology, the Fordson model was produced in increasing quantities as it continued to dominate the market during the period 1918–1928. Yet during the same period its technical specifications remained virtually unchanged, notwithstanding the fact that it suffered from poor stability characteristics (Sahal, 1978). This is not to say that the experience acquired in the production of the Fordson was in vain. The point is that it proved to be valuable in the long run rather than immediately. As another example, while the output of steam locomotives continued to increase for several decades past the middle of the nineteenth century, their design remained relatively unchanged from 1850 to 1864 (Bruce, 1952, pp. 26, 39). Further, many innovations that originate in the experience of the past compel a reduction in the current level of output because new production techniques may be required that cannot be introduced at once. A good example of this is the considerable delay in the delivery of the new 360 series of IBM computers from the time of its planned introduction (Abernathy and Wayne, 1974). These specific examples are illustrative of a very general phenomenon. Technical progress and the acquisition of production experience do not always go hand in hand. The association between the two constitutes a *long-term* relationship.

It seems worthwhile to consider two other formulations of the general hypothesis concerning the role of accumulated experience in the process of technological innovation. First, as discussed above, the learning by doing version of this hypothesis assumes that the relevant experience is acquired in the *production* process. It is plausible, however, that at least some of the useful know-how is acquired in the *utilization* of technology. For example, it may well be that in some cases the possibilities for the modification and refinement of a technique are first identified by its users. Indeed, many of the functional defects in the performance of a technique may never become apparent except during the course of its actual operation. Manufacturers may subsequently introduce the requisite changes in the new versions of the technology on the basis of feedback received from users. Frequently, learning is necessitated by failures in the actual use of technology, especially as a consequence of an unwarranted increase in its scale. The intensive R&D in marine propulsion technology during the postwar years was undertaken as a matter of simple necessity, to find out why American war ships cracked. This painstaking investigation took nearly a decade. As Mostert (1975, p. 101) points out, it is going to take much longer to find out the true limits of supertankers. Likewise, a great deal of current R&D in aircraft technology has been in response to the much more frequent tire blowouts with the wide-body jet. Indeed, some of the most important lessons in the development of new techniques are learnt from failures in the course of their utilization. In the circumstances, it is evident that technical progress is dependent on the extent of familiarity with the operation and maintenance of the new technique. The relevant variable in the explanation of the innovation process is then cumulated utilization of technology (i.e., capital stock) rather than cumulated production volume. This suggests what may be called a "learning via diffusion" hypothesis of technological innovation. In its essence, the increased adoption of a technology paves the way for improvement in its characteristics. The hypothesis is obviously complementary to the previously proposed "diffusion by learning" hypothesis (Chapter 5) that improvement in the characteristics of a technology enhances the scope of its adoption.

Second, it is apparent that the technical change process is not always a matter of "learning" or accumulation of relevant experience. In certain cases it is also subject to "nonlearning," or the acquisition of misleading information. Specfically, the proposed hypothesis is well demonstrated by an important exception to it—the oft-cited view that pioneers in the field of technological development suffer a disadvantage relative to newcomers. The reasons are many, such as resistance to change, the effect of sunk costs, and the inherent difficulty of introducing new techniques insofar as they do not conform to the specifications of the existing plant and equipment (Frankel, 1955). Thus it is sometimes claimed that the industrialized countries, such as Britain, are handicapped in their plans to modernize technology by their early start.

This suggests what may be called a hypothesis of "disadvantage of beginning." In its operational form, this hypothesis can be taken to imply that the younger the age of capital stock, the better are the prospects for technical progress. That is, the possibilities of technological innovation become increasingly limited as capital stock grows older. The age variable (i.e., "oldness") in turn can be measured as a ratio of capital stock to gross investment. Finally, it is apparent that if this hypothesis holds, the alternative learning via diffusion hypothesis must be rejected as untenable.

In summary, the foregoing considerations point to three different versions of the first main proposition concerning the role of accumulated experience in the process of innovation: the learning by doing hypothesis, the learning via diffusion hypothesis, and the disadvantage of beginning hypothesis.

The second main proposition of our theory has its origin in the observations that the characteristics of the environment surrounding the use of a technology play an important role in its long-term development. As an example, consider the three distinct types of farm tractor: the track type, the wheel type, and the garden variety. We may well ask why it was necessary to develop the technology along different lines in the first place. A detailed account of the history of the farm tractor is provided later in Section 5. Briefly, however, the answer is that the development of the different tractor types was necessitated by the technical and physical requirements of farming. Among other things, although the characteristics of relatively uneven terrain favored the track type of tractor, it was unsuitable for use elsewhere. Thus it was necessary to develop the wheel type of tractor (including the garden variety) for use on relatively uniform terrain. As another example, the development of locomotive technology has led to the emergence of three different types of machinery to serve different functions: yard switching, road freight, and road passenger service. The distinguishing features of these different types of locomotive should be evident. The design of the switching service locomotive is characterized, among other things, by a relatively short driving wheel base and the absence of a car body superstructure. In contrast, the design of the other two types of locomotive is characterized by a relatively bigger driving wheel base along with a full-width car body and a streamlined nose. The former design is best suited to the requirements of low-speed operation, the latter two types to those of high-speed operation. These specific examples illustrate a very general point: The evolution of a technology often proceeds along more than one pathway so as to meet the requirements of its task environment. This is reflected in the fact that there exist many different types of the same technology, at least during the initial stages of its evolution. Moreover, as indicated by the following case studies, the scope of innovation within each of the different types of a technology is also governed by the nature of the task environment.

Conversely, the process of development is often retarded due to lack of adaptation between technology and the larger system of its use. Historically, the mismatch between technology and its task environment has caused much irreparable damage. A striking example of this is the change, during the British occupation of Egypt in the years 1882–1914 (Richards, 1978), from the old "basin" method of irrigation to perennial irrigation, and of the crop-rotation system within perennial irrigation. The older system, dating back to the time of the pharaohs, consisted of the canals dug through the highland along the banks of the Nile, while the remainder of the land was divided into basins by dikes. The canals made it possible to control the flood waters, while the dikes trapped the sediment-laden water. Under this system, not only was there little use for fertilizers, since the flooding supplied the basins with nutrients every year, but also very little land preparation was needed for crops sown after the flood. Further, the occasional flooding of higher fields as well as the annual washing of the basin lands prevented the deposit of salts in the soils. The new perennial irrigation made it possible to raise the level of water in the Nile and the canals during the dry preflood summer months by means of various dikes, dams, and canals. This enabled both expansion of the arable land and its intensified use via switching from a three- to a two-year crop rotation. However, the benefits that accrued from increased crop yields proved temporary at best, and the long-term consequences were nothing short of disastrous. Not only did the soil begin to deteriorate rapidly as a result of the shortened fallow, but the rise in the water table had the effects of suffocating the deep roots of the plants and salinating the soil. The intensified use of the land further contributed to the rapid growth of insect pests, leading to quite frequent and severe infestations. Such examples of the adverse consequences of technological change are not infrequent. In modern times, the "green revolution" has failed to materialize in certain parts of the world because high-yield plant varieties have not been adapted to conditions where there is lack of water control and little diversity of crops (Mynt, 1972). Likewise, transport technology has failed to take root in many developing countries for want of necessary improvements in the road standards and communications network. All this is to be expected, since the successful development of a technology depends on how well it dovetails with the larger system of its use.[2]

In its influence on the process of innovation, one salient system characteristic is its scale. As has already been discussed at length, the specialization of many vital activities of an organization generally depends on its scale (Chapter 4). For example, the division of labor in cities in terms of both the number and diversity of service establishments, manufacturers, and retail stores depends on their sizes measured in terms of their populations (Zipf, 1949, p. 376). So also the number of occupational specialities as well as the number of occupational types in a "primitive" society are governed by the size of the settlement (Naroll, 1956). Similarly, it is plausible that the development of a technology is a function of the scale of the organization designed to secure its utilization. There are several more specific reasons for this.[3]

First, the very adoption of a technology might depend on a certain minimum efficient size below which (or a miximum size above which) its use may not be optimal. Typically, the introduction of special-purpose machinery becomes feasible only beyond a certain scale. The introduction of the reaper on family-sized farms in the United States provides something of a classic example of this (David, 1975). Although the reaper had been invented in the early 1830s, its use made little headway until the next 20 years. One main reason for this was simply that farm sizes were not sufficiently large for labor savings made possible through the use of the reaper to compensate for the initial cost of its acquisition. It was only when the price of reapers fell relative to the wage rate and the farm sizes increased that widespread use of the reaper became feasible by the mid 1850s. Thus a certain scale of task may be a prerequisite to the adoption of a new technique.

Second, insofar as characteristics of the capital stock employed by large- and small-scale organizations differ from one another, this difference may have a bearing on the diffusion of an innovation. Both age and the size of capital stock are relevant to the phenomenon under consideration. Thus, recall that the former is important since it determines the scope of investment, the latter because it determines the possibilities of learning in the utilization of technology. The two variables may affect the process of technical change in alternative ways. Conceivably, the older the stock, the greater is the speed with

---

[2] For a somewhat similar viewpoint see Reynolds (1966) and Solo (1966).

[3] Note that some of the following considerations are clearly relevant to the *diffusion* of new process techniques. However, they also have a bearing on the process of *innovation* in view of the previously stated learning via diffusion hypothesis.

which a new technique is adopted. But it is equally possible that the larger the stock, the slower is the pace of diffusion of an innovation. Regardless of which of these two variables predominates, however, it is apparent that the scale variable enters in an essential way in the phenomenon under consideration.

Third, it is very generally true that change in the scale of a system beyond a certain point increases the complexity of its structure (cf. Chapter 4). This, in turn, may *warrant* the development of a new technology. An example of this is the installation of dial phones against the background of increase in the volume of telephone calls during the 1950s. As another example, it is often necessary to substitute methods of flow production for batch production with increase in the volume and rate of input utilization. Equally important, insofar as changes of scale are accompanied by an increase in the number of physical and technical operations involved, they may *induce* the use of an advanced technique.

Fourth, in certain circumstances, the adoption of an innovation may depend on the extent to which it can be tried out on a limited basis. The trialability of the innovation is in turn governed by the scale of utilization. Thus, according to one investigation in this area, maximum innovations were found to occur when industries were operating at about 75% of capacity (Mansfield, 1964, p. 27). New techniques could not be introduced when production was at its maximum apparently because of the implied restriction on the scope of experimentation. In sum, it is plausible that the adoption of an innovation is determined not only by the absolute but also by the relative scale of technological utilization.

The fifth reason is somewhat related to the third and the fourth. In many cases, the requisite criteria of stability and reliability cannot be satisfied unless changes in technology are in keeping with the scale of the larger system within which it is embedded. Thus, electric utilities have been observed to obey the rule of thumb that the maximum capacity of a generating unit should be no more than 7–10% of the total capacity of the generating system (Kirchmayer et al., 1955). In this manner, a single unit's failure can be compensated by the utility's normal 15–25% reserve capacity, and the repercussions from the forced outage of a single unit are largely avoided. Actually, the scale effect under consideration is even stronger at a more aggregated level. A striking instance of this is that the maximum capacity of turbogenerators has increased in a fixed proportion to the total installed capacity over a period of nearly five decades (see Chapter 8 for further details). What is more, the scope of innovative activity in other areas such as petrochemicals has been determined by the scale of the overall system in an analogous manner.

Finally, in many cases, the technology of a system must depend on the scale of its utilization due to economic reasons, the origin of which in turn can be traced to certain fundamental factors of a physical nature. For example, many of the technological advances in electricity generation have been made possible by an increase in the scale of the electricity transmission network. The reason for this is simple: Whereas capacity increases with the square of the voltage, the capital and operating costs of transmission facilities increase in direct proportion to voltage (Meek, 1972, p. 74). In consequence, growth in the efficiency of power systems has been made possible, in no small measure, by transmission over greater and greater distances.

Together these considerations point to what may be called a "specialization via scale" hypothesis of technological innovation. This should not be taken to imply that advances in technology necessarily depend on a trend toward "bigness" or "smallness" of the system scale. For example, on the one hand, it may well be that the adoption of a new technique becomes feasible only after a certain increase in the scale of input utilization. On the other hand, the large scale itself may be a barrier to change in technology insofar as it requires a great many alterations in the existing organization of a specialized plant and equipment. According to the proposed hypothesis, variations of scale in either direction affect the course of innovative activity. Within the framework outlined here, the relevance of scale to innovation processes is due mainly to its bearing on the *systemic* nature of technical progress. If there is a crucial variable in determining whether or not a new technique and the prevailing setup of technical operations together form a coherent whole, it is the scale of input utilization. Our proposition is therefore simply that the overall scale of the system governs the state of its technology component.

The above proposition may be readily contrasted with the commonly held viewpoint that it is the state of technology that governs the overall scale of the system. To cite some of the more obvious examples, the substitution of railways for horse carriages extended the scale of transportation far beyond what was previously possible. The development of the diesel locomotive in turn is supposed to have significantly increased average route length. The advances in aircraft technology opened new routes of their own. The advent of the jet aircraft, for example, is believed to have affected the scope of air transportation in much the same way as the diesel locomotive did the extent of ground transportation. Indeed, the list of such examples can be almost indefinitely multiplied.

While there is an element of truth in the conventional viewpoint, it does not give a complete picture of technical progress. Consider the case of advances in ground transportation. All that the available evidence indicates in support of the conventional viewpoint is that the initial breakthrough in the design of the locomotive opened many new routes where none had existed before. A close examination of the facts and figures further indicates that past the initial stage of technical advances, the design of the locomotive was influenced by the existing route structure to a far greater extent than was the route structure by the design of the locomotive. For example, the diesel electric locomotive came into widespread use in Europe by 1925 and it was produced in commercial quantities in the United States at about the same time. However, the technology lay dormant until 1934 simply because the initial design of the locomotive was ill suited to the long-distance route characteristics of American transportation and there was little operating experience to draw upon. A close examination of advances in air transportation reveals an essentially similar phenomenon. The origin of the DC-3 aircraft—an innovation of major significance in the history of piston-type aircraft—lay in the necessity to serve the already existing mail carriage and low-density passenger routes. The advent of jet aircraft did lead to the inauguration of new long-range routes. However, the initial form of the jet was uneconomical for use on short- and medium-range routes. A number of innovations in its design, such as the installation of a fan-type engine, eventually made it possible to use jet aircraft on a wide variety of routes. The history of these developments is more fully discussed in Section 5 of this chapter. The main point here is that a great many advances in transportation subsequent to the introduction of a new technique were necessitated by the requirements of the existing route structure, while the technology in turn had comparatively limited influence on the scale of operations.[4] The lesson to be drawn is that innovations in a technology past the infant stage of its development depend much more on the size of the overall system than does the system size on advances in technology.

To a certain extent, then, the relationship between a technology and the scale of its operations may well be of a mutual causal nature. As indicated by the foregoing discussion, however, we can justifiably assume that technology is more susceptible to the influence of scale than scale is to the influence of technology. Broadly put, there are two important reasons for this. First, technology is only one among a host of factors underlying the observed variations in scale. In general, determinants of the scale tend to be very many and the effect of each relatively small. For example, the increased use of tractors is only one factor in determining farm size. Other factors are types of crops produced, irrigation facilities, topography, relative factor prices, and so on. Put another way, scale cannot be changed in one stroke because it involves a wide variety of constraints. Thus, new canals and waterways cannot be built just because the technology is available. The advent of an innovation may well be a spur to the expansion of navigational facilities. However, this cannot be done at will because there exist numerous other obstacles, such as those posed by geography and demography. For the purpose of analysis, therefore, we can

---

[4] For a somewhat similar viewpoint in the context of plant mechanization see Baldamus (1953).

assume that the causality runs from the overall scale of the system to its technology component.

The second reason is related to the first. The time constants of technological change and scale change differ vastly from each other in their respective magnitude. One key characteristic of the size distribution is that it cannot change rapidly. Witness the fact that modern-day tank ships follow virtually the same route structure as in the past. This allows them to utilize, where possible, many of the same phenomena that assisted navigation from time immemorial, such as the southwest monsoon and its drift, and the Mozambique and Agulhas currents. This in turn makes possible considerable saving in fuel and time (see, e.g., Mostert, 1975, pp. 35–36). Thus, very often it is easier to change the technology in relation to scale than it is to change scale in relation to technology. The causal direction of the relationship between scale and technology will be further discussed on a case-by-case basis in the following empirical investigations. Suffice here to note that it is not entirely without justification to regard scale as a predetermined variable in its relation to technology.

Finally, one important question remains: What is a suitable measure of the system scale? While the scale variable is frequently employed in empirical investigations reported in a wide variety of disciplines including biology, economics, and sociology, it lacks a conceptual underpinning (see Cantley and Glagolev, 1978; Dullemeijer, 1974; and Kimberly, 1976, for a survey of the relevant literature). Typically, the term is used in an *ad hoc* manner and in widely different ways. Without undertaking a detailed discussion of the problem, our own attempt at measurement of scale is as follows. First, it seems that the essence of the theoretical propositions outlined above is best captured when the scale of a system is conceived in terms of its physical aspects. Second, different types of technical operations are best considered in terms of different measures of scale. That is, appropriate measures of scale are different in the case of different situations.

Our proposal then amounts to the following. With regard to advances in agricultural technology, the relevant measure of overall scale concerns some areal dimension of the system (e.g., acreage per farm). There are two reasons for this. First, a number of unrelated investigations such as studies of investment in farm machinery have repeatedly pointed to the importance of this type of scale variable (see, e.g., Fox, 1966; Sahal, 1977). Second, according to the a priori evidence presented in Section 5, farm size emerges as an important determinant of technological innovation in its own right.

As regards advances in transportation technology, one relevant measure of overall system scale is a distance factor or some related variable such as route miles, or number of ports. As indicated by studies in regional science, there exist significant regularities associated with the variation in distance factors (Israd, 1956, Chapter 3). The pattern of commodity flow is a case in point: shipments fall off fairly regularly with distance. In general, the distance factor is found to play a highly significant role in a variety of activities, including not only transportation per se but also family migration and communication.

With regard to technological innovations in electricity generation units, one appropriate measure of overall system scale is plant size in, say, megawatts. At the next higher level, consider an electricity generation system. In this case, one relevant measure of the scale of the larger system is the transmission circuit capability in, say, kilovolts. As for technical advances in electricity transmission, one appropriate measure of the overall system scale is the capability of the distribution system, for example, in terms of the number of subnetworks embraced by the bulk power network or the number of distribution points in it.

In sum, different types of technology are best characterized in terms of different scale variables. There is no universally applicable measure of scale.

3.   ALTERNATIVE HYPOTHESES OF TECHNOLOGICAL INNOVATION

It seems worthwhile to consider certain alternatives to the theoretical propositions regarding technological innovation advanced in the preceding section. Among the various viewpoints expounded in the literature on the subject, two theoretical propositions have commanded widespread belief. One is the hypothesis of demand for technology. The other is the theory of induced innovations. A discussion of each follows.

3.1   The Hypothesis of Demand for Technology

The hypothesis of demand for technology was originally put forth by Schmookler. Its basic idea can be very simply expressed: The inventive output of an industry varies in direct relation to the volume of its sales. In the author's statement (Schmookler, 1962, p. 17):

> To the degree that inventions are made either by producers or consumers of a commodity, more money will be available for invention when the industry's sales are high than when they are low. Increased sales imply that both the producing firms and their employees will be in a better position than before to bear the expenses of invention. Large purchases of a product also suggest that its buyers are better able to finance invention. Finally, the current business practice of setting research budgets at a fixed percentage of sales tends to assure, in recent years, the [validity of the] relation . . . .

Further, the relationship between technological innovation and demand is postulated to hold in both the long run and the short run (Schmookler and Brownlee, 1962, p. 165).

Schmookler (1966) himself provides some convincing evidence in support of the view that invention processes are governed by market demand forces. Concerning the railroad industry, for which extensive data are available, his analysis indicates that there is a close relationship between demand for railroad equipment and inventive activity measured in terms of the number of patents issued. What is more, this relationship is found to hold for a period of more than a century. The hypothesis under consideration is also supported by the results from his analysis of cross-sectional data on a large number of industries in the years both before and after the Second World War. Specifically, his findings indicate that capital goods invention tends to be distributed among industries in relation to the corresponding distribution of investment. However, as Schmookler himself recognized, these results must be qualified in one important respect: they depend on a classification of inventions according to the industry of use, not according to the industry of their origin.

Since Schmookler's original work, a host of investigations have attempted to examine the "demand-pull" hypothesis of technological innovation in one form or the other. It is impossible to discuss these studies here in detail because of limitations of space (see Mowery and Rosenberg, 1979, for a critical review). It is sufficient to point out that there is some evidence in support of the theory. However, it is by no means clear-cut.

According to the critiques, there are two main objections to the theory of demand for technology. First, innovative activity is inherently full of uncertainty. Being an unknown quantity, a priori, demand for a new product cannot be a determining factor in its development. As Nelson has put it (1972, p. 42):

> The assumption of a well-perceived demand curve for product or supply curve for input is plausible only if one can describe mechanisms whereby these curves in fact get well perceived. This would seem to imply considerable experience on the part of the firms in the industry in the relevant environment of demand and supply conditions. This clearly cannot be assumed in an environment of rapid change in either demand or supply conditions. In particular, it seems completely implausible in considering the demand for a major innovation.

Second, if the detailed case studies of the history of technology are any guide, it seems that innovations generally originate either in an attempt to overcome some limitation or bottleneck in the design of a technique (Usher, 1954) or when new technical opportunities present themselves in the light of previous experience. The demand for a product may well be one of the main stimuli for innovation, but this does not guarantee that the solutions of technical problems can, in fact, be found. Rosenberg (1972, p. 51) thus casts his criticism of the theory of demand for innovations in terms of the following quotation from scene 1 of *Henry IV, Part I*:

GLENDOWER: I can call the spirits from the vasty deep.
HOTSPUR: Why, so can I, or so can any man; but will they come when you do call for them?

In summary, numerous arguments have been raised both for and against the theory. In our own opinion the exact status of the theory must be regarded as unsettled pending futher empirical evidence. Anticipating some of the results of the following investigations, however, we acknowledge that the truth seems to lie somewhere between the extreme positions on the subject.

3.2   The Hypothesis of Induced Innovations

The hypothesis of induced innovations was originally set out by Hicks (1932) and further developed by, among others, Fellner (1961) and Kennedy (1964). In its essence, it posits that as labor becomes dearer relative to capital, firms are compelled to adopt labor-saving techniques. This may involve selection from among a set of available techniques (i.e., substitution of one factor of production for another) or the search for entirely new techniques with the explicit objective of saving labor. The gist of the hypothesis is the latter implication. It is as if it were easier to design and produce a technique that saves one factor of production relative to another than a technique without any such bias.

In one important and thoughtful essay, Salter (1969) has undertaken a critical examination of this hypothesis. As he points out, the notion of induced innovation is subject to the serious objection that the basic aim of a firm using a technique is to minimize total costs, and it is a moot point whether this minimization is accomplished by saving labor or capital. In his own statement (Salter, 1969, pp. 43–44):

The entrepreneur is interested in reducing costs in total, not particular costs such as labour costs or capital costs. When labour costs rise any advance that reduces total cost is welcome, and whether this is achieved by saving labour or capital is irrelevant. There is no reason to assume that attention should be concentrated on labour-saving techniques, unless, because of some inherent characteristic of technology, labour-saving knowledge is easier to acquire than capital-saving knowledge.

Nevertheless, as Salter is quick to note, there may be other reasons for a labor-saving bias in technical advances. One reason of a historical nature is that the initial stages of technological development consisted of the progressive mechanization of manual operations. By its very nature such a phenomenon is characterized by a labor-saving bias. We would like to know whether this is still true now that mechanization has reached an advanced stage.

The hypothesis of induced innovation has been much debated in the literature on the subject; limitations of space preclude its further discussion here. However, the interested reader will find the details in the references cited above. For the purpose of the present investigation, it will suffice to examine whether the hypothesis in fact holds when confronted with actual data on technical progress.

4.   FORMAL SPECIFICATION OF THE THEORETICAL PROPOSITIONS

The following attempt at formal specification of the proposed hypotheses is based on theoretical considerations advanced in Chapter 4. To recapitulate briefly, there are two fundamental features of technological evolution. First, the development of a technology inevitably reaches certain dead ends with little prospect for further advances in its capability. This is a short-term process of equilibration. Second, however, such periods of crisis in innovative activity often prove to be temporary. That is, stagnation of technology seldom lasts forever. This is a long-term process of disequilibration.

The phenomenon of equilibration can be formulated in many different ways. One way is to consider the well-known Gompertz function

$$Y = K \exp[-\exp(A - Bt)] \tag{6.1}$$

where Y is a measure of technology, K is an equilibrium stage of development, and A and B are constants. From (6.1) it is evident that the function ranges from zero at t equal to minus infinity to the upper limit K at t equal to plus infinity. One central feature of the proposed formulation is that the relative growth rate $(1/Y) \times (dY/dt)$ decreases exponentially with time, since

$$\frac{dY}{dt} = KB \exp(A - Bt) \exp[-\exp(A - Bt)] = BY \exp(A - Bt)$$

or

$$\frac{1}{Y} \frac{dY}{dt} = B \exp(A - Bt). \tag{6.2}$$

Further, the curve is not symmetrical around its inflection point. Differentiating again, we have

$$\frac{d^2 Y}{dt^2} = B^2 Y \exp(A - Bt)[\exp(A - Bt) - 1]. \tag{6.3}$$

Thus, it can be readily verified that the inflection point occurs at $t = A/B$ where $Y = K/e$.

Letting $P = Y/K$ and upon taking successive logarithms, we can rewrite Eq. (6.1) as

$$\log(-\log P) = A - Bt.$$

Further, if we let $A = \log(-\log a)$ and $B = -\log b$, then

$$\log(-\log P) = \log(-\log a) + t(\log b)$$

or

$$\log P = (\log a)(b^t)$$

or

$$P = a^{b^t}.$$

We therefore obtain an integral form of the Gompertz equation:

$$Y = Ka^{b^t}. \tag{6.4}$$

Equation (6.4) can be rewritten as

$$\log Y = \log K + b^t \log a. \tag{6.5}$$

The derivative of Eq. (6.5) with respect to time is given by

$$\frac{d \log Y}{dt} = b^t \log a \log b. \tag{6.6}$$

Substituting Eq. (6.5) into (6.6), we have

$$\frac{d \log Y}{dt} = \gamma(\log K - \log Y), \qquad \gamma = -\log b. \tag{6.7}$$

According to Eq. (6.7), the rate of growth of Y depends on the parameter γ. Specifically, the greater the value of γ, the greater is the rate of growth of Y.

After approximation of the derivative of log Y by its difference log Y_t − log Y_{t-1}, (6.7) reduces to

$$\log Y_t - \log Y_{t-1} = \gamma(\log K_t - \log Y_t). \qquad (6.8)$$

However, since long-term development is a process of disequilibration, the equilibrium level of growth, K, in (6.8) cannot be assumed to remain constant. According to the two principal theoretical propositions advanced in Section 2, the long-term development of technology (consisting of variations in K) is governed by accumulated experience of a practical nature (X) and the scale of the larger system (Z) within which a given technology is embedded. One simple way to express these relationships is in the forms

$$\log K_t = \alpha_1 + \beta_1 \log X_t \qquad (6.9)$$

and

$$\log K_t = \alpha_2 + \beta_2 \log Z_t. \qquad (6.10)$$

By combining Eqs. (6.8), (6.9), and (6.10), the following two principal models of technological development can be obtained:

$$\log Y_t = \alpha_1(1-\lambda) + \beta_1(1-\lambda)\log X_t + \lambda \log Y_{t-1} \qquad (6.11)$$

and

$$\log Y_t = \alpha_2(1-\lambda) + \beta_2(1-\lambda)\log Z_t + \lambda \log Y_{t-1} \qquad (6.12)$$

where

$$\lambda = 1/(1+B) \qquad (6.13)$$

since

$$\gamma = -\log b = B \qquad (6.14)$$

and

$$0 < \lambda < 1 \qquad (6.15)$$

since B is essentially a positive quantity.

According to these models, the coefficient itself of any given explanatory variable, say X, is an estimate of its effect on the chosen measure of technology, Y, in the short run. The estimate of the long-run effect of X on Y can be obtained by dividing the observed coefficient of X by $1-\lambda$. The parameter λ may be called a coefficient of disequilibration: The lower λ is, the higher is the speed with which innovations within any given area make it possible for the technology to advance from the old to the new equilibrium.

The two other hypotheses of technological innovation discussed in Section 3 can be specified in the same way. That is, both of them can be expressed in the general form of Eq. (6.11) or (6.12) because both demand and the relative price ratio variable are relevant to an explanation of long-term technological development.

It is interesting to note that the proposed models can also be derived from an alternative premise. As discussed in Section 2, one characteristic feature of the role played by accumulated experience in the process of technological development is that it involves significant time lags. In the simplest case, suppose that the state of present technology is determined by the total stock of previously accumulated know-how; that is

$$\log Y_t = c_0 + d_0 \log X_t + d_1 \log X_{t-1} + d_2 \log X_{t-2} + \cdots$$

$$= c_0 + \sum_{j=0}^{\infty} d_j \log X_{t-j}. \tag{6.16}$$

Not all past experience is equally relevant, however. This is taken into account here by means of a mathematical scheme due to Koyck (1954). The essential features of this scheme are well known. In the present context they imply that the role of past experience in technical progress declines geometrically; that is,

$$d_i = \eta^i d_0, \qquad 0 < \eta \leqslant 1. \tag{6.17}$$

By combining Eqs. (6.16) and (6.17) we obtain

$$\log Y_t = c_0 + d_0 \log X_t + (\eta d_0) \log X_{t-1} + (\eta^2 d_0) \log X_{t-2} + \cdots \tag{6.18}$$

and upon lagging by one period

$$\log Y_{t-1} = c_0 + d_0 \log X_{t-1} + (\eta d_0) \log X_{t-2}$$

$$+ (\eta^2 d_0) \log X_{t-3} + \cdots. \tag{6.19}$$

Multiplying Eq. (6.19) throughout by η and subtracting from Eq. (6.18) we obtain

$$\log Y_t = c_0(1-\eta) + d_0 \log X_t + \eta \log Y_{t-1}. \tag{6.20}$$

Equation (6.20) has the same general form as Eq. (6.11) or (6.12).

However, while the final forms of the two models under consideration are identical to each other, their stochastic structures are not. This can be shown as follows. Let u be a disturbance term. As is customary, it is assumed to be a real random variable that is normally distributed with zero mean and constant variance; that is,

$$u_i \sim N(0,\sigma_u^2). \tag{6.21}$$

Further, it is assumd that the values of u_i are independent of the values of any other u_j; that is,

$$E(u_i u_j) = 0 \quad \text{for} \quad i \neq j, \tag{6.22}$$

and every disturbance term u_i is independent of the explanatory variables; that is,

$$E(x_{ji} u_i) = 0 \quad \text{for} \quad j = 1, 2, \ldots, k; \, i = 1, \ldots, n. \tag{6.23}$$

If stochastically specified, Eqs. (6.8) and (6.9) are given by the equations

$$\log Y_t - \log Y_{t-1} = \gamma(\log K_t - \log Y_t) + \upsilon_{1t} \tag{6.24}$$

and

$$\log K_t = \alpha_1 + \beta_1 \log X_t + \upsilon_{2t}, \tag{6.25}$$

respectively, where the υ's are disturbance terms with the properties outlined above. By combining Eqs. (6.24) and (6.25), we obtain

$$\log Y_t = \alpha_1(1-\lambda) + \beta_1(1-\lambda) \log X_t + \lambda \log Y_{t-1}$$
$$+ (\upsilon_{1t} + \gamma\upsilon_{2t}). \tag{6.26}$$

With regard to the alternative derivation of the model, the stochastic forms of Eqs. (6.18) and (6.19) are given by

$$\log Y_t = c_0 + d_0 \log X_t + (\eta d_0) \log X_{t-1}$$

$$+ (\eta^2 d_0) \log X_{t-2} + \cdots + \varepsilon_t \qquad (6.27)$$

and

$$\log Y_{t-1} = c_0 + d_0 \log X_{t-1} + (\eta d_0) \log X_{t-2}$$

$$+ (\eta^2 d_0) \log X_{t-3} + \cdots + \varepsilon_{t-1}, \qquad (6.28)$$

respectively, where ε is another disturbance term whose properties are also given by Eqs. (6.21), (6.22), and (6.23). By combining Eqs. (6.27) and (6.28), we obtain

$$\log Y_t = c_0(1-\eta) + d_0 \log X_t + \eta \log Y_{t-1}$$

$$+ (\varepsilon_t - \eta \varepsilon_{t-1}). \qquad (6.29)$$

Comparison of Eq. (6.26) with (6.29) makes it apparent that the two models in fact differ from each other in their stochastic structures. For example, the disturbance term $(\upsilon_{1t} + \gamma \upsilon_{2t})$ in Eq. (6.26) does not involve an autoregressive scheme, as does the disturbance term $(\varepsilon_t - \eta \varepsilon_{t-1})$ in Eq. (6.29). Moreover, the lagged dependent variable is not independent of the disturbance term in the latter case. Thus in contrast with Eq. (6.26), Eq. (6.29) cannot be estimated by means of ordinary least squares. In the following empirical investigation we have chosen to employ the former rather than the latter model. The choice will be defended on the grounds of simplicity.

5.   CASE STUDIES

This section presents a number of case studies of technological innovation in farm tractors, transportation equipment, digital computers, and electric power plants over the course of time. In each case, a brief account of the history of the technology is provided first. This description draws heavily on the references cited below where the reader will find further details. Against the background of a priori information, the theoretical propositions are then applied to the data on the various cases of technological innovation under consideration. Finally, an overview of the results from the empirical analysis is given at the conclusion.

The chosen measures of technology and the explanatory variables employed in these case studies have been dictated by empirical necessity. In all cases they represent some relevant functional property of the technology under

consideration. As discussed at length in Chapter 2, one main advantage in using functional measures of technology is that they readily provide us with an
objective and appropriately weighted index of both major and minor innovations. For example, the following analysis of technological advances in the
farm tractor in terms of the evolution of the horsepower-to-weight ratio over
the course of time makes it possible to consider a wide variety of innovations
such as the substitution of aluminum for cast iron in the construction of
pistons, the use of pneumatic tires and the three-point hitch and control
system. Whenever the data are available, we have, of course, considered
several alternative measures of advances in one and the same technology.
(Thus, technological innovations in farm tractors, for example, have been further analyzed in terms of the evolution of fuel-consumption and mechanical
efficiencies.) In sum, the investigations reported here are concerned with explaining both major and minor innovations in the long-term development of
new techniques.

Another aspect of the following analyses deserves attention. In the case of
farm tractors and digital computers, the available data on technical change are
based on gross investment in the product. In the case of transportation equipment they are based on the existing stock of products in use. Thus, there are
important differences between these two types of data. In economic terminology, the former case concerns advances in *best-practice* technology, the
latter concerns advances in *general-practice* technology. It is the former case
that comes closest to a true representation of innovative activity.

In both of the above two cases, we have employed the mean values of the
technological variables. Thus, for example, the focus of our analysis is on the
average level of best-practice technology. An investigation of advances in best-
practice technology per se is of no less importance. However, given the
specialized nature of this topic, it is examined separately in Chapter 8. Here we
are solely concerned with the more general aspects of long-term technology
development.

Further, it is customary to make a distinction between product and process innovation. The two are, of course, related: An improvement in production
methods constitutes a process innovation, while an improvement in the
resulting outcome is an instance of product innovation. Put in a somewhat extreme form, the former consists of new techniques of making old things, the
latter of old techniques of making new things. As discussed earlier, the two
types of change frequently take place together. A somewhat related distinction
is between embodied and disembodied technological change. The former consists of innovations in the form of new equipment, the latter in the form of improved methods and organization that increase the productivity of both old
and new techniques. The majority of the following analyses are concerned

with the more general and complex case of product innovation (in farm tractors, transportation equipment, and digital computers). One of the analyses is concerned with both product and process innovations (in electric power plants).

All together, the various cases under consideration thus provide a wide spectrum of long-term technological development processes.

5.1   Technological Innovation in the Farm Tractor (1920–1968)

5.1.1   A BRIEF HISTORY OF THE TECHNOLOGY

The farm tractor of today had its origin in a design very similar to that of the self-propelled steam traction engine of the late nineteenth century. By 1908, two principal forms of gasoline-powered tractor designs had been developed; one was of the wheel type, the other of the track type. Further advances in the technology were made possible by a number of minor and major innovations both in the product itself and in the system of tooling and production (Gray, 1954, 1958; Sahal, 1978; Worthington, 1966). The development of a "frameless" type of wheel tractor in 1913 allowed the belt pulley gears, final drive, and clutch housing to be combined with the engine crankcase into a unitary construction. This assemblage was largely made possible by adoption of various precision techniques of manufacturing leading to the gradual transformation of the production setup from what was essentially a foundry into something of a machine shop. The development of the power takeoff in 1918 made possible the transmission of power from the tractor engine to the various drawn implements.

One of the most successful early tractor models was the Fordson, first marketed in 1917 through Ford car dealerships. According to the results of tractor tests by the Agricultural Engineering Department of Ohio State University, its maximum horsepower was 8.9 and its weight was 2700 lb. Its ploughing speed was 3 miles per hour. The construction of the Fordson was of the frameless type and it was made of cast iron rather than boiler plate steel. Moreover, it was a product of straight-line assembly, which resulted in a substantially lowered cost of production. In 1918 it had already captured about 25% of the market. By 1925, its share of the market had risen to about 70%.

By 1920, a certain basic form of tractor design had emerged which was to remain essentially the same for the next two decades. The subsequent advances in technology were made possible by gradual modifications of this basically invariant pattern of machine design. In the statement of Gray and Dieffenbach (1957, p. 393):

> From the standpoint of development, the 1920 tractor, taken collectively, embodied fundamental engineering and designing found perhaps in more refined

form in tractors of the present day. The one piece cast-iron frame, replaceable wearing parts, force-feed and pressure-gun lubrication, enclosed-transmission, carburetor manifolding, air cleaner, electric lighting and starting, the high-tension magneto ignition with impulse starter, enclosed cooling system, antifriction bearings, alloy and heat-treated steels, and the power take-off had all been introduced, and some experiments had been made with rubber tires. The light-weight low-price tractor had been designed and widely accepted, and several fair-ly successful motor cultivator-type units were on the market.

However, while the reliability of the tractor was considerably improved by this time, its utility was limited to preharvest drawbar operations. Indeed, a tractor was hitherto no more than a "land locomotive"—a pulling machine. It was unsuitable for most farm operations except plowing. The versatility of the machine was gradually improved by a number of technical advances in the following years. One milestone of progress in this area was reached with the successful development of the general-purpose tractor in 1924, the culmination of nearly a decade's efforts. This form of the machine featured high rear-axle clearance and adjustable front-wheel spacing which enabled it to traverse be-tween the rows for cultivation. The introduction of the mechanical lift in 1929 made it possible to lift the implements off the ground when turning at the end of a row. These developments jointly made possible the use of tractors to harvest not only nonrow but also row crops such as corn and cotton.

Since the early 1920s, a number of attempts had been made to use rubber blocks on tractor wheels. However, these attempts were thwarted by the very large size of early tractor drive wheels. Over the years it gradually became possible to reduce the diameter of a tractor's drive wheels so as to secure max-imum traction on weak soils. This reduction paved the way for the successful adoption of rubber tires on tractors in 1934. A new tire design consisting of widened rims with changing outside diameters was introduced in 1938. In con-sequence, the lateral stability of tractors was considerably improved. After much experimentation, it was also found that the tires could be filled with water to provide needed ballast and cushion. The water was eventually re-placed by a solution of calcium chloride because of its desirable antifreezing properties. As a consequence of these developments, the share of tractors equipped with rubber tires increased from barely 14% in 1935 to nearly 95% in 1940.

By this time, the form of tractor design had been taxed to its limit. The development of technology had reached a plateau due mainly to the increase in complexity resulting from the continual modification of a design that was otherwise unchanged in its fundamental aspects. Judging by the growth of various performance measures such as the horsepower-to-weight ratio, there was little prospect for additional improvement. Further advances in technology were nevertheless made possible by increased adoption of the three-point hitch for control of integrated implements that was developed a few years ago. In the statement of Baker (1970a, p. 32)

> Historically speaking the tractor ha(s) evolved around an essentially unchanged configuration. The only true innovation ha(s) been the three-point linkage and control system. Further development (i)s likely to be along established lines . . . .

Although we might question whether the three-point hitch and control system was the only unique development in the design of the farm tractor, it was undoubtedly a major innovation. Its significance lay in its integration of two hitherto largely independent technologies: the tractor and the implement. The resulting simplification of the form of the combined system effectively paved the way for further advances in both technologies. Like virtually all other major innovations in tractor technology, the successful development of the implement linkage and control system was made possible by many seemingly modest advances in the design process. An integral tractor plow supported by a three-point hitch was first incorporated in the Fordson model in 1917. However, it took 17 more years of experimental work to develop its relatively perfected form in 1938. Even so it was plagued by a number of defects. Among other things, it lacked control over the working depth of heavy implements when it was used on light soils. A solution of these problems had to wait for another 15 years. Others are being tackled to this date.

In parallel, numerous other advances in tractor and implement control were taking place. The introduction of hydraulic remote control in 1947 made possible improved control of large drawn implements. The development of the continuous running power takeoff (PTO) in the same year allowed the tractor's clutch to be disengaged without impeding power to the implements. This was soon followed by the introduction in 1950 of the 1000-rpm PTO, which made possible the transmission of higher power. A few years later, in 1953, power steering was introduced. The all-weather cab was widely accepted by this time.

A review of developments since the war reveals two somewhat distinct trends. First, the average belt or PTO horsepower of the tractor has more than doubled, from about 27 hp to 69 hp, during the period 1948–1968. Second, so-called garden tractors of relatively low horsepower have been developed for the very small farm. There has been a rapid increase in the use of these lighter tractors alongside machines of higher horsepower. Other innovations in recent years have been the introduction of dual rear wheels in 1965 and that of four-wheel drive in 1967. They have permitted increased drawbar pull under adverse soil conditions.

In retrospect it is apparent that seemingly minor modifications of the basic design did much to enhance the capability of the machine. Equally important, the major innovations, such as power takeoff, pneumatic tires, and three-point linkage for control of integrated implements, were themselves founded on myriad refinements in the early forms of their design. Thus it may be said that virtually all advances in farm tractor technology including both major and minor innovations appear to have been made possible through the accumulation of design and production experience over the course of time.

Of significance is the fact that the thrust of technical advances in the farm tractor past the initial stage of its evolution has been toward meeting the physical and technical requirements of farms. If the goal of design efforts during period 1900–1920 was to seek a reliable form of the machine, ever since then it has been to improve its versatility. A case in point is the development, noted earlier, of the general-purpose tractor. It was prima facie an outcome of various attempts to adapt the tractor to a far greater range of *farm* operations than in the past. In the words of one perceptive observer of the contemporary scene (Baker, 1931, p. 207),

> When one analyzes the overall design and functioning of the general purpose tractors, one notes at once how the design of the mechanical engineer has been adapted to the requirements of agriculture by the agricultural engineer. The laying out of a successful general purpose tractor calls for a familiarity with crops, soils and topography. Such factors influence height, clearance, tread, center of gravity, turning radius, attachment points ahead, behind and beneath.
>
> There were many efforts to design general purpose tractors before a successful one was produced. The handicap of the early designers was inadequate knowledge of farming. They had not sufficient agricultural engineering training. Only the farm equipment industry could originate the general purpose tractor.

A wide variety of innovations in tractor technology were likewise necessitated by the system of farming. The use of staggered high cast-iron (chilled) lugs in the steel wheels of tractors in the early 1930s was a consequence of the attempt to reduce flotation resulting from using the tractor in sand and similar light soils. The use of rubber tires in tractors was in part necessitated by the requirements of working in weak soils. The development of the three-point hitch and control system was primarily an outcome of the attempt to solve the persistent problem of the integration of tractors with other farm implements. The incorporation of four-wheel drive was necessitated by the poor traction conditions in wet soils. In summary, there is considerable evidence that design of the tractor has been largely shaped by the exigencies of farming.

Moreover, it is striking that the majority of advances in tractor technology have come about through changes in the design of the machine rather than in the system of farming. As an example, in 1927 Peere & Co. introduced a four-wheel cultivating machine intended for three-row planting and cultivation. However, the three-row idea simply did not make any headway and eventually the machine had to be replaced by a tricycle model adapted to both two- and four-row cultivation. The history of the farm tractor is replete with such examples. This is not to imply that advances in technology, in turn, had no influence on the system of farming. It is rather that *certain* characteristics of the system of farming, such as cropping practices and the scale of production units, were the product of evolution over a much longer period than the development of the technology. More often than not, it was therefore a case of adaptation of technology to the farm system rather than change of the farm system in response to technology. Further, many of these characteristics of the farming system were determined by such variables as the mode of irrigation, load-bearing capacity of the soil, insect and pest population, crop mix, and topographical conditions. These variables, in turn, could be changed only within a certain narrow range. Consequently, while advances in technology were largely governed by certain technical and physical characteristics of the farm organization, the former had little influence on the latter.

In terms of its bearing on the nature and direction of technical progress, one of the most salient characteristics of farm organization appears to have been the size distribution of the production units. This is evidenced by the divergence in the evolution of tractor technology that led to two distinctly different designs—the track type and the wheel type, as noted earlier. The crawler or the track type of tractor was especially suited to the requirements of large farms and uneven terrain but could not be efficiently used under other conditions. Thus it was necessary to develop the wheel type of tractor for use on small farms and more or less uniform terrain. Further changes in the design of each of these two main types of tractor are also attributable to the existing composition of the farm size structure. Thus the development of the garden version of the wheel-type tractor has evidently been necessitated by the requirements of very small and much less specialized farms.

The causal relationship between technological innovations in the tractor and farm size distribution as outlined here is further supported by evidence from other unrelated studies of the subject. According to Sargen (1975, pp. 229–231),

A distinct feature of the process [of tractorization] in the Northern part of the U.S. is that the farm size distribution remained fairly stable over time. While there was a tendency for a decline in the number of farms over time (and a concomitant increase in the average farm size), the variance of the farm size distribution in the North was basically unaltered by "tractorization". This suggests that, to a significant extent, tractor technology adapted to the existing farm structure, and it is relevant for us to consider why this was the case.

The role that the tractor building industry played in developing a wide variety of models and tractor sizes was certainly a critical factor. A quite discernable pattern is evident on the part of U.S. tractor manufacturers. The gasoline tractors that were marketed extensively in the 1910's and 1920's were large machines that were exclusively suited for the highly specialized grain farms in the Mid-West, where the topography was generally quite flat. With the development of the general purpose tractors in the 1920's, along with the change from steel wheels to rubber tires, the potential market for tractor services in effect was extended to small, less specialized farms.

. . .What is significant is that tractor producers were able to find technical solutions to preserve the existing farm size distribution while allowing mechanization to go forward to smaller farms. This at least raises the possibility that technical solutions may also be developed to overcome the barrier of scale in developing countries, such that the benefits of mechanization on agriculture productivity can be dispersed in a uniform fashion throughout the agricultural sector.

It should be noted that in contrast with the northern part, there was considerable change in the existing farm size distribution in the southern part of the United States. However, there is no conclusive evidence to indicate that this was mainly a consequence of tractorization itself. Moreover, the changes in farm size distribution in both North and South were confined to the portion of the size distribution under 100 acres (Sargen, 1975, p. 232). Above this size, the dispersion of the overall farm size distribution remained by and large unchanged over the course of time.

The predominantly one-way dependence of technological innovation on farm structure has undergone little change over the course of time. It is even likely to persist in the forseeable future. As one spokesman of the industry has put it (Baker, 1970b, p. 391),

The future of tractors appears most vulnerable to new methods of land preparation, planting and cultivating. If an alternative to high horsepower requirement for land preparation is found practical, then the present evolutionary trend in tractor design could alter drastically.

The general picture that emerges from these considerations is that certain predetermined technical characteristics of farm organization have played an important role in the phenomenon under consideration. Specifically, the long-term development of tractor technology appears to have been governed in no small measure by the existing farm size structure.

In summary, the a priori information does not refute the role of learning and scale factors in the development of farm tractor technology. In what follows we will therefore proceed to examine whether the theory is also supported by the results of quantitative analysis.

5.1.2 RESULTS OF EMPIRICAL ANALYSIS

The empirical analysis reported in this section is concerned with changes in three chosen measures of technology during the period 1920–1968: average fuel-consumption efficiency measured in terms of drawbar horsepower-hours per gallon of fuel used (Y_1); average mechanical efficiency measured in terms of the ratio of drawbar horsepower to belt or PTO horsepower (Y_2); and average belt or PTO horsepower per 1000 lb of unballasted tractor weight (Y_3). The horsepower variable in the chosen measures (Y_2, Y_3) refers to belt horsepower during the period 1920–1958 and to the PTO horsepower during the period 1959–1968. This is simply a reflection of the fact that modern tractors no longer have integral belt pulley power takeoffs. Further, the tractors were first designed to be ballasted in 1941. The data on these variables have been obtained from laboratory tests of tractor performance under both ballasted (Y_1, Y_2) and unballasted conditions (Y_3). In all three cases the data pertain to tractors marketed in the United States.

The advances in fuel-consumption efficiency have been due to the successful use of pneumatic tires, improvement in the quality of fuel, the shift from gasoline to propane and diesel fuel, and the use of more durable valve, piston, and ring alloys, which have allowed, among other things, an increased compression ratio, thereby improving the utilization of fuel. It should be noted, however, that the chosen measure of advances in technology has received very different priorities by manufacturers in different countries. The improvement in fuel-consumption efficiency has been one of the main lines of development in the case of tractors built in Europe but not of those produced in the United States. Particularly in the postwar period, 30- to 110-hp tractors have been manufactured in Europe even by U.S. multinational corporations. Machines of over 110 hp have been produced mainly in the United States, with many machines of over 250 hp being derivatives of construction and earth-moving vehicles. Accordingly, the data employed in the following analysis consist of two populations with machines of less than 110 hp being fuel efficient and those of more than 110 hp not being so. Admittedly, as the average horsepower of tractors sold in the United States has approached this 110-hp benchmark, there has been a varying bias in the data base. In sum, the chosen measure of fuel-consumption efficiency is sensitive to the differing conditions under which tractors have been produced and sold in different countries. There does not seem to be any easy way in which this problem can be circumvented. When interpreting the results of analysis of the growth in fuel-consumption efficiency it will therefore be useful to keep in mind that they are possibly subject to a certain limitation.

At first glance, mechanical efficiency may not seem to be a very important measure of the technology under consideration. Hypothetically, if it were a highly rated technical parameter, tractor producers and farmers would opt for crawlers and not wheeled machines. However, there are several reasons to regard it as a significant variable in the analysis of technological change. To begin with, mechanical efficiency is defined as the ratio of drawbar horsepower to belt or power takeoff horsepower. It is therefore an appropriately *normalized* measure that is largely independent of changes in the absolute size of the technology. Moreover, as discussed below, it does reflect a great many changes in the design and performance characteristics of the tractor over the course of time. It is also a proxy measure of many qualitative changes such as the improved maneuverability of the machine.

The observed advances in mechanical efficiency have been largely made possible by the development of the power takeoff and the successful use of pneumatic tires. Further improvements in the chosen measure of efficiency have been due to the development of the three-point hitch for control of integrated implements, and more recently, to the use of four-wheel drive.

The use of pneumatic tires on tractors deserves further explanation. On the surface of things, it might seem to be merely a by-product of tire manufacturers' development programs, which were mainly devoted to other industrial applications. As discussed earlier, however, this was hardly the case. In reality, there always existed a pressing need for the use of rubber tires—indeed, the earliest attempts to use rubber blocks on steam traction engines for farm use date back to 1871—and tire manufacturers were ever ready to oblige. Yet it took nearly a decade of systematic effort before the use of rubber tires on tractors could make any headway at all. The final breakthrough in 1932 came about not so much as a result of the efforts of tire manufacturers as it did from the attempts of tractor manufacturers to reduce the size of the drive wheels. Subsequent improvements leading to more effective use of rubber tires were likewise primarily an outcome of learning in the tractor manufacturing industry.

The horsepower-to-weight ratio is perhaps the most stringent of all three measures of technology. This is because in spite of weight-transferring innovations, tractors are essentially counterbalanced machines. To a certain extent, any reduction in the specific weight of the tractor must in turn be compensated by adding more and more front weights. Thus the observed growth in the horsepower-to-weight ratio over the course of time has been somewhat sluggish, although far from negligible. The advances have been due to an increase in the working speed of tractors made possible by the use of pneumatic tires, the improved utilization of tractor weight made possible by the three-point hitch and control system, and the increase in the operating speed by various means such as the use of aluminum instead of cast-iron pistons. Further improvements in the horsepower-to-weight ratio during recent years have been

due to the use of high-speed, turbocharged engines, although they were not
necessarily introduced for this reason.

In summary, advances in the three chosen measures of technology reflect
a wide variety of innovations in the farm tractor over nearly five decades from
the time of its genesis.

The results from the test of the learning by doing hypothesis of
technological innovation are presented in Table 6.1. As can be seen from the
values of the coefficients of determination and the corresponding values of the
F ratio, the theoretical relationship performs extremely well in the explanation
of changes in all three chosen measures of technology. Because of the presence
of the lagged dependent variable in these equations, the Durbin–Watson test
statistic is biased toward 2 and should not be regarded as evidence against correlation in the residuals. Its estimate is provided here as a test for the presence
rather than the absence of serial correlation (This interpretation of the
Durbin–Watson test applies throughout to all the results presented in this
chapter.) In this regard, the performance of the model is highly satisfactory.
The coefficients of the cumulative production variable are significant and are
of positive sign in all three cases, in keeping with the theoretical expectation.

Table 6.1:   Test of the Learning by Doing Hypothesis of Technological
Innovation in the Farm Tractor (1920–1968)

Equation	Estimated relationship[a]
(1)	$\log Y_{1(t)} = -0.05 + 0.07 \log X_t + 0.68 \log Y_{1(t-1)}$
	(0.08) (0.04) (0.12)
	$R^2 = 0.93,\ S = 0.03,\ F = 247.6,\ N = 42,\ \text{D–W} = 2.24$
(2)	$\log Y_{2(t)} = 0.42 + 0.05 \log X_t + 0.64 \log Y_{2(t-1)}$
	(0.14) (0.02) (0.12)
	$R^2 = 0.89,\ S = 0.02,\ F = 153.84,\ N = 42,\ \text{D–W} = 1.74$
(3)	$\log Y_{3(t)} = -0.07 + 0.09 \log X_t + 0.56 \log Y_{3(t-1)}$
	(0.09) (0.03) (0.14)
	$R^2 = 0.76,\ S = 0.05,\ F = 61.52,\ N = 42,\ \text{D–W} = 1.85$

[a]Definitions: The variables Y_1, Y_2, and Y_3 denote fuel-consumption efficiency,
mechanical efficiency, and horsepower-to-weight ratio, respectively. The explanatory
variable, X, is cumulated tractor production quantities in hundreds; R^2 is the coefficient
of determination, F the ratio of variance explained by the model to unexplained
variance; D–W the Durbin–Watson test statistic; and N the number of observations corresponding to the time periods 1920–1941, 1948–1968. Standard errors of coefficients
are given in parentheses.

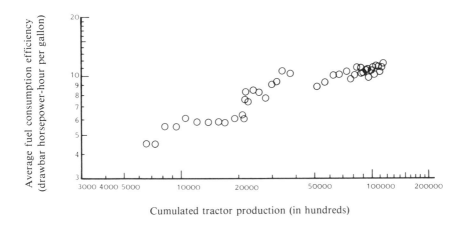

Fig. 6.2   Relationship between fuel-consumption efficiency of tractors and cumulated tractor production, 1920–1968.

Further, there exists a well-behaved pattern of long-term learning in the process of technological innovation. This is shown in Figure 6.2 for the illustrative case of advances in the fuel-consumption efficiency of farm tractors. The observed pattern of advances can be roughly divided into three stages. In the first stage, during the period 1920–1931, growth in the volume of cumulated tractor production has relatively little influence on fuel-consumption efficiency. Past the initial stage, however, performance of technology rapidly picks up with the accumulation of relevant experience. The beginning of this second stage of technological evolution over the period 1932–1941 is clearly attributable to the use of pneumatic tires in farm tractors. The advances during this stage further illustrate a point noted earlier: Major innovations frequently undergo numerous modifications resulting from the accumulation of relevant experience until their potential is fully exploited. Finally, advances in technology can be seen to level off in the third stage of evolution during the period 1948–1968. This slowdown is attributable to the increased use of power-consuming accessories in these years. The systematic nature of the observed pattern is particularly striking inasmuch as it holds over a period of nearly four decades.

The overall fit of the theoretical relationship to the data is further shown here in a diagrammatic form in Figure 6.3. As is evident, its performance is generally very good. In conclusion, a wide variety of the innovations under consideration are demonstrably attributable to learning in the design and production of farm tractors over the course of time.

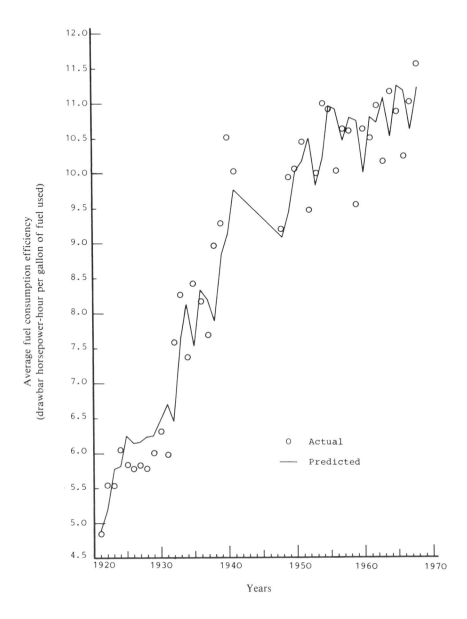

Fig. 6.3   Performance of the learning by doing hypothesis of technological innovation in predicting the fuel-consumption efficiency of farm tractors, 1921–1968.

The results from the test of the two alternative versions of the learning hypothesis are presented in Tables 6.2 and 6.3. It can be seen that the perform-ance of the learning via diffusion hypothesis in the explanation of techno-

Table 6.2:   Test of the Learning via Diffusion Hypothesis of Technological
Innovation in the Farm Tractor (1920–1968)

Equation	Estimated relationship[a]
(1)	$\log Y_{1(t)} = \begin{array}{c} 0.04 \\ (0.04) \end{array} + \begin{array}{c} 0.07 \log X_t \\ (0.04) \end{array} + \begin{array}{c} 0.69 \log Y_{1(t-1)} \\ (0.12) \end{array}$ $R^2 = 0.93,\ S = 0.03,\ F = 246.1,\ N = 42,\ \text{D-W} = 2.25$
(2)	$\log Y_{2(t)} = \begin{array}{c} 0.46 \\ (0.16) \end{array} + \begin{array}{c} 0.05 \log X_t \\ (0.02) \end{array} + \begin{array}{c} 0.66 \log Y_{2(t-1)} \\ (0.12) \end{array}$ $R^2 = 0.89,\ S = 0.02,\ F = 151.6,\ N = 42,\ \text{D-W} = 1.75$
(3)	$\log Y_{3(t)} = \begin{array}{c} 0.05 \\ (0.07) \end{array} + \begin{array}{c} 0.09 \log X_t \\ (0.04) \end{array} + \begin{array}{c} 0.56 \log Y_{3(t-1)} \\ (0.14) \end{array}$ $R^2 = 0.76,\ S = 0.05,\ F = 61.83,\ N = 42,\ \text{D-W} = 1.85$

[a]Definitions: The explanatory variable, X, is tractor stock, i.e., number of tractors on farms in thousands. For the remaining definitions see footnote a, Table 6.1.

Table 6.3:   Test of the Disadvantage of Beginning Hypothesis of Technological
Innovation in the Farm Tractor (1920–1968)

Equation	Estimated relationship[a]
(1)	$\log Y_{1(t)} = \begin{array}{c} 0.13 \\ (0.05) \end{array} + \begin{array}{c} 0.03 \log X_t \\ (0.02) \end{array} + \begin{array}{c} 0.87 \log Y_{1(t-1)} \\ (0.05) \end{array}$ $R^2 = 0.92,\ S = 0.03,\ F = 240.9,\ N = 42,\ \text{D-W} = 2.51$
(2)	$\log Y_{2(t)} = \begin{array}{c} 0.35 \\ (0.13) \end{array} + \begin{array}{c} 0.03 \log X_t \\ (0.01) \end{array} + \begin{array}{c} 0.81 \log Y_{2(t-1)} \\ (0.07) \end{array}$ $R^2 = 0.88,\ S = 0.02,\ F = 148.9,\ N = 42,\ \text{D-W} = 1.86$
(3)	$\log Y_{3(t)} = \begin{array}{c} 0.13 \\ (0.07) \end{array} - \begin{array}{c} 0.0007 \log X_t \\ (0.03) \end{array} + \begin{array}{c} 0.85 \log Y_{3(t-1)} \\ (0.09) \end{array}$ $R^2 = 0.72,\ S = 0.05,\ F = 49.7,\ N = 42,\ \text{D-W} = 2.11$

[a]Definitions: The explanatory variable, X, is the ratio of the number of tractors on farms to the number of tractors sold. For the remaining definitions see footnote a, Table 6.1.

logical change is nearly as good as that of the previously tested learning by doing hypothesis (Table 6.2). What is more, the estimated coefficients of these two relationships are virtually identical to each other. Next, consider the results from the test of the disadvantage of beginning hypothesis (Table 6.3). In the explanation of changes in both fuel-consumption efficiency and mechanical efficiency, the age variable (i.e., the ratio of capital stock to gross investment) enters with a positive sign. As such, this is an eminently plausible result because it implies that the older the tractor stock, the greater the speed with which farmers purchase new machines and hence the greater the rate of technological advance. However, this interpretation contradicts the theoretical proposition under consideration. In the explanation of change in the horsepower-to-weight ratio, the sign of entry of the age variable is negative, but its coefficient is insignificant even though there is little evidence of collinearity. In conclusion, the disadvantage of beginning hypothesis must be rejected for the case at hand. The reasons for the failure of the hypothesis are not difficult to see. *Inter alia*, the hypothesis might at best hold in those cases where experience acquired in the *diffusion* of technology plays no role in the process of innovation. However, in any instance where technological change is a product of learning via diffusion—as in the present case—the hypothesis must be ruled out by definition.

In summary, the results of the empirical analysis do not refute the role played by the accumulation of practical experience in the process of technological innovation. They also indicate that in the case of tractor technology the relevant experience has been acquired in both the machine-building and machine-using sectors.

Table 6.4 presents a test of the "specialization via scale" hypothesis of technological innovation. As indicated by the coefficients of determination and the corresponding values of the F ratio, the relationships perform quite well in an explanation of observed changes in all three chosen measures of technology. The parametric estimates of the farm size variable in the first two equations are insignificant. However, this insignificance is attributable to the high correlation between the explanatory variables themselves. The sign of the farm size variable is positive, indicating that observed advances in chosen measures of technology have been made possible by the increase in the average scale of the production units over the course of time.

Further, there exists a systematic pattern of association between changes in technology and farm size. This is shown in Figure 6.4 for the illustrative case of changes in fuel-consumption efficiency in relation to farm size. It can be seen that the pattern is very nearly S-shaped. The whole pattern may be divided into two stages of technological evolution. In the first stage, during the period 1920–1941 there is considerable advance in the fuel-consumption efficiency of tractors resulting from a relatively small increase in farm size. In the second stage, during the period 1948–1968, there is considerable increase in farm size but relatively little change in the efficiency of tractors. Obviously, increase in the use of relatively big tractors on ever increasing farm sizes during

Table 6.4:   Test of the Specialization via Scale Hypothesis of Technological
Innovation in the Farm Tractor (1920–1968)

Equation	Estimated relationship[a]
(1)	$\log Y_{1(t)} = \begin{array}{ccc} 0.015 & + & 0.05 \log X_t & + & 0.86 \log Y_{1(t-1)} \\ (0.09) & & (0.06) & & (0.07) \end{array}$
	$R^2 = 0.92,\ S = 0.03,\ F = 228.7,\ N = 42,\ \text{D–W} = 2.49$
(2)	$\log Y_{2(t)} = \begin{array}{ccc} 0.21 & + & 0.048 \log X_t & + & 0.82 \log Y_{2(t-1)} \\ (0.10) & & (0.042) & & (0.09) \end{array}$
	$R^2 = 0.87,\ S = 0.02,\ F = 136.29,\ N = 42,\ \text{D–W} = 1.82$
(3)	$\log Y_{3(t)} = \begin{array}{ccc} -0.36 & + & 0.37 \log X_t & + & 0.37 \log Y_{3(t-1)} \\ (0.14) & & (0.10) & & (0.15) \end{array}$
	$R^2 = 0.79,\ S = 0.05,\ F = 73.72,\ N = 42,\ \text{D–W} = 1.83$

[a]Definitions: The explanatory variable, X, is the average acreage per farm. For the remaining definitions see footnote a, Table 6.1.

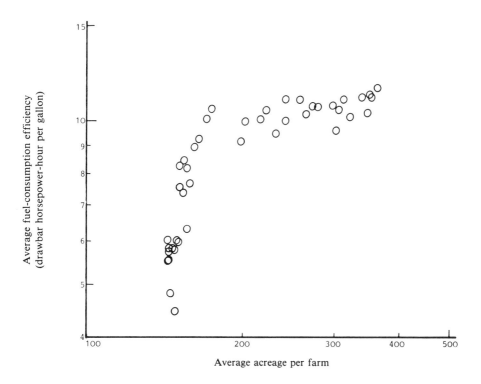

Fig. 6.4 Relationship between the fuel-consumption efficiency of farm tractors and average farm size, 1920–1968.

the postwar years has been at the cost of little saving in their fuel consumption. Apparently, there is an optimal size in the use of technology. As is equally apparent, however, actual changes in technology have not been in keeping with this; there is "overshooting" involved.

The overall fit of the theoretical relationship to the data is also shown in a diagrammatic form in Figure 6.5. As is evident, its performance is generally very good. In conclusion, the observed instances of innovative activity under consideration seem to be in keeping with the manufacturers' attempt to adapt the tractor to changes in the average farm size.

In contrast to the learning and scale factors, a priori, the demand for farm tractors appears to have played relatively little *direct* role in the observed advances in technology. In certain cases, innovations in technology were made available in advance of any widespread demand for them. In other instances, the desired advances in technology did not materialize despite significant demand simply because the solutions of the relevant technical problems could not be found immediately. This was true of both major and minor innovations. The development of the "frameless" type of wheel tractor, as noted earlier, was made possible without any manifest demand for the new technology. In contrast, although from the very outset there existed widespread demand for tractors capable of higher field and road speeds, several decades had to elapse before the successful introduction of pneumatic tires. In postwar years, for example, increased demand appears to have been maintained almost solely by an increase in tractor horsepower. Thus certain innovations may have been in response to the demand, but the great majority of them seem to be due mainly to a technological learning process.

The results from the test of the hypothesis are presented in Table 6.5, where demand is measured in terms of gross investment in tractors by numbers. It might be felt that the demand variable should somehow be adjusted for "quality change" (e.g., by multiplying the tractor sales figures by the average horsepower). However, we have chosen not to do so, for two reasons. First, for the problem at hand, adjustment for quality change in the independent variable is likely to introduce a spurious correlation in the estimated relationships, since the dependent variable is itself a measure of technology. Second, "number of tractors sold" is an important variable in its own right. The inference drawn from the a priori information is generally confirmed by the evidence from the empirical analysis. In the explanation of changes in both fuel-consumption efficiency and the ratio of drawbar horsepower to belt horsepower, the demand variable enters with a negative sign, thereby contradicting the theoretical proposition under consideration. The demand variable is significant and positive in its relationship to the change in the horsepower-to-weight ratio of the tractor, but the explanatory power of the

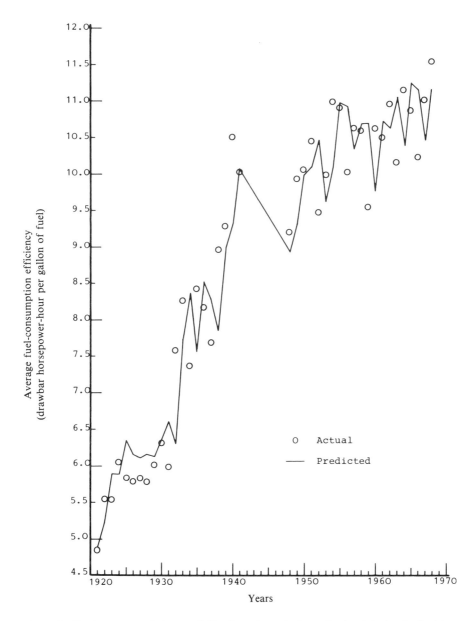

Fig. 6.5 Performance of the specialization via scale hypothesis of technological innovation in predicting the fuel-consumption efficiency of farm tractors, 1921–1968.

equation is relatively poor. In conclusion, the relevance of the hypothesis of demand for technology may well be limited to the special case of cost-reducing innovations. It does not seem to be very generally applicable.

Table 6.5:   Test of the Hypothesis of Demand for Technology in the
Farm Tractor (1920–1968)

Equation	Estimated relationship[a]
(1)	$\log Y_{1(t)} = \begin{array}{l} 0.11 - 0.016 \log X_t + 0.94 \log Y_{1(t-1)} \\ (0.05) \quad (0.02) \qquad\qquad (0.05) \end{array}$
	$R^2 = 0.92,\ S = 0.03,\ F = 227.9,\ N = 42,\ \text{D–W} = 2.64$
(2)	$\log Y_{2(t)} = \begin{array}{l} 0.16 - 0.006 \log X_t + 0.92 \log Y_{2(t-1)} \\ (0.10) \quad (0.013) \qquad\qquad (0.06) \end{array}$
	$R^2 = 0.93,\ S = 0.02,\ F = 131.8,\ N = 42,\ \text{D–W} = 1.94$
(3)	$\log Y_{3(t)} = \begin{array}{l} - 0.01 + 0.07 \ \log X_t + 0.74 \log Y_{3(t-1)} \\ (0.09) \quad (0.03) \qquad\qquad (0.09) \end{array}$
	$R^2 = 0.75,\ S = 0.05,\ F = 58.5,\ N = 42,\ \text{D–W} = 2.13$

[a]Definitions: The explanatory variable, X, is gross investment in tractors, i.e., the number of tractors sold each year (in hundreds). For the remaining definitions see footnote a, Table 6.1.

The test of the induced innovation hypothesis in the present case raises a number of difficulties. Especially problematic is the conceptualization of the relative price ratio (i.e., the price of capital in relation to the price of labor). According to the theory, the marginal product of a technique should be equated to its annalized cost, and this cost includes both the annalized capital cost and the variable cost of operation. The computation of the latter is complicated by the fact that it depends on the rate of utilization of the machine. Moreover, even if the appropriate cost of capital can be calculated, we should perhaps further allow for changes in the working capacity and efficiency of technology. This is because part of the increase in tractor prices is the result of improvement in their performance. Additionally, the functional form of the relationship between technological innovation and relative price ratio needs to be considered in its own right.

In the present investigation, it has not been possible to estimate the cost of capital in accordance with the theory because the relevant data are lacking. Rather, the tractor prices have been employed as such in relation to the wage rate. This is a serious limitation which must be taken into account in evaluating the following results from the test of the theoretical proposition. We have not adjusted the relative price ratio (i.e., the ratio of tractor price to labor) for changes in the performance characteristics of the technology since the very objective of our analysis is to explain how and why such changes have occurred. Moreover, the (unadjusted) relative price ratio is a quite meaningful variable as such. Finally, the causal association between changes in technology and relative price ratio is best regarded as a long-term relationship for it is unlikely to be operative on a day-to-day basis. Consequently, the same form of the model as the one utilized in the test of other hypotheses is applicable.

The results from the test of the model are presented in Table 6.6. The explanatory power of the model is generally very good and the results from the Durbin–Watson test are satisfactory. On the other hand, as can be seen from a comparison of the results in Table 6.6, with those presented in Tables 6.1 and 6.4, the performance of the theory of induced innovations is somewhat inferior to that of the learning by doing and specialization via scale hypotheses of technological change. This is particularly true in the explanation of the growth of the tractor horsepower-to-weight ratio. The coefficients of relative price ratio are significant in all the estimated relationships except Eq. (1), where the independent variables themselves are highly correlated. Most important, however, in all three cases of technological change in farm tractors, the relative price ratio enters with a positive sign, thereby contradicting the theoretical proposition under consideration. This may be partly because, as noted earlier, the chosen measure of relative price ratio falls short of the theoretical prescription. Nevertheless, the fact remains that more than any other industry, farming offers considerable scope for factor substitution.

Table 6.6:   Test of the Hypothesis of Induced Innovations in the
Farm Tractor (1920–1961)

Equation	Estimated relationship[a]
(1)	$\log Y_{1(t)} = 0.05 + 0.06 \log X_t + 0.73 \log Y_{1(t-1)}$
	$\phantom{\log Y_{1(t)} =} (0.05) \quad (0.04) \qquad\quad (0.13)$
	$R^2 = 0.92,\ S = 0.03,\ F = 183.9,\ N = 35,\ \text{D–W} = 2.25$
(2)	$\log Y_{2(t)} = 0.30 + 0.03 \log X_t + 0.77 \log Y_{2(t-1)}$
	$\phantom{\log Y_{2(t)} =} (0.16) \quad (0.019) \qquad\quad (0.12)$
	$R^2 = 0.91,\ S = 0.02,\ F = 161.6,\ N = 35,\ \text{D–W} = 1.79$
(3)	$\log Y_{3(t)} = 0.10 + 0.09 \log X_t + 0.51 \log Y_{3(t-1)}$
	$\phantom{\log Y_{3(t)} =} (0.08) \quad (0.038) \qquad\quad (0.15)$
	$R^2 = 0.68,\ S = 0.05,\ F = 34.16,\ N = 35,\ \text{D–W} = 1.81$

[a]Definitions: The explanatory variable, X, is the ratio of tractor price to labor. The number of observations corresponds to the time periods 1921–1941, 1948–1961. For the remaining definitions see footnote a, Table 6.1.

Herein we would expect the theory of induced innovations to hold the best. This expectation is not borne out by the empirical results from the test of the theory.

It would nevertheless be wrong to reject the basic hypothesis under consideration. According to the earlier results, one important determinant of observed advances in technology has been change in farm size. However, the observed variations in relative price ratio over time reveal that the increase in farm size may well be at least partly attributable to an increase in real wages (see Figure 6.6). This seems to be particularly true during the period 1933–1941 (compare Figure 6.4 with 6.6). Indeed, increasing the operated land area is typically one of the few ways of increasing labor productivity. Following this logic, the theory of induced innovations may still be relevant to the case at hand, albeit in a different guise.

One possible shortcoming of the preceding analysis is that it considers the various hypotheses in isolation from each other. In reality, however, signifi-

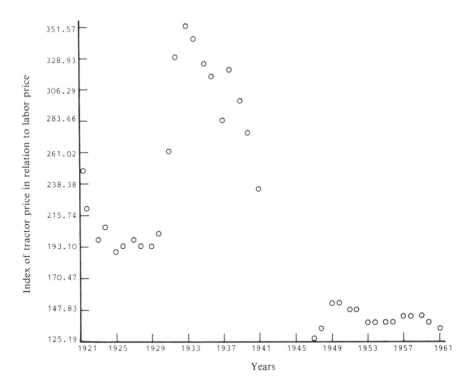

Fig. 6.6 Temporal variations in the index of tractor price in relation to labor price, 1921–1961.

cant interactions may well be involved. For example, farm size and tractor stock variables may exert a *combined* effect on innovative activity insofar as large farms require a larger tractor stock, which then induces additional technological advances. It is therefore of considerable interest to combine the various theoretical propositions into a single "hybrid" model of technological innovation.

The parametric estimates of the proposed model are presented in Table 6.7. Equations (1) and (2) correspond to an explanation of improvement in the fuel-consumption and mechanical efficiency of the farm tractor during the periods 1921–1941 and 1948–1961. The standard errors of the estimated coefficients are generally very high due to multicollinearity between the explanatory variables. The sign of the relative price ratio is wrong in both (1) and (2). The scale variable enters with a positive sign, indicating that the main consequence of increase in farm size has been greater specialization in the use of new machinery rather than greater efficiency in the use of existing stock. The age variable also enters with a positive sign, indicating that a portion of the observed changes in the technology has undoubtedly occurred via replacement investment. Equation (3) offers a substantially different explanation of technological change in terms of improvement in the horsepower-to-weight ratio over the course of time. The sign of entry of the relative price ratio is theoretically correct in this relationship. The age variable enters with a negative sign, pointing to the depressing influence of the existing stock of machinery on the pace of technological change. However, the sign of the lagged dependent variable is negative, indicating that the model is misspecified. In summary, the proposed hybrid model does not offer any additional insights into the nature of technological innovation.

Finally, the performance of the various hypotheses under consideration may be compared by testing how well they predict the turning points in the development of technology. This can be easily determined by examining runs in the time sequence of residuals from the fitted model in each case. The results are presented in Table 6.8. Briefly, changes in fuel-consumption efficiency are best explained in terms of the learning by doing hypothesis of technological change. The changes in mechanical efficiency are best accounted for in terms of the specialization via scale hypothesis of technological change. Finally, the growth of the horsepower-to-weight ratio is best explained in terms of the hypothesis of demand for technology. Thus different hypotheses are peculiarly suited to the explanation of different aspects of technological change. In two out of three cases, however, the performance of *either* of the two proposed hypotheses of learning and scaling in technological innovation is superior to that of the alternative hypotheses.

Table 6.7: Parametric Estimates of a Hybrid Model of Technological Innovation in the Farm Tractor (1921–1961)

Equation	Estimated relationship[a]	R^2	S	F	N	D-W
(1)	$\log y_{1t} = -0.49 + 0.11\ \log x_{1t} + 0.19\ \log x_{2t} + 0.008\ \log x_{3t}$ $\quad\quad\ \ (0.56)\ \ (0.09)\quad\quad\ \ (0.19)\quad\quad\ \ (0.03)$ $\quad\quad\quad\quad\quad\quad\quad\quad + 0.81\ \log y_{t-1}$ $\quad\quad\quad\quad\quad\quad\quad\quad\ \ (0.10)$	0.92	0.03	89.9	35	2.45
(2)	$\log y_{2t} = \quad 0.03 + 0.05\ \log x_{1t} + 0.12\ \log x_{2t} + 0.017\ \log x_{3t}$ $\quad\quad\ \ (0.23)\ \ (0.05)\quad\quad\ \ (0.10)\quad\quad\ \ (0.016)$ $\quad\quad\quad\quad\quad\quad\quad\quad + 0.76\ \log y_{t-1}$ $\quad\quad\quad\quad\quad\quad\quad\quad\ \ (0.11)$	0.92	0.02	84.1	35	1.79
(3)	$\log y_{3t} = \quad 0.17 - 0.26\ \log x_{1t} + 0.56\ \log x_{2t} - 0.07\ \log x_{3t}$ $\quad\quad\ \ (0.51)\ \ (0.11)\quad\quad\ \ (0.14)\quad\quad\ \ (0.03)$ $\quad\quad\quad\quad\quad\quad\quad\quad - 0.06\ \log y_{t-1}$ $\quad\quad\quad\quad\quad\quad\quad\quad\ \ (0.17)$	0.83	0.04	36.28	35	2.13

[a]Definitions: The variable y_1 is average fuel-consumption efficiency in horsepower-hours per gallon, y_2 is average mechanical efficiency (ratio of drawbar horsepower to belt horsepower), and y_3 is average belt horsepower per 1000 lb of unballasted tractor weight; x_1 is the ratio of price of tractor to labor, x_2 is average acreage per farm, and x_3 is average age of tractor stock. For remaining definitions see footnote a, Table 6.1.

Table 6.8: Relative Performance of Various Hypotheses of Technological Innovation:

Test of Runs in Time Sequence of Residuals

Measure of technology	Learning by doing hypothesis		Specialization via scale hypothesis		Induced innovation hypothesis		Hypothesis of demand for technology		Hypothesis of disadvantage of beginning	
	Z	P_z	Z	P_z	Z	P_z	Z	P_z	Z	P_z
1. Fuel-consumption efficiency	0.156	0.437	0.798	0.212	1.17	0.119	0.798	0.212	0.798	0.212
2. Mechanical efficiency	0.798	0.212	0.296	0.383	0.692	0.244	−0.413	0.339	0.296	0.383
3. Belt horsepower per 1000 lb of unbal-lasted weight	0.533	0.297	−1.39	0.08	−0.647	0.258	−0.156	0.437	1.737	0.04

Definitions: The variable Z is a unit normal deviate defined as (expected number of runs of signs − observed number of runs of signs)/expected standard deviation of run distribution; P_z is the probability of obtaining a value of Z equal to or less than the computed value.

5.2 Technological Innovation in the Locomotive (1904–1967), Tank Ship (1914–1970), and Aircraft (1932–1965)

5.2.1 A BRIEF HISTORY OF TECHNOLOGICAL CHANGE IN TRANSPORTATION EQUIPMENT

THE DEVELOPMENT OF LOCOMOTIVE TECHNOLOGY.   The history of the modern-day locomotive dates back to 1815 when George Stephenson built an experimental engine for the Stockton and Darlington Railway in the United Kingdom. This engine was essentially an outcome of his attempts to adapt Watt's steam engine to the needs of locomotion over a period of some 35 years. In 1829, Stephenson built the locomotive Rocket, the locus classicus of the modern-day steam locomotive inasmuch as it contained all the essential elements of the subsequent designs (Bruce, 1952; Snell, 1971). It incorporated a horizontal boiler about 3.3 ft in diameter and 6 ft long and it operated at a pressure of nearly 30 pounds per square inch (psi). The firebox was 3 ft by 2 ft and the boiler tubes were approximately 6 ft long. Each of the two front wheels (the drivers) and the two trailer wheels had a diameter of 3.7 ft and 2.5 ft, respectively. The weight of the engine was nearly 9500 lb. It could haul a weight of some 90 tons at a maximum speed of about 16 miles per hour.

One of the earliest American locomotives was the Best Friend of Charleston, placed in operation on the South Carolina Railroad in 1831. The salient features of this locomotive included a vertical boiler capable of generating steam at a pressure of 50 psi, two cylinders, and four wheels (all drivers), each having a diameter of 54 inches. Its weight was 4.5 tons and it had a tractive effort of 400 lb.

In the following year, Baldwin succeeded in building a locomotive model, Old Ironsides, that was to remain in service for more than 20 years. The success of this model paved the way for the establishment of the Baldwin Locomotive Works. The Baldwin output steadily increased during the next three decades, reaching nearly 130 locomotives a year by 1864.

In parallel, a number of important advances were being made in manufacturing techniques. Early locomotives were necessarily crude for want of reliable and accurate means of producing and machining cylinder casting and fabricating boiler plates. By the turn of the century, techniques employed in the foundries were considerably improved and the machine tools for accurate fabrication of the components had become available. In this way it became possible to improve the reliability and efficiency of the locomotive. At the same time the introduction of the automatic air brake (invented by Westinghouse in 1872) and car coupler (invented around 1875) made possible regular long-train operation.

Availability of improved material was another major factor enhancing the capability of technology. One development of particular significance was the substitution of steel rails for iron rails during the Civil War period. First introduced by the Pennsylvania Railroad, the steel rails soon came into widespread use because they could withstand much higher loads and were nearly ten times more durable than iron rails. In 1890 they accounted for nearly 80% of the total track mileage. About a decade later, it became possible to replace iron castings with steel castings in the construction of the component parts of the locomotive.

By this time the innovation potential of technology was virtually exhausted. One main constraint to the evolution of technology lay in the fact that the size of locomotives was limited by the hand-fired engines, which had an intake of about 5000 lb of coal per hour. This in turn restricted the continuous locomotive output to about 1500 hp (Bruce, 1952). Thus locomotives of this era were designed for maximum tractive effort at low speed rather than maximum horsepower.[5] They were best suited to the requirements of the freight service.

In 1906, the Baldwin Locomotive Works introduced the articulated compound engine, an outcome of more than 15 years of prior research work devoted to improving the application of the older form of articulated engines so that they could meet the requirements of high-speed operation. This engine further improved the tractive effort of the locomotive. The Baldwin model was well adapted to the American railroad characteristics and its successful use for freight service ensured its widespread adoption in subsequent years.

A practical form of the mechanical stoker was first used by the Pennsylvania Railroad as early as 1904. By 1912 the first mechanical stoker to use the steam-jet overfeed system of coal distribution was perfected. Another important advance in this area was the substitution, beginning in 1913, of a pneumatically operated power reverse gear for the hand lever. The introduction of the unit drawbar and radial buffer in 1916 eliminated the need for a safety chain in coupling the engine and tender together. These developments jointly made it possible to eliminate the bottleneck in the evolution of technology.

Simultaneously, a number of improvements were being made in the construction of freight cars. Together they made it possible to nearly double the ratio of the capacity to the deadweight tonnage (DWT) of the cars during the

---

[5]Tractive effort is a measure of pull based on boiler pressure and the dimensions of the cylinder and driving wheels. It must overcome three types of train resistance: on curves, on grades, and during acceleration. A locomotive is more and more likely to slip as tractive force exceeds 15–25% of the weight on the driving wheels.

period from 1870 to 1910. While seemingly minor in nature, these changes had profound implications for locomotive technology. In the statement of Fishlow (1966, p. 641),

> Had the powerful twentieth century engines been developed without that simultaneous remarkable advance in freight-car construction, much more of the increased power would have been dissipated in the nonproductive task of hauling dead weight. A higher ratio of dead weight requires either more or heavier trains to deliver the same payload, both involving additional expense. If 1910 tonnage had to be moved in 1870 freight cars, it would have required about 3.3 of them to equal one 1910 car, and at twice the weight. With identical load factors under both technologies, the same loads would have been carried in four trains of identical weight (but with 3.3 times as many cars) as were actually transported in three.

Further advances in manufacturing techniques made it possible to produce the engine-bed frame as an integral casting by the mid 1920s. In 1933, the all-welded boiler was introduced with considerable success.

In the meantime a great deal of progress was being made in the maintenance and service requirements of railroads (Barger, 1951). The introduction of chemically treated ties and the use of improved ballast and tie-plates during the 1920s considerably extended the useful life of the equipment. The adoption of cast-steel frames integral with the cylinder, the chemical treatment of the locomotive boiler water supply, and the introduction of roller bearings during 1929–1939 reduced the frequency of repair work. Together these developments reduced the manpower requirement for maintenance work by nearly 40% over a decade.

The continuous modification of the steam locomotive with reciprocating engine over the course of time had been necessarily accompanied by an increase in the complexity of the machine design. By the mid 1940s, its capabilities had been fully exploited. This made it necessary to develop the diesel-electric locomotive.

One of the earliest attempts to combine the diesel engine and an electric transmission to propel a rail vehicle dates back to 1913, when a diesel-electric rail car was built by the General Electric and the Atlas Diesel companies in Sweden. Some 30 such cars using 150-hp engines and two axle-hung motors were manufactured in the following years for use in Sweden, Denmark, and France. Soon thereafter Sulzer Brothers in Switzerland built a diesel-electric car, five of which were put to use on the Prussian and Saxon Railways in 1915 (Jewkes et al., 1970). The first diesel-electric locomotives in the United States were built in 1917 with GM-50 engines and Lemp controls; three of these locomotives were sold for commercial use (Pinkepank, 1966). However, the success of these early attempts was relatively limited. One main problem lay in the comparatively high weight-to-power ratio of the diesel engine. Moreover, a suitable means for torque conversion to obtain maximum engine power over a wide range of speed was lacking.

The experience gained from the widespread use of diesel engines in submarines during the First World War made it possible gradually to reduce the weight-to-power ratio. In the meantime, the substitution of electric for mechanical transmission had become a real possibility with the development of the direct-current (dc) generator, dc traction motor, and a practical axle drive system, devices originally designed for application in street railways. These developments laid the basis for the application of the internal combustion engine and electric transmission on rail vehicles, first for switching and subsequently for main-line operations.

In 1923 General Electric and Ingersoll-Rand built an experimental diesel-electric switching locomotive. It incorporated a four-stroke diesel using solid injection and capable of developing 300 brake horsepower at 600 rpm. It weighed nearly 60 tons, with the engine and the dc generator weighing 17,000 lb and 97,500 lb, respectively. The successful demonstration of this model on a number of railroads was quickly followed by the production of what is widely regarded as the first commercially successful diesel locomotive, the Jersey Central Line's No. 1000, in 1925. By this time the diesel-electric locomotive had come into widespread use in Europe. However, major spadework lay ahead in adapting it to the long-distance route characteristics of the American transportation system.

In 1928 Charles Kettering of General Motors initiated a systematic program of R&D in the design and production of a two-cycle diesel (Addie, 1977). In 1930 General Motors acquired the Winton Engine Company, a manufacturer of gasoline and distillate engines, and the Electro-Motive Company, which had close ties to Winton. This acquisition set the stage for a concerted attack on many hitherto unsolved engineering problems. Among the various achievements of this program were the development of a unit injector integrating the pumping and fuel-metering functions into a single device; an application of the engine-driven positive-displacement blower for scavenging; and a welded lightweight steel crankcase.

In 1934, a landmark was reached with the development of a combined system of a General Electric dc generator and a redesigned version of the 201A diesel engine initially built for submarines. The success of this model in trial runs established a number of advantages in the use of diesel locomotives. It became apparent that the diesel locomotive could be used on long runs with good performance (e.g., reduced wear on the track at high speeds because of the absence of reciprocating components) and economy owing to its much higher thermal efficiency in comparison with the steam locomotive. This paved the way for the introduction of the diesel-electric locomotive in main-line passenger service.

In the following year, General Motors launched a program for mass production of diesel locomotives. This was an important innovation in its own right, since it marked a departure from the job shop tradition of the past. In 1937, General Motors expanded its manufacturing facilities for integrated production of all major locomotive components including both the diesel engine and electric transmission. The 201A diesel engines were not entirely satisfactory, however, in railroad application. They lacked both power and durability. This led General Motors to embark upon the design of a narrow, V-type two-cycle engine. The 12-cylinder version of this engine was capable of developing 1100 brake horsepower and a traction rating (i.e., horsepower to main generator) of 1000 hp at 800 rpm. In 1938, the Electro-Motive Division of General Motors introduced the E-3 2000-hp streamlined passenger locomotive, powered by two of these 12-567 diesel engines. In the following year, it introduced the FT 5400-hp freight locomotive, consisting of an assembly of four separate FT units rated at 1350 hp each. Between 1940 and 1947, a total of 1097 model FT's were manufactured. The era of streamlined diesel locomotives had thus begun.

In 1941, Alco and General Electric introduced a multiduty diesel locomotive incorporating a 1000-hp engine and an offset cab. This was a predecessor of what later came to be called a general-purpose locomotive combining the yard switching and rail freight operations into a single unit. By 1943, the share of diesel in the locomotive stock had risen to nearly 10% of the total.

In 1949, General Motors introduced the GP-7 1500-hp general-purpose locomotive. Lacking a full-width car body, it was suitable for yard operations inasmuch as it provided good visibility and ready access to components of the propulsion system. At the same time, it could also be utilized in freight operations, where the relatively low speed requirements made a streamlined nose unnecessary. This type of locomotive proved to be highly successful, as is evidenced by the fact that between 1949 and 1954 a total of 2619 GP-7 locomotives were manufactured and sold. Here we can hardly avoid noticing a striking parallel between the evolution of the locomotive and of farm tractor technology in that both have led to the development of a general-purpose design.

These and numerous other innovations in the design of diesel engines paved the way for its widespread adoption. By 1959, the diesel locomotive had virtually displaced the steam locomotive in railroad operation.

In retrospect, it is striking that a great many advances in locomotive technology have been made possible by gradual modification of the basic design.

As in the case of farm tractors discussed earlier, new techniques reached a state of readiness only after elaborate changes. We look in vain for a case where a new technique was developed in a full-fledged form from the very outset. The piston valve, for example, had been used as early as 1890. However, it took another 12 years before it could be fully incorporated in locomotive design. The mechanical stoker required some eight years of refinement before it became practicable. The compound articulated steam locomotive was first developed in Europe in 1888. However, it took another 15 years to increase its speed to acceptable standards. The resulting problem of vertical instability at higher speeds required some three decades of further engineering effort before it was satisfactorily resolved in 1940 by the application of a flat bearing on the boiler support. The diesel engine was first applied in rail vehicles in 1913. However, it took nearly two decades of experimental work before it made any headway in 1934. It took yet another decade to make numerous modifications in the design until the requisite level of performance and reliability was attained. Thus it is apparent that advances in locomotive technology have been made possible by a host of modifications in the basic design through the accumulation of relevant experience.

The general thrust of the advances in locomotive design, as in so many other cases, appears to have been toward greater effectiveness in the role of technology in railroad operations. We have already referred to the emergence of the three distinct types of locomotive to perform the different tasks involved in yard switching, road freight, and passenger service. Beyond that, individual innovations are likewise attributable to requirements of the task environment. The success of the first 2662 articulated compound engine built in 1906 in the United States by the Baldwin Locomotive Works was in no small measure the result of its adaption to the American road requirements. Conversely, the American locomotive's 2882 articulated engine with four cylinders, introduced in 1911, failed to make any headway despite its outstanding design characteristics because it did not meet the operational requirements of the time: it was too powerful for the draft gears of many of the freight cars. Similarly, recent years have witnessed the development of several variants of the general-purpose locomotive so as to meet the specialized requirements of heavy drag operations (which demand a high level of tractive effort in a single unit), of secondary-line operations (where train size and speeds are lower), etc. Likewise, the possible substitution of the electric locomotive for diesel motive power in the future seems to be conditional upon the growth of high traffic density lines.

THE DEVELOPMENT OF TANK SHIP TECHNOLOGY.   The notion of carrying oil in bulk against the ship's plating was originally conceived by Ludwig Noble in 1876. The first oceanic vessel embodying this idea was a steamer,

Zoroaster, constructed in 1878 at the yards of Lindholmen-Motala in Sweden for use on the Caspian Sea. Its engines were amidships and it had a carrying capacity of 250 tons of kerosene in 21 vertical cylindrical tanks. Subsequently, the tanks were removed and the oil carried to the skin. Steam was obtained by burning fuel oil (Lisle, 1936). In the following year, another such vessel, *Buddah*, was completed to ship oil from the Baku fields in Russia to continental Europe. Soon thereafter a number of wooden sailing ships were converted so as to carry oil in their holds. Initially, there were widespread doubts about the feasibility and safety of these vessels. However, the converted ships carried oil so much less expensively than by any other means that bulk transportation of oil continued to grow throughout the following decade.

Hitherto it had been common to experience leakage in rivets and plates. Moreover, the gas in the void space between the tanks or in the double bottoms was a fire hazard. A recognition of these problems led to important changes in tanker design best illustrated by the tank ship *Charlois* and its successors *Era* and *Noka*, built in 1887. The *Charlois* had an oil capacity of nearly 3500 tons in six large tanks and its expansion tanks were big enough so that, with minor modifications, it could carry ordinary cargoes. Equipped with two powerful pumps, it could discharge the entire cargo in about 15 hours. It did not have a double bottom. The design of the *Era* and *Noka* was largely patterned after that of the *Charlois*. Among their prominent features was the ability to carry oil against the bottom as well as the side and deck plating. The ends of the tank space were fitted with cofferdams. The design of these ships was an important landmark in the history of tanker technology. In the statement of Frear (1945, p. 137),

> The "Charlois", "Era" and "Noka" brought to an end a chapter in the evolution of the marine transport of oil and established a precedent in design which has undergone no fundamental change except in equipment and details of construction up to the present time. The financial advantages of carrying oil in bulk had finally become fully realized. Whereas in 1886 there were only about 12 such ocean-going vessels in existence, there were, in 1891, from 70 to 80 running from Batoum to European ports alone.

The bulk transportation of oil had become a flourishing business by the turn of the century. In 1885, virtually all of the cargo from the United States to Europe was carried in barrels. By 1906, however, this practice had been all but discarded in favor of carrying the oil in bulk.

The widespread use of tankers was nevertheless beset by a number of problems. Most new tankers had their machinery aft, which frequently led to failures in the decks due to an increase in the compressive stress. Matters were not helped when in 1981 Lloyd issued a recommendation that the tanks in the middle of the ship be filled with water ballast. A number of tankers with ballast amidships broke into pieces in the course of their voyage.

Hitherto there was considerable resistance to the passage of tankers through the Suez Canal because of the seeming danger this posed to other ships in the restricted waterway. In 1892, the tanker *Murex* was especially built to overcome the apparent difficulty. It had large cofferdams or ballast tanks between its cargo tanks. They were filled with water so that if the ship were to go astray its weight could be reduced by discharging water rather than oil. It is interesting to note that modern-day tankers follow much the same principle.

By 1906, it became possible to employ longitudinal framing in the construction of tankers. One prominent feature of this system of construction was the use of stiffeners connected to the bulkheads or to each other at the ends by brackets. The Isherwood system, as it came to be called, was first incorporated in the ship *Paul Paix*, built in 1908. It permitted an increase in the deadweight capacity and strengthening of the deck against compressive stresses while reducing the material requirements and checking the leakage from the bulkhead of the system. The engines of the *Paul Paix* were placed amidships. However, this system subsequently made it possible for the machinery of large tankers to be placed aft. Its successful use marked the end of the earlier so-called transverse system of construction. The new system went through a series of improvements leading to a "bracketless system," which was first used in the ship *British Inventor*, constructed in 1926.

In parallel, a number of improvements were being made in marine propulsion technology. Louis Hunter (1949) provides one of the most perceptive accounts of the early developments in the history of technology. While his description concerns the evolution of the steamboat on western rivers in the antebellum period, it merits quotation because it is very generally applicable (pp. 121–122):

> The history of the steamboat is also the history of foundry and machine-shop practice, of metalworking techniques and machine tools, and of the practical art of steam engineering. The story is not, for the most part, one enlivened by great feats of creative genius, by startling inventions or revolutionary ideas. Rather, it is one of plodding progress in which the invention in the formal sense counted far less than a multitude of minor improvements, adjustments, and adaptations. The heroes of the piece were not so much such men as Watt, Nasmyth, and Maudslay, Fulton, Evans, and Shreve—although the role of such men was important—but the anonymous and unheroic craftsmen, shop foremen, and master mechanics in whose hands rested the daily job of making things go and making them go a little better. The story of the evolution of steamboat machinery in the end resolves itself in large part into such seemingly small matters as, for instance, machining a shaft to hundredths instead of sixteenths of an inch, or devising a cylinder packing which would increase the effective pressure a few pounds, or altering the

design of a boiler so that cleaning could be accomplished in three hours instead of six and would be necessary only every other instead of every trip. Matters such as these do not often get into the historical record, yet they are the stuff of which mechanical progress is made, and they cannot be ignored simply because we know so little about them.

By the middle of the nineteenth century, the steamship was perfected to a considerable extent. However, as discussed earlier (Chapter 5), it took another three decades for the steamship to prove its superiority over the sailing ship. By 1865, it became possible to substitute the screw propeller for the paddle wheel. The simple engine in turn was displaced by the compound engine about a decade later. It was undoubtedly a development of major significance. As Graham puts it (1956, pp. 82–83),

> Generally speaking, in the 1830's the ordinary steam pressure in marine boilers averaged 5 lb.; in the 1840's, 10 lb., and in the 1850's, with the introduction of the tubular boiler, (patented by Lord Dundonald in 1848) 20 lb. The first major advance from this elementary stage towards 50 and 60 lb. pressure came in the (late) 1860's with the gradual adoption of compound cylinders, accompanied by circular boilers of sufficient stoutness to take the increased pressure and surface condensers that allowed the continual use of fresh instead of sediment-bearing salt water. The introduction of the compound engine marked a notable advance in marine engineering; by passing steam from the first cylinder (where the initial pressure was great) to a second cylinder of greater bore, where there was naturally less pressure per unit of area, the amount of power from a given amount of steam was not doubled, but it was considerably increased. A reduction of almost 60 per cent fuel consumption was effected, which made possible by 1870 a profitable tea trade to China. Had the compound engine been perfected twenty years earlier, there would still have been sailing ships on the ocean, but there would have been no great age of sail as represented by the sixties and seventies; and again, had John Napier and his fellow engineers of Glasgow been able to produce in 1870 pressures of 125 lb. or more to the square inch instead of 60 or thereabout, then and then only would the completion of the Suez Canal have marked the major turning-point in the life of sail.

By 1885, triple-expansion engines came into general marine use. The engines designed during this period were not, however, powerful enough to propel the relatively large vessels. This made it necessary to equip them with twin screws, a practice that became commonplace by the turn of the century.

Steam turbines were used in ships for the first time around 1900. Their main advantage lay in the elimination of vibrations, which made their use particularly attractive in battleships. However, there were no suitable means to use them efficiently at relatively low speeds. By 1910 it became possible to circumvent this problem by means of double helical reducing gear. However, use of the geared turbine was beset by a number of problems (in matching the high-speed turbine to a more efficient propeller speed). Its widespread adoption was therefore delayed until after the Second World War.

The diesel engine was first incorporated in a tanker in 1903 (Goldbeck, 1970; Lisle, 1936; Qvarnström, 1970). Its reliability was considerably improved during the first World War owing to its use in submarines. This paved the way for its widespread adoption in other marine applications. In recent decades, the use of diesel-powered ships has grown rapidly in comparison with that of ships equipped with steam or gas turbines. (For a comparison of their relative advantages and disadvantages see, e.g., Meurer, 1969.) One main reason for this is undoubtedly that the former are considerably more fuel saving than the latter. To be sure, the fuel-consumption efficiency of the turbine engine has significantly improved over the course of time as a consequence of steady increases in operating pressures and temperatures (Robinson et al., 1948). The use of advanced steam data and reheat cycles beginning in the 1960s has led to further reduction in the fuel requirement. However, advances in the diesel engine have generally outpaced those in turbines. For example, the application of turbocharging has made it possible to utilize the energy of exhaust gases, thereby making possible improvement in the horsepower-to-weight ratio of marine diesel engines. These advances, in turn, have made it possible to increase the cargo-carrying capacity of tankers while reducing their crew requirements. In this manner the size of the diesel engine operating on a single shaft has continued to grow over the course of time (Al-Timimi, 1975). In 1963, the maximum capability of a single-screw engine consisted of propelling a 130,000-DWT vessel. Currently, a single screw is capable of generating as much as 50,000 shaft horsepower in order to propel a tanker of 300,000 DWT. Finally, it is now possible to utilize the same grades of heavy fuel oil in running the diesel engine as were once used in steam turbine ships. In recent years many large tankers have been retrofitted with diesels in an attempt to reduce the costs over steam plants.

The average size and speed of tankers have increased considerably over the course of time. In 1900, the average size and speed of the worldwide stock of tankers (of 2000 gross tons and over) were little more than 5000 DWT and 9 knots, respectively. By 1940, the average DWT had nearly doubled and the average speed had increased by 2 knots. At the time it was thought that tanker dimensions had reached their upper limit as dictated by the considerations of maximum practical draft given the available port facilities. However, tanker size increased sharply after the closing of the Suez Canal in 1967. Currently, the average capacity and speed figures stand in excess of 50,000 DWT and 15.5 knots, respectively. The concept of a large tanker itself has changed against the background of rising demand for oil (Keith, 1968). In 1956, the size of the largest tanker, the *Sinclair Petrolore*, was 56,089 DWT. Ten years later, the largest tanker, *Idemitsu Maru*, had a size of 206,000 DWT. The increase in the size of other merchant ships has been less dramatic but still substantial. Further, there has been a rapid growth in the use of specialized ships during the postwar years. Included in this category are liquefied natural gas carriers, pellet ships, container ships, barge carrying vessels, roll-on–roll-off vehicle transporters, etc.

These changes in technology have been made possible in no small measure by various changes in the nature of the land–sea interface and in terminal facilities, including the use of offshore cargo-handling equipment. As Albu puts it (1976, p. 516),

> The concept of sea transport has been undergoing a qualitative change leading to a systems approach in which more attention is paid to the ship and shore. These new ideas are equivalent to the mass production of transport services involving mechanical handling at the ports and technical advances in ship design; the object being to smooth and speed up the flow of goods from factory or plant to the consumers.

Other recent advances in the design of tankers have included the development of the bulbous bow (first tried in the *Cimarron*, ordered by the Standard Oil Company in 1938) and the cylindrical bow, both incorporated for the first time in Japanese vessels in 1967. The former has made it possible to increase the speed and horsepower-to-weight ratio of the tanker while the latter has contributed toward increased stability and reduction in weight.

As in so many other instances of product innovation, a great deal of the technical progress in tankers is attributable to improvement in construction materials and advances in production technology. By the middle of the nineteenth century, iron had been substituted for wood as the main material employed in the construction of ships. Iron, in turn, was displaced by steel around 1885. The bulk transportation of oil on the Caspian Sea was nevertheless being carried out mainly by wooden sailing ships as late as in 1907. One main reason was that sailing vessels were generally regarded as safer than steamers for this purpose. The use of steel in the construction of tankers gradually increased as the number of modifications in tanker design enhanced their seaworthiness. The introduction of the fluted bulkhead in 1939 brought about a considerable saving of steel as well as of maintenance time while improving drainage and resistance against corrosion. In parallel, advances in welding techniques eliminated the need for riveting. However, completely welded tankers were not entirely successful. Many of them developed serious cracks in their plating. Thus, it became necessary to employ a combination of riveting and welding. Welded connections, first used in a ship in 1920, came into general marine use in the 1930s. In the wake of the Second World War, there were two important developments in the construction of ships: the assembly of prefabricated components under one cover (first tried in 1907) and the use of flow production methods for building standardized ships. More recently, a number of new methods of steel-plate marking and cutting have been adopted in the construction of ships of all types (cf. Chapter 5). They have included the use of the optical lofting technique and the photoelectrically controlled cutting machine. Simultaneously, the use of high-tensile-strength steel in the construction of tankers has made possible considerable savings in weight and cost. During the First World War a number of concrete cargo ships were built in the United States. Their hulls were highly resistant to sea stress damage and they could be built in a relatively short time. Currently, the possibility of building a reinforced concrete hull for large liquefied natural gas tankers is being reconsidered.

THE DEVELOPMENT OF AIRCRAFT TECHNOLOGY.   The beginning of civil aviation in the United States is generally traced to the Air Mail and Air Commerce Acts of 1925–1926. The infant state of aircraft technology is best illustrated by the Ford Trimotor, one of the most popular of the early planes, which first entered service in 1929. The Ford 5-AT-C weighed nearly 13,500 lb, with a wing loading of 73 lb per square foot, and was capable of developing 1350 hp with its three engines. It had a seating capacity of 15 and typical cruise speed of about 113 miles per hour. By the end of 1932, the U.S. domestic fleet employed more than 400 planes (Phillips, 1971a). The majority of them—some 60%—were of the single-engine type with an average seating capacity of 10. However, the potential for improvement of the single-engine type of plane was necessarily limited. Thus much of the subsequent technological development centred on multiengine aircraft.

One milestone in technical progress was the DC-3 aircraft, which first entered service in 1936. It incorporated an all-metal structure with two engines and an unbraced monoplane wing (i.e., without struts and wires, as in case of the earlier biplanes). Its other features included small wings with flaps, a variable-pitch propeller (making it possible to change the angle of the blades to the airstream during flight), and a number of drag-reducing devices such as retractable landing gear. Its construction was stressed skin (in which the outer skin of the structure is used to withstand both primary structural loads and air pressure). In all, its aerodynamic and propulsive efficiencies were considerably higher than those of its nearest rivals, including the Boeing and Ford Trimotors and other earlier models of airplanes. Above all, it set the stage for considerable further technical progress in civil aviation.

Among the various sources of improvement in the performance of aircraft with propellers and piston engines were increased size; higher wing loadings due to the use of more efficient wing flaps; the availability of higher-grade fuels; the use of stronger alloys; and improved methods of construction. The use of slotted flaps made it possible to increase wing loading from 24 lb per square foot on the DC-3 to 31 lb per square foot on the DC-4E without significantly increasing the takeoff run. Yet higher wing loadings were made possible by a reduction in skin area in relation to capacity due to a change in the form of the fuselage. The improved quality of fuels made possible an increase in supercharger pressures (i.e., increased compression ratios). Beginning in 1940 it became possible to equip aircraft with pressurized cabins. By 1949, the capacity of aircraft with pressurized cabins was nearly 5% of the total capacity in terms of seat miles per hour.

Together these developments enabled aircraft to operate more efficiently at higher altitudes, thereby making it possible to attain substantially higher air speed than in the past. This is best illustrated by the performance and design characteristics of the DC-6B, which first entered service in 1951. It was capable of developing 10,000 hp with its four engines while weighing 107,000 lb. Its allowable wing loading was 73 lb per square foot and it had a cruise speed of nearly 300 miles per hour with a typical seating capacity of 66. Thus, in less than a quarter of a century since the introduction of the Ford Trimotor, the speed of aircraft had increased 165% and the typical seating capacity grew by a factor of more than 4.

One overriding feature of these advances in aircraft technology stands out: They were made possible by a series of refinements in existing techniques rather than by any radical breakthrough. This is evidenced by the fact that the basic structure of the DC-3 design remained intact long after its sales were discontinued. As discussed in Chapter 2, the DC-3 was a product of a great deal of prior development effort (Phillips, 1971b). In turn, it became a focal point for much subsequent innovative activity. Thus, the essential features of the DC-6 were much the same as those of the DC-3. The difference between the two lay in details rather than in the fundamental aspects of design. While this generation of aircraft went through numerous modifications over the course of time, it remained true to its initial conception. Individually, many of the changes in technology were modest in scope. Collectively, however, their impact was substantial. As Miller and Sawers put it (1968, p. 128):

> What did not happen to airliner design is more interesting than what did in the quarter-century between the introduction of the DC-2 and that of the big jets in 1958. Apart from the first turbine-powered transports in Britain, no major innovations were made in airliner design, though the design of military aircraft was being revolutionized by the introduction of the jet engine. Airliners changed from the DC-2 mostly in size, number of engines and power; and these alterations sufficed to increase their cruising speed from 170 m.p.h. to 310–330 m.p.h. and their range with capacity payload from 600 miles to 4760 miles . . . . The enormous

growth in air travel between the 1930's and 1950's was thus not the result of any great improvement in the design of the airliner, though it was helped by its higher speed and longer range which made international air travel practicable. All the efficiency that made the airliner a cheap enough means of travel to attract passengers in a significant number depended on the innovations of the early 1930's.

By the 1950s, the technology of piston-type aircraft had been stretched to its limit. The time was ripe to exploit the hitherto dormant jet-power aircraft first successfully flown in 1939. However, a number of obstacles stood in the way of the jet engine's use in aircraft. Its fuel consumption was higher than that of the piston engine. Its reliability could not be easily ascertained due to the use of new alloys and the higher temperature at which it operated. Moreover, its capability to develop thrust at low speeds was limited. Nevertheless the combination of a gas turbine with a jet propulsion system did hold considerable promise for the future. In the piston engine, the compression ratio was, of course, limited by the detonation of the fuel charge in the cylinder. No such limit existed in the case of the turbine. In principle, it was possible to attain a higher compression ratio. Most important of all perhaps was the fact that the design of the turbine engine was simpler than that of the piston engine because of the absence of rubbing surfaces. At any rate, Boeing's decision to build the 707 prototype paved the way for replacement of the propeller and piston engine by the turbojet.

The fuel consumption of jet aircraft was gradually reduced with the development of axial flow compressors during 1948–1957. This in turn made possible a reduction in the size and weight of the airplane. Certain other structural and aerodynamic problems arose with the increase in speed brought about by the use of the jet engine in aircraft. Smoothness of skin became an important factor, as did buckling under stress. The solution consisted of making the wing skin out of thicker sheet. This, however, increased the weight. Moreover, the requisite form could not be produced by shaping sheet metal. These problems were eventually resolved by machining the stiffeners for the skin out of the same slab of metal as the skin. The increased weight of the aircraft was in turn offset by the use of higher wing loading through the use of more efficient wing flaps. In 1958, leading-edge flaps with changes in the wing section were introduced, making it possible to reduce drag at high speeds. These innovations expedited the widespread adoption of jet aircraft. By 1960, jet planes provided nearly 49% of the total capacity of the airliner fleet in terms of seat miles per hour.

The installation of the fan type of engine in the early 1960s was another major source of improvement in aircraft technology. Turbofan engines provide greater propulsive efficiency than the conventional turbojet by increasing the total compressor airflow and reducing the net velocity of the jet by directing the additional airflow through the fan rather than through the combustion chambers and turbine wheels. In this way it becomes possible to increase power and fuel-consumption efficiency. Early turbofan engines reduced specific fuel consumption by about 25%. The fuel requirement of the fan-type engine was further reduced by raising the bypass ratio (the proportion of the airflow that passes through only the fan). By 1966, the jet-powered airliner was fully established as the most economical means of transportation for flights of more than 200 miles carrying 50 or more passengers.

5.2.2   RESULTS OF EMPIRICAL ANALYSIS

Before proceeding to the test of theoretical propositions, certain aspects of the available data should be noted. First, the data on technical change in all three cases—locomotives, tank ships, and aircraft—are based on the stock of equipment in use and they are best described by a roughly exponential trend. This has the implication that the values of λ, the coefficient of the lagged dependent variable, are expected to be relatively high. More important, with regard to the specialization via scale hypothesis, it is very likely that the regression relationship concerns proper estimation of technological change as a function of scale rather than the converse because scale will be subject to much greater abrupt shifts than the chosen measures of technology. This is analogous to a classical procedure in agricultural economics: it is customary to assume that abrupt shifts in supply enable identification of relatively stable demand function. To elaborate, the size of the observed residuals in the technology function is expected to be much smaller than that of the scale function since the determinants of the latter are excluded.

Second, however, the available data on accumulated experience in the development of the tank ship and of aircraft are somewhat inadequate. As regards tankers, the data on average service speed and stock are based on worldwide statistics, whereas the data on cumulated production are based on U.S. statistics. The worldwide stock of tankers is perhaps the only sensible measure of cumulated experience in the *utilization* of technology, given that oil companies frequently choose a flag of convenience in registering their vessels. Strict compatibility between different variables then requires that the cumulated *production* experience be measured in terms of the worldwide output of tankers. However, the paucity of historical data does not allow for this. In the case of aircraft, it is well known that some of the important technical advances in civil aviation originated in the military development effort. However, there is no obvious way in which the available data on cumulated production of commercial aircraft can be adjusted to take into account the fact that a part of the relevant experience was acquired in the production of military aircraft. In view of these limitations of the data, it is apparent that the learning by doing hypothesis of technological innovation cannot be adequately tested in the case of tank ships and aircraft.

The results from the actual test of the learning by doing hypothesis in all three cases are presented in Table 6.9. The first thing to be said about these results is that the coefficients of the cumulated production variable are insignificant in the reported equations. However, this is attributable to the high correlation between the two explanatory variables themselves. Second, the coefficients are further sensitive to the origin of the scale (i.e., the time period from which the production figures are cumulated). Nevertheless, the performance of the theoretical model is excellent in all three cases. Thus, the estimated equations leave a mere 1% variance unexplained in the data. This conclusion must be somewhat qualified for the equations concerning technical change in tank ships and aircraft because the results from the Durbin–Watson test are inconclusive. The possible serial correlation in the residuals of these equations is attributable to the limitation of the available data on cumulated production of tank ships and aircraft, as discussed earlier. In all, the R^2 values are nevertheless so high that confidence in the theoretical model is justified.

Table 6.9:   Test of the Learning by Doing Hypothesis of Technological Innovation in Transportation Equipment

Equation	Estimated relationship[a]
(1)	$\log Y_{1(t)} = \quad 0.17 \ + \ 0.0016 \log X_t \ + \ 0.96 \log Y_{1(t-1)}$ $\qquad\qquad (0.03) \quad (0.005) \qquad\qquad (0.01)$ $R^2 = 0.99, \ S = 0.004, \ F = 19333.8, \ \text{D–W} = 1.65, \ N = 42 \ (1904–1945)$
(2)	$\log Y_{2(t)} = \quad 0.007 \ + \ 0.003 \log X_t \ + \ 0.99 \log Y_{2(t-1)}$ $\qquad\qquad (0.01) \quad (0.002) \qquad\qquad (0.01)$ $R^2 = 0.99, \ S = 0.005, \ F = 7093.3, \ \text{D–W} = 1.35, \ N = 57 \ (1914–1970)$
(3)	$\log Y_{3(t)} = \ - \ 0.11 \ + \ 0.007 \log X_t \ + \ 0.99 \log Y_{3(t-1)}$ $\qquad\qquad (0.09) \quad (0.01) \qquad\qquad (0.05)$ $R^2 = 0.99, \ S = 0.008, \ F = 989.04, \ \text{D–W} = 1.23, \ N = 19 \ (1947–1965)$

[a]Definitions: The variables Y_1, Y_2, and Y_3 denote the average tractive effort of the locomotive, average service speed of the tank ship, and average speed of the aircraft, respectively; while the explanatory variable X represents cumulated production quantities in corresponding cases; R^2 is the coefficient of determination, S the standard error of the estimate, F the ratio of variance explained by the model to unexplained variance, D–W the Durbin–Watson test value, and N the number of observations from the time period shown in parentheses. The standard errors of the regression coefficients are given in parentheses.

This conclusion is further substantiated by the fact that the relationship between chosen measures of technology and the respective measures of cumulated production quantities is rather well behaved and systematic in all three cases under consideration (Figures 6.7–6.9). With regard to the case of

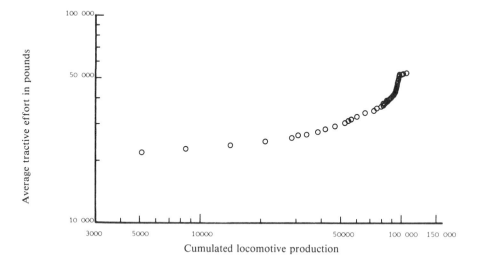

Fig. 6.7   Relationship between average tractive effort of locomotives and cumulative production, 1903–1945.

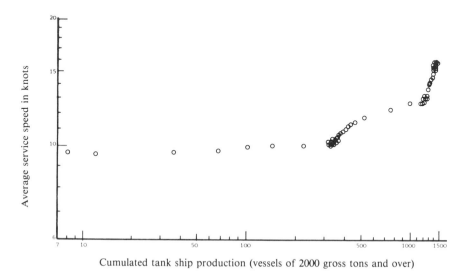

Fig. 6.8   Relationship between average service speed of the tank ship and cumulative production, 1914–1970.

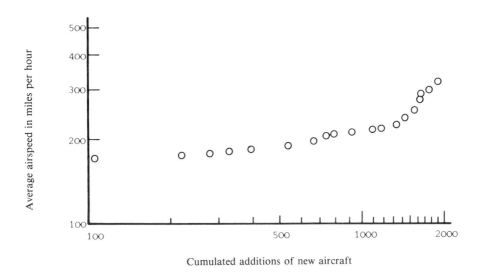

Fig. 6.9   Relationship between average airspeed and cumulative additions of new aircraft, 1947–1965.

locomotives, three distinct stages of technological development may be identified (Figure 6.7). The first stage of development, over the period 1903–1907, may be explained in terms of advances in manufacturing techniques and availability of improved materials for the construction of locomotives. The second stage, over the period 1908–1920, is associated with the development of the mechanical stoker and power reverse gear as well as numerous improvements in the superheater, the feed-water heater, and the force-feed lubricator. Finally, the third stage, over the period 1921–1943, is attributable to the last phase of advances in the steam engine and the gradual exploitation of the diesel engine's potential.

As regards the case of tankers, again, three stages of technological development can be identified (Figure 6.8). In the first stage, during the period 1914–1920, growth in the volume of cumulated tank ships has relatively little influence on service speed. Past this initial stage, the performance of technology rapidly picks up. The second stage of technological evolution, over the period 1921–1942, is attributable to the use of the welded hull and turbine engines in tankers. The development then levels off in the last years of this stage. Finally, the third stage of development, during 1943–1973, is attributable to such recent advances as turbocharging, the use of advanced steam data and reheat cycles, and the use of high-tensile steel in the construction of ships. In summary, while the first phase of development is primarily associated with advances in production technology (strutural engineering), the second stage has to do with improvement in the product itself via improvement in the power plant and the materials of construction.

In the case of aircraft, two stages of technological development can be identified (Figure 6.9). The first stage, over the period 1947–1958, corresponds to the introduction of the pressurized cabin in piston-type aircraft. The second stage of development, over the period 1959–1965, obviously pertains to the evolution of jet aircraft.

In view of the systematic nature of the observed patterns, it is evident that the learning by doing hypothesis is operative in all three cases of technological development.

The results from the test of the specialization via scale hypothesis are presented in Table 6.10. The explanatory power of the relationship is very good, although the results from the Durbin–Watson test are inconclusive in the case of the residuals of the equation concerning aircraft. This seems to be due to the fact that growth of airports and landing fields does not capture all the essential aspects of changes in average route density over the course of time. The coefficients of the scale variable are not well determined, due to multicollinearity between the explanatory variables. This is evidenced by the fact that the growth of technology and of the scale of larger systems for its use are otherwise highly correlated, as shown here in the case of aircraft (Figure 6.10). The observed pattern may be neatly divided in two stages, corresponding to the development of piston-type and turbojet-type aircraft.

The overall fit of the specialization via scale hypothesis to the data is further shown in a diagrammatic form for locomotive technology in Figure 6.11. As is evident, its performance is quite good.

Table 6.10: Test of the Specialization via Scale Hypothesis of Technological Innovation in Transportation Equipment

Equation	Estimated relationship[a]
(1)	$\log Y_{1(t)} = \begin{array}{c} 0.11 \\ (0.23) \end{array} + \begin{array}{c} 0.009 \log X_t \\ (0.045) \end{array} + \begin{array}{c} 0.97 \log Y_{1(t-1)} \\ (0.01) \end{array}$
	$R^2 = 0.99, S = 0.01, F = 4484.1, \text{D–W} = 2.31, N = 64 \ (1904–1967)$
(2)	$\log Y_{3(t)} = \begin{array}{c} 0.03 \\ (0.06) \end{array} + \begin{array}{c} 0.02 \log X_t \\ (0.02) \end{array} + \begin{array}{c} 0.96 \log Y_{3(t-1)} \\ (0.05) \end{array}$
	$R^2 = 0.98, S = 0.01, F = 825.9, \text{D–W} = 1.18, N = 33 \ (1933–1965)$

[a]Definitions: The explanatory variables, X, in Eqs. (1) and (2) are total railroad mileage and the number of airports and landing fields, respectively. For the remaining definitions see footnote a, Table 6.9.

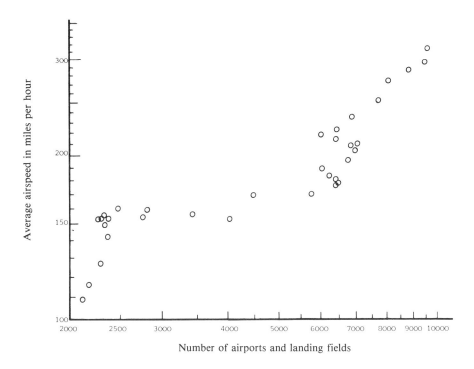

Fig. 6.10   Relationship between average speed of aircraft and number of airports and landing fields, 1932–1965.

The results from the test of the learning via diffusion hypothesis are presented in Table 6.11. The R^2 values are uniformly as high as in the test of learning by doing hypothesis. However, as earlier, the Durbin–Watson test results are inconclusive in two cases. The coefficients of the stock variables in the relationships concerning both the tank ship and aircraft are significant. However, the stock variable enters with a negative sign in the explanation of innovation in locomotive technology. Although this is plausible in that the already existing stock may be an obstacle to innovation, it contradicts the basic hypothesis under consideration.

Next, the hypothesis of the disadvantage of beginning is tested, with the results as presented in Table 6.12. The age variable is measured as the ratio of stock of machines to annual production (i.e., due to lack of data, gross investment is approximated in terms of production). The coefficient of age variable is insignificant in the explanation of change in the tractive effort of the locomotive. However, this is attributable to the problems of multicollinearity. In the explanation of technological change in aircraft and the tank ship, the coefficients of the lagged dependent variables are greater than unity, in violation

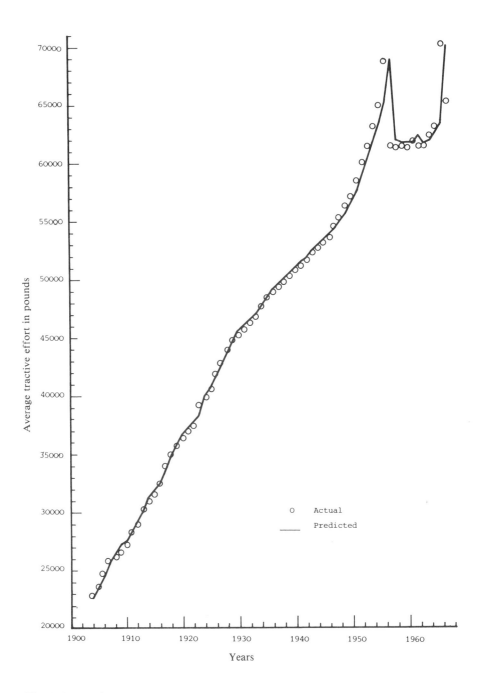

Fig. 6.11   Performance of the specialization via scale hypothesis of technological innovation in predicting the tractive effort of locomotives, 1904–1967.

Table 6.11:   Test of the Learning via Diffusion Hypothesis of Technological Innovation in Transportation Equipment

Equation	Estimated relationship[a]
(1)	$\log Y_{1(t)} = \underset{(0.14)}{0.17} - \underset{(0.017)}{0.002} \log X_t + \underset{(0.01)}{0.96} \log Y_{1(t-1)}$ $R^2 = 0.99,\ S = 0.01,\ F = 4481.8,\ \text{D–W} = 2.32,\ N = 64\ (1904\text{–}1967)$
(2)	$\log Y_{2(t)} = \underset{(0.008)}{0.007} + \underset{(0.002)}{0.004} \log X_t + \underset{(0.01)}{0.98} \log Y_{2(t-1)}$ $R^2 = 0.99,\ S = 0.004,\ F = 13848.7,\ \text{D–W} = 1.25,\ N = 74\ (1900\text{–}1973)$
(3)	$\log Y_{3(t)} = \underset{(0.006)}{0.07} + \underset{(0.013)}{0.029} \log X_t + \underset{(0.03)}{0.93} \log Y_{3(t-1)}$ $R^2 = 0.99,\ S = 0.01,\ F = 935.05,\ \text{D–W} = 1.16,\ N = 33\ (1933\text{–}1965)$

[a]Definitions: The explanatory variables, X, represent cumulated utilization of technology in terms of stock, i.e., the number of machines in service. For the remaining definitions see footnote a, Table 6.9.

Table 6.12:   Test of the Disadvantage of Beginning Hypothesis of Technological Innovation in Transportation Equipment

Equation	Estimated relationship[a]
(1)	$\log Y_{1(t)} = \underset{(0.03)}{0.14} - \underset{(0.0015)}{0.002} \log X_t + \underset{(0.007)}{0.97} \log Y_{1(t-1)}$ $R^2 = 0.99,\ S = 0.003,\ F = 20132.15,\ \text{D–W} = 1.63,\ N = 42\ (1904\text{–}1945)$
(2)	$\log Y_{2(t)} = \underset{(0.004)}{0.0014} - \underset{(0.0009)}{0.0036} \log X_t + \underset{(0.004)}{1.01} \log Y_{2(t-1)}$ $R^2 = 0.99,\ S = 0.004,\ F = 38055.26,\ \text{D–W} = 1.72,\ N = 52\ (1914\text{–}1970)$
(3)	$\log Y_{3(t)} = - \underset{(0.06)}{0.16} - \underset{(0.009)}{0.0013} \log X_t + \underset{(0.02)}{1.08} \log Y_{3(t-1)}$ $R^2 = 0.99,\ S = 0.008,\ F = 963.17,\ \text{D–W} = 1.27,\ N = 19\ (1947\text{–}1965)$

[a]Definitions: The explanatory variables, X, refer to the age of the capital stock defined as the ratio of the number of machines in service to annual production (Eq. (1), (2)) or the ratio of the number of machines in service to annual sales (Eq. (3)). For the remaining definitions see footnote a, Table 6.9.

of the theoretical specification. In the case of the tank ship, however, this may be due to multicollinearity. Finally, the coefficient of age variable is insignificant in its relationship with change in the speed of aircraft even though there is no evidence of significant correlation between the explanatory variables themselves.

Above all, however, the age variable consistently enters with a negative sign, in accordance with the theoretical proposition under consideration. Consequently, the learning via diffusion hypothesis is rejected. Together with the results previously presented, this has the important implication that the origin of learning *in all three cases* of technological change seems to lie in the machine-producing rather than the machine-using sector. There is considerable a priori evidence that technological change in these cases is, by and large, attributable to the initiative of the equipment manufacturers rather than the equipment users. Historically, the railroads have conducted very little R&D of their own. Thus, it is apparent that "much of the innovative drive in the transport sector comes from suppliers to transportation companies" (Gellman, 1971, p. 189). This is likewise true of technological innovation in aircraft. The initiative for advanced airplanes has almost always come from the manufacturers rather than from airlines. Attempts to adhere too closely to the specifications of buyers have frequently proved to be failures (Miller and Sawers, 1968, pp. 259–260). Moreover, there are perhaps only eight major airlines with any leverage to dictate the design and performance characteristics of a new aircraft type. Thus the orders placed by individual airlines do not generally affect the basic design (Williams, 1964, p. 242). In summary, the results of our analysis are in keeping with other qualitative information.

Finally, the results from the test of the hypothesis of demand for technology are presented in Table 6.13. In the explanation of technological change in the locomotive and tank ship, it has been necessary to employ the production variable as an approximation of demand since data on the latter are not available. The production variable enters with a positive sign in both cases. It is insignificant in relation to the tractive effort of locomotives but this may well be due to multicollinearity. The explanatory power of the relationship is very high in both cases. However, as can be seen from the estimated relationship concerning the service speed of tank ships, the coefficient of the lagged dependent variable is equal to unity, in violation of the theoretical specification even though there is no evidence of multicollinearity. The theoretical specification is similarly violated in the estimated relationship concerning the speed of aircraft, despite the fact that the explanatory variables are hardly correlated with one another. Thus, the theoretical proposition must be regarded as somewhat inadequate in its explanation of technological change in both tank ships and aircraft.

Table 6.13:   Test of the Hypothesis of Demand for Technology
in Transportation Equipment

Equation	Estimated relationship[a]
(1)	$\log Y_{1(t)} = \quad 0.12 \; + \; 0.002 \log X_t \; + \; 0.97 \log Y_{1(t-1)}$ $\qquad\qquad\quad (0.03)\quad (0.0014)\qquad\quad (0.007)$ $R^2 = 0.99,\ S = 0.003,\ F = 20299.4,\ \text{D-W} = 1.64,\ N = 42\ (1904\text{–}1945)$
(2)	$\log Y_{2(t)} = -\ 0.001 \; + \; 0.004 \log X_t \; + \; 1.00 \log Y_{2(t-1)}$ $\qquad\qquad\quad (0.004)\quad (0.0009)\qquad\quad (0.003)$ $R^2 = 0.99,\ S = 0.004,\ F = 41257.8,\ \text{D-W} = 1.84,\ N = 52\ (1914\text{–}1970)$
(3)	$\log Y_{3(t)} = -\ 0.16 \; + \; 0.001 \log X_t \; + \; 1.07 \log Y_{3(t-1)}$ $\qquad\qquad\quad (0.06)\quad (0.008)\qquad\quad (0.02)$ $R^2 = 0.99,\ S = 0.008,\ F = 963.01,\ \text{D-W} = 1.27,\ N = 19\ (1947\text{–}1965)$

[a]Definitions: The explanatory variables, X, refer to annual production quantities (Eqs. (1), (2)) or sales (Eq. (3)). For the remaining definitions see footnote a, Table 6.9.

These results are consistent with a priori evidence. As Phillips (1971a, pp. 148-149) has pointed out in the case of aircraft, demand did play a role in stimulating the development of new airframes in the 1920s. During the 1930s most improvements in technology, such as the development of radial engines, came from funds provided by the government in the early 1920s for military purposes. Commercial aircraft after 1930 were almost exclusively based on the R&D conducted by government agencies, especially NACA. The technology of turbojet and turbofan engines was also developed independently of the demand for new commercial airplanes (Schlaifer and Heron, 1950, pp. 156-198, 332-508). Thus it is hardly surprising to find that technical change in aircraft seems not to be strongly related to demand for innovations. If anything, there is overwhelming evidence to indicate the opposite. That is, demand was, in fact, created by the innovations.

In summary, the theory of demand for technology does hold in an explanation of the improvement in the tractive capability of locomotives. Its performance is much less satisfactory, however, in explaining change in the service speed of tank ships and aircraft over the course of time.

5.3   Technological Innovations in Digital Computers (1944–1967)

5.3.1   A BRIEF HISTORY OF TECHNOLOGY

The invention of the first mechanical device capable of performing arithmetic operations in a digital manner dates back to the time of Pascal and Leibniz in the seventeenth century (Goldstine, 1972). Following these pioneering attempts, Charles Babbage successfully built a small working model of a "difference engine" (a special-purpose machine) around 1822. In the following year, he succeeded in receiving government support for what was to have been the biggest such machine ever built. However, the production technology of the time was far behind the construction requirements of the machine and it was incomplete as late as in 1833, by which time the government support had been withdrawn. Babbage was not, however, deterred in his efforts and he began work on an "analytical engine" (a general-purpose machine) which was to occupy him until his death in 1871. Although this machine was never constructed, its design had all the essential aspects of a modern-day computer. In the statement of Serrel et al. (1962, p. 1042, reference to the figure deleted),

Babbage's design had all the elements of a modern general-purpose digital computer; namely: memory, control, arithmetic unit, and input/output. The memory was to hold 1000 words of 50 digits each, all in counting wheels. Control was to be by means of sequences of Jacquard punched cards. The very important ability to modify the course of a calculation according to the intermediate results obtained—now called conditional branching—was to be incorporated in the form of a procedure for skipping forward or backward a specified number of cards. As in modern computer practice, the branch was to be performed or not depending upon the algebraic sign of a designated number. The arithmetic unit, Babbage supposed, would perform addition or subtraction in one second while a 50×50 multiplication would take about one minute. Babbage spent many years developing a mechanical method of achieving simultaneous propagation of carries during addition to eliminate the need for fifty successive carry cycles. Input to the machine was to be by individual punched cards and manual setting of the memory counters; output was to be punched cards, printed copy, or stereotype molds. When random access to table of functions—stored on cards—was required, the machine would ring a bell and display the identity of the card needed. Although Babbage prepared thousands of detailed drawings for his machine, only a few parts were ever completed.

Except for a few intermittent attempts, little progress was made in this area until the Second World War. One of the first breakthroughs came in 1944 when an electromechanical computing machine, the IBM Automatic Sequence Controlled Calculator (nicknamed Mark I) was successfully made operational at Harvard's Cruft Laboratory. Another noteworthy product of the wartime effort was ENIAC, the Electronic Numerical Integrator and Computer successfully completed in 1946 at the Moore School of the University of Pennsylvania. It is generally agreed that the construction of these machines heralded the dawn of large-scale automatic computation.

The main function of ENIAC was the computation of ballastic trajectories and firing tables requiring integration of a system of ordinary differential equations. Of necessity, it was a special-purpose machine. The concept of a general-purpose, stored-program computer was then proposed in 1945 by John Mauchly, J. P. Eckert, and J. von Neumann. This led to the design of EDVAC, the Electronic Discrete Variable Automatic Computer, which was built during 1947–1950. It contained nearly 5900 vacuum tubes along with 12,000 semiconductor diodes and utilized the binary number system. Its design became the basis for a number of subsequent design efforts. One of the most outstanding of these efforts was UNIVAC I, built in 1951 by Remington Rand. Although it lacked many peripheral devices, it was the first machine embodying the concept of direct recording onto magnetic tape from a typewriter keyboard. Both EDVAC and UNIVAC I utilized mercury acoustic delay lines for high-speed memory and vacuum tubes for logic circuits. Four years later, in 1955, IBM introduced its computer model 702 utilizing electrostatic memory storage devices (cathode-ray tubes) for commercial data processing.

The IBM 702 was, however, subject to a number of limitations (Rosen, 1969). It was too slow—it took 115 microseconds to read a standard five-character instruction. Moreover, its electrostatic memory was far less reliable than the mercury delay line storage of other machines such as UNIVAC I. Magnetic drum storage systems were first conceived in 1933 and were developed in a practicable form by ERA (later Remington) in 1946. Their obvious advantage over other storage systems, including both mercury delay line and electrostatic memory, lay in their potential to substantially reduce the price of computing machines. However, the earlier versions of magnetic drum computers were not very satisfactory; they had to rely on various means such as minimum access-time coding to obtain satisfactory performance. Eventually, their design was perfected. In 1955, IBM began delivery of its highly successful model 705 with its faster and more reliable magnetic core memory.

In parallel, a number of attempts were being made to develop transistors suitable for economical use in computers. Their advantage over the vacuum tube obviously lay in both their small size and their greater reliability. However, the switching speeds of the earliest transistorized computers were relatively slow. One main difficulty was the production of transistors of uniform quality. One important advance in this area was the development of the surface barrier transistor in 1954. This paved the way for widespread use of transistors for high-speed computers.

The initial design of high-speed transistorized computers is best illustrated by LARC, the Livermore Atomic Research Computer, and the STRETCH machine, built by Remington Rand and IBM, respectively, under a contract from the Livermore and Los Alamos research laboratories of the U.S. Atomic Energy Commission. While remarkably fast, they failed to live up to their performance expectations due primarily to their large size. Neither was a commercial success. Nevertheless, their design provided much impetus to further successful efforts in this area. By 1959, transistors had virtually displaced vacuum tubes in nearly all types of computers.

In 1964, IBM announced six new models of its System 360 with the objective of standardizing various computer characteristics such as character codes, instruction codes, and modes of arithmetic. This was soon followed by the development of solid-state hybrid circuits incorporating discrete but very small transistors. The 360 series has had a tremendous impact on computer technology inasmuch as it has been sold in huge quantities and many of its characteristics have been adopted by other manufacturers. Most recently, solid-state medium-scale integration (MSI) and large-scale integration (LSI) units have been developed. They are characterized by the production of entire subsystems as monolithic units. Their examples include ILLIAC IV and the Navy AADC, All Application Digital Computer.

The history of computer technology is often described in terms of new developments in machine components and in the computers themselves (including changes in the architecture and software). There is some disagreement as to what constitutes a new phase of development, or a new "generation," as it is commonly called. However, the following classifications have been more or less accepted (Joseph, 1972; OECD, 1969; Turn, 1974).

CLASSIFICATION OF COMPONENT GENERATIONS

1.   Generation 0 (up to 1953). This generation was characterized by the use of relays and vacuum tubes in building specialized computers such as ENIAC.
2.   Generation 1 (1951–1958). This generation was characterized by the development of commercial computers such as UNIVAC 1, IBM 701, IBM 704, and IBM 709.
3.   Generation 2 (1958–1969). This generation was characterized by the use of transistors. Examples include the IBM 7090 and CDC 6600 as well as such supercomputers as STRETCH and LARC.
4.   Generation 3 (1967 to the present). This generation is characterized by solid-state hybrid circuits used in the IBM 360 series, UNIVAC 1108, etc.
5.   Generation 4. This generation has been called one of solid-state medium-scale integration (MSI) and large-scale integration (LSI). It is characterized by the development of monolithic circuits.

CLASSIFICATION OF COMPUTER GENERATIONS

1.   Generation 1 (1951–1952). This generation comprises special-purpose computers for commercial and scientific use. It is charac-

terized by single job operation, machine language, and
subroutines.

2. Generation 2 (1958–1960). This generation comprises general-
    purpose computers and is characterized by independent and simul-
    tanously operating input–output, high-speed main memory, batch
    processing, higher-level languages, and macroassemblers.

3. Generation 3 (1963–1965). In this generation computer systems
    (i.e., families of computers) were developed for general informa-
    tion processing. These systems are characterized by multiprogram-
    med and time-shared operation, remote-terminal interactive and
    job-entry systems, modular programs, and conversational systems.

4. Generation 4 (1970–1972). This generation consists of networks of
    computer systems for on-line information processing. It is charac-
    terized by direct higher-order language processing, extendable
    languages, meta-compilers and microprogrammable computers.

Each new generation of computer technology has led to important advances in performance and design characteristics. Joseph (1972) estimates that
on average, each new component generation has effected a 10-fold increase in
speed, a 20-fold increase in memory capacity, and a 10-fold increase in reliability, while reducing the component cost 10-fold and the system cost
2.5-fold.

In summary, while advances in computer technology have been truly spectacular, they have been founded on a succession of seemingly minor improvements. Knight (1967, p. 493), commenting on his own extensive research in this
area, states that "most of the developments in general-purpose digital computers resulted from small, undetectable improvements, but when they were
combined they produced the fantastic advances that have occurred since
1940." In conclusion, it is fair to say that the development of so-called science-
oriented, high-technology items is not immune from learning.

5.3.2   RESULTS OF EMPIRICAL ANALYSIS

Due to limitations of the available data, it is not possible to test any
proposition except the learning by doing hypothesis of innovation in the development of computers. The dependent variable employed in the following
analysis is homogeneous computing power measured in operations per second.
It is based on an elaborate model developed by Knight (1966). As noted in the
Appendix, this variable is based on the measurement of and interaction between numerical capabilities (e.g., time to perform a standardized number of
additions), memory capacity, and input–output capabilities (e.g., speed of input–output units).

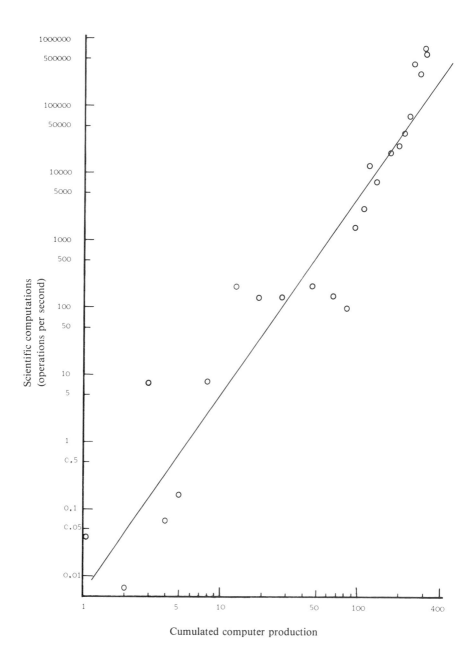

Fig. 6.12 Relationship between computing power and cumulated computer production, 1944–1967.

Table 6.14:   Test of the Learning by Doing Hypothesis of Technological
Innovation in the Digital Computer (1944–1967)

Equation	Estimated relationship[a]
(1)	$\log Y_{1(t)} = -$ 1.75 $+$ $2.43 \log X_t$ $+$ $0.17 \log Y_{1(t-1)}$ (0.73) (0.71) (0.22) $R^2 = 0.95$, $S = 0.75$, $F = 86.56$, D–W $= 1.86$, $N = 23$
(2)	$\log Y_{2(t)} = -$ 1.61 $+$ $2.48 \log X_t$ $+$ $0.07 \log Y_{2(t-1)}$ (0.69) (0.68) (0.22) $R^2 = 0.92$, $S = 0.82$, $F = 60.30$, D–W $= 1.97$, $N = 23$

[a]Definitions: The variables Y_1 and Y_2 represent the "homogeneous computing power"
of the machines designed for scientific and commercial use, respectively, and X is the
cumulated production quantities; S is the standard error of estimate, F the ratio of the
variance explained by the model to the unexplained variance, D–W the Durbin–Watson
test statistic, and N the total number of observations corresponding to the period
1944–1967. The standard errors of estimated coefficients are given in parentheses.

The results from the test of the theoretical proposition are presented in
Table 6.14. The explanatory power of the relationships is very good, as indicated by the high R^2 values. This is also illustrated in Figure 6.12 for the case
of advances in scientific computation. For purposes of comparison, the actual
observations are plotted against a trend based on a simple power function. A
close scrutiny of the data shows that deviations from this fitted trend seem to
be in correspondence with certain failures in the development of new systems.
Nevertheless, some such "errors" in turn appear to have provided important
"lessons" in the learning process, and the estimated relationship in Table 6.14
is close to but significantly different from a simple power function relationship. In general, the pattern of advances is remarkably well behaved considering the tremendous increase in computing power over the course of time. The
coefficients of the explanatory variables are highly significant in both relationships and the Durbin–Watson test values are satisfactory. In all, the performance of the theoretical model is excellent.

5.4   Technological Innovation in the Generation of Electricity (1920–1970)

5.4.1   A Brief History of Technological Change

At present, electricity is generated in three main types of plants: (1) steam-
powered plants, which may be either fossil fueled or nuclear plants; (2) hydroelectric plants; and (3) internal-combustion plants, including gas turbines and
diesel engines which are used on a limited basis often in emergency start-up

operations. The main fossil fuels used in steam-powered plants are coal, natural gas, and oil. Historically, coal has been the main fuel used for the generation of electricity. Although its use declined somewhat in the 1960s (it still accounted for nearly 64% of the total output of electricity by steam plants in 1967), it is beginning to regain some lost ground.

The steam-powered electricity generating unit consists of four major components: a boiler or steam generator; a steam turbine; an electrical generator; and various auxiliary equipment. Since, however, the turbine is directly connected to a single electrical generator and in unit construction it receives input from its own boiler, it is common to speak of a boiler–turbine–generator complex.

The principal measure of technological change employed in the following analyses is the fuel-consumption efficiency of plants measured in kilowatt-hours per unit of fuel. All three types of fuel (i.e., coal, oil, and natural gas) are considered. A detailed history of innovations in electrical turbogenerators is presented in Chapter 8. Suffice here to note that the observed advances in plant efficiency are due mainly to the increase in steam pressure and temperature made possible by advances in metallurgy, the use of double reheat units (in which the partially expanded steam from the turbine is fed back to the boiler for reheating up to the throttle pressure and temperature, whereupon it is returned to the turbine for further expansion), and various improvements in the integrated system of man–machine interactions, such as the use of process simulation techniques to optimize both the design and the operation of plants.

In recent years, the rate of improvement in the efficiency of power plants has slowed down. To some extent, this situation is a reflection of the deterioration in the quality of fuel and of constraints imposed by environmental considerations. There are, however, three other main reasons: First, increased steam temperature requires the use of more costly alloys, which in turn entail maintenance problems of their own. The benefits gained in efficiency from their use are considerable. At times, however, these benefits have been outweighed by the increased cost of using these alloys. Thus there has been a decrease in the maximum throttle temperature from 1200 °F in 1962 to about 1000 °F in 1970. Second, there has been lack of motivation to increase the efficiency in the use of gas in both steam-powered and internal-combustion plants because of the artificially low price of fuel due to the Federal Power Commission's wellhead gas price regulation. Finally, as Fisher (1974) points out, there has been a slowdown in generation efficiency due to the heavy use of low-efficiency gas turbines necessitated by delays in the construction of nuclear power plants.

5.4.2   RESULTS OF EMPIRICAL ANALYSIS

Due to limitations of the available data, only two hypotheses concerning the role of learning and scaling in technological innovation are tested in the present case. The advances in the fuel-consumption efficiency of plants are due to improvement in the use of existing systems of plants and equipment, as well as to the development of new techniques. Obviously, any relevant measure of learning must take into account both of these activities. However, paucity of data does not allow for this. Thus, accumulated experience in innovative activity is simply measured here in terms of cumulative net production of electrical energy in kilowatt-hours.

Plant size is commonly measured in terms of either the availability or the capacity factor. The former may be defined, for example, as the ratio of production time to total time in a given period. The latter may be defined, for example, as the ratio of kilowatt-hours of production to the licensed design power level. Among other things, the capacity factor is determined by cyclical fluctuations in the demand for electricity and forced outage. In a sense, the availability factor has to do with the maturity, the capacity factor with the newness, of the system. Historical data on the availability factor are not readily accessible. We have therefore measured the scale of plant utilization in terms of both plant capacity and the ratio of net total production of electrical energy to total number of plants.

The results from the test of the learning by doing hypothesis are presented in Table 6.15. The model performs extremely well in its explanation of variations in all three chosen measures of fuel-consumption efficiency of plants over the course of time, leaving a mere 1% of the variance unexplained. This conclusion must be qualified in the case of variations in the efficiency of steam-powered plants (Eq. (1)) since, as indicated by the Durbin–Watson test value, there is clear evidence of positive autocorrelation in the residuals of the model. The implication is that while the parametric estimates of the model are unbiased, its explanatory power has been overestimated. The standard error of the coefficient of the cumulated production variable is relatively high. However, this is attributable to the collinearity between the two explanatory variables themselves. Further, one main reason for the presence of serial correlation is that the available data are unadjusted for progressive degradation in the quality of fuel. In particular, the data do not take into account the shift from the use of high-sulfur coal to low-sulfur coal during the 1960s as a consequence of various environment protection regulations. Thus, the results from the Durbin–Watson test of the residuals from the recomputed version of the model for the period 1921–1957 turn out to be quite satisfactory. The perform-

Table 6.15   Test of the Learning by Doing Hypothesis of Technological
Innovation in Electricity Generation (1920–1970)

Equation	Estimated relationship[a]
(1.)	$\log Y_{1(t)} = -0.04 + 0.007 \log X_{1(t)} + 0.91 \log Y_{1(t-1)}$ $\qquad\qquad (0.08)\quad (0.012)\qquad\qquad (0.05)$ $R^2 = 0.99,\ S = 0.009,\ F = 5390.2,\ \text{D–W} = 0.74,\ N = 50$
(2.)	$\log Y_{2(t)} = \quad 0.11 + 0.04 \log X_{2(t)} + 0.72 \log Y_{2(t-1)}$ $\qquad\qquad (0.01)\quad (0.01)\qquad\qquad (0.08)$ $R^2 = 0.99,\ S = 0.01,\ F = 1913.6,\ \text{D–W} = 1.91,\ N = 50$
(3).	$\log Y_{3(t)} = -0.20 + 0.013 \log X_{2(t)} + 0.87 \log Y_{3(t-1)}$ $\qquad\qquad (0.12)\quad (0.011)\qquad\qquad (0.06)$ $R^2 = 0.99,\ S = 0.01,\ F = 2871.9,\ \text{D–W} = 1.43,\ N = 50$

[a]Definitions: The variables Y_1, Y_2, and Y_3 are average fuel-consumption efficiency of
fossil-fuel electric power plants in kilowatt-hours per pound of coal, per gallon of oil,
and per cubic foot of natural gas, respectively; X_1 and X_2 are cumulated net production
of electrical energy (in millions of kilowatt-hours) by steam-powered and internal-
combustion plants, respectively; R^2 is the coefficient of determination, S the standard
error of estimate, F the ratio of the variance explained by the model to the unexplained
variance, D–W the Durbin–Watson teşt value, and N the number of observations corresponding to the period 1920–1970. The standard errors of the coefficients are shown
in parentheses.

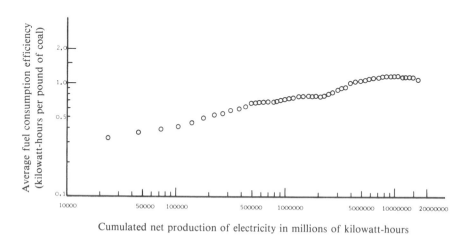

Cumulated net production of electricity in millions of kilowatt-hours

Fig. 6.13 Relationship between fuel-consumption efficiency of steam-powered
electricity-generating plants and the cumulative production of electricity, 1920–1970.

ance of the theoretical model is further illustrated in Figure 6.13. It is apparent that the relationship between growth in fuel-consumption efficiency and cumulated production has remained fairly consistent for a period of more than five decades. At the $\alpha = 0.01$ level, the results from Durbin–Watson tests of residuals are satisfactory in the remaining two cases (Eqs. (2), (3)). The parameters of the estimated relationship are also generally well determined.

The overall fit of the learning by doing hypothesis to the data on the growth of fuel-consumption efficiency is illustrated in Figure 6.14 for the case

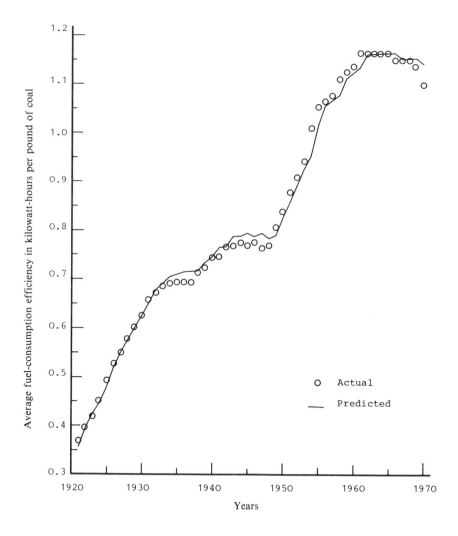

Fig. 6.14 Performance of the learning by doing hypothesis of technological innovation in predicting the fuel-consumption efficiency of steam-powered plants, 1921–1970.

of steam-powered plants. Evidently, the performance of the model is excellent.

Table 6.16 presents the results from the test of the specialization via scale hypothesis. The explanatory power of the model is very good, as indicated by uniformly high R^2 values. As before, however, the residuals of the model are autocorrelated in the explanation of improvement in the fuel-consumption efficiency of steam-powered plants. With regard to the results from the application of the model to improvement in the fuel-consumption efficiency of

Table 6.16:   Test of the Specialization via Scale Hypothesis of Technological Innovation in Electricity Generation (1920–1970)

Equation	Estimated relationship[a]
(1)	$\log Y_{1(t)} = -0.05 + 0.02 \log X_{1(t)} + 0.85 \log Y_{1(t-1)}$ $\quad\quad\quad\quad (0.02)\quad\ (0.008)\quad\quad\quad\ (0.03)$ $R^2 = 0.99,\ S = 0.009,\ F = 6177.03,\ \text{D–W} = 0.85,\ N = 50$
(2)	$\log Y_{2(t)} = \quad 0.15 + 0.03 \log X_{2(t)} + 0.84 \log Y_{2(t-1)}$ $\quad\quad\quad\quad (0.03)\quad\ (0.01)\quad\quad\quad\ (0.03)$ $R^2 = 0.99,\ S = 0.014,\ F = 1947.56,\ \text{D–W} = 2.20,\ N = 50$
(3)	$\log Y_{3(t)} = -0.17 + 0.03 \log X_{2(t)} + 0.86 \log Y_{3(t-1)}$ $\quad\quad\quad\quad (0.03)\quad\ (0.009)\quad\quad\ (0.03)$ $R^2 = 0.99,\ S = 0.011,\ F = 3512.14,\ \text{D–W} = 1.72,\ N = 50$
(4)	$\log Y_{1(t)} = -0.11 + 0.032 \log C_{(t)} + 0.92 \log Y_{1(t-1)}$ $\quad\quad\quad\quad (0.06)\quad\ (0.01)\quad\quad\quad (0.01)$ $R^2 = 0.99,\ S = 0.008,\ F = 5771.28,\ \text{D–W} = 0.87,\ N = 50$
(5)	$\log Y_{2(t)} = -0.006 + 0.028 \log C_{(t)} + 0.91 \log Y_{2(t-1)}$ $\quad\quad\quad\quad (0.08)\quad\ (0.028)\quad\quad\ (0.021)$ $R^2 = 0.98,\ S = 0.015,\ F = 1713.58,\ \text{D–W} = 2.11,\ N = 50$
(6)	$\log Y_{3(t)} = -0.187 + 0.029 \log C_{(t)} + 0.92 \log Y_{3(t-1)}$ $\quad\quad\quad\quad (0.09)\quad\ (0.022)\quad\quad\ (0.016)$ $R^2 = 0.99,\ S = 0.012,\ F = 2894.90,\ \text{D–W} = 1.55,\ N = 50$

[a]Definitions: The variables Y_1, Y_2, and Y_3 are average fuel-consumption efficiency of fossil-fuel electric power plants in kilowatt-hours per pound of coal, per gallon of oil, and per cubic foot of natural gas, respectively; X_1 and X_2 are average scale of utilization of steam-powered and internal-combustion plants, respectively, measured in terms of the ratio of net production of electrical energy in millions of kilowatt-hours to number of plants; C is the capacity factor measured in terms of the ratio of production to installed power in kilowatt-hours. For the remaining definitions see footnote a, Table 6.15.

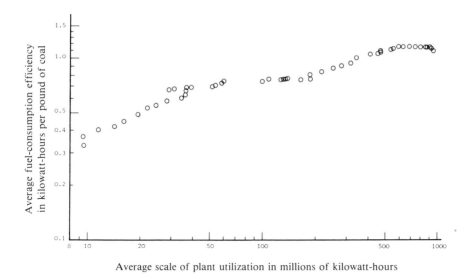

Fig. 6.15   The relationship between fuel-consumption efficiency and scale of utilization of the steam electric power plants, 1920–1970.

internal-combustion plants, there is no evidence of serial correlation in the residuals. The parametric estimates of the model are highly significant. In general they are not very sensitive to chosen measure of scale.

The performance of the theoretical model is further illustrated in Figure 6.15 for the case of improvement in the efficiency of steam-powered plants. Evidently, there exists a well-behaved relationship between growth of efficiency and scale of plant utilization.

6.   DIFFERENTIAL LEARNING, SCALE INVARIANCE, AND TECHNOLOGICAL INSULARITY IN INNOVATION PROCESSES

Two parameters of the relationships proposed in the preceding section are of special interest. One is the long-term coefficient of learning (β_1) or scaling (β_2). The other is the coefficient of disequilibration (λ), which is an inverse measure of the rate of technological evolution, B. Specifically, $B = 1 - \lambda/\lambda$. Thus, the lower the λ, the higher the rate of technical advances. The estimates of these parameters for various cases are presented in Tables 6.17 and 6.18.

Consider the long-term coefficients of learning and scaling. It is evident that they differ from case to case, which implies that the pace of learning or scaling differs from one technology to the other. This raises the question whether the observed differences are real or merely due to, say, differences in the units of measurement of different variables.

According to the modern theory of dimension, the notion of a "true" dimension of a variable is by and large meaningless because a variable can be defined in terms of any group of scale transformations (see Chapter 9). The relevant question is, what choice of dimensions is of maximum usefulness? That is, the dimensional specification of a variable is necessarily based on pragmatic but otherwise arbitrary grounds.

For the problem in hand, it seems that both the cumulated production quantity variable (X) and (any) chosen measure of technology (Y) are appropriately regarded as having the dimension of time. The rationale for this is discussed in Chapter 9. Suffice here to note that the long-term coefficient of learning is a nondimensional variable. In contrast to this, it seems that scale variables in different cases cannot be usefully defined in terms of a common dimension. Rather, as regards the various applications of the model to technological change in farm machinery, transportation systems, and electricity generation, the scale variable (Z) is appropriately regarded as having the dimension of L^2 (area), L (distance), and ML^2T^{-2}, respectively, where M denotes mass, L length, and T time. In summary, while the significance of observed differences in the long-term coefficients of scale can be determined only on an intraindustry-wide basis, there does not seem to be any such restriction when comparing the long-term coefficients of learning.

One noteworthy feature of the long-term coefficients of learning presented in Table 6.17 is that they closely correspond to the generally held view on the technical progressivity of different industries. At the two extremes, the learning coefficients are highest in the case of digital computers and aircraft, and lowest in the case of the farm tractor and locomotive. Thus it is reassuring that the a priori information does not refute the results of the quantitative analysis.

The next question is whether there is a significant difference between the learning coefficients in various cases. This is determined here by analysis of variance for want of any better method. (The same procedure is utilized in the other comparisons that follow.) The analysis is based on the classification of various cases into four groups of relatively similar technologies:

1. farm tractors and locomotives,
2. tank ships and aircraft;
3. computers for scientific and commercial use;
4. steam-powered and internal-combustion types of electric power plants.

Table 6.17: Coefficients of Long-Term Learning and Disequilibrium in the Process of Technological Innovation

Case	Source Table	Source Equation	Technology	Measure of technology	Coefficient of learning (β_1)	Coefficient of disequilibration (λ)
1	6.1	(1)	Farm tractor	Fuel-consumption efficiency	0.22	0.68
2	6.1	(2)	Farm tractor	Mechanical efficiency	0.14	0.64
3	6.1	(3)	Farm tractor	Horsepower-to-weight ratio	0.20	0.56
4	6.9	(1)	Locomotive	Tractive effort	0.04	0.96
5	6.9	(2)	Tank ship	Service speed	0.30	0.99
6	6.9	(3)	Aircraft	Air speed	0.70	0.99
7	6.14	(1)	Computers for scientific use	Homogeneous computing power	2.93	0.17
8	6.14	(2)	Computers for commercial use	Homogeneous computing power	2.67	0.07
9	6.15	(1)	Steam-powered electricity-generating plants	Fuel-consumption efficiency (kw-hr/lb of coal)	0.08	0.91
10	6.15	(2)	Internal-combustion electricity-generating plants	Fuel-consumption efficiency (kw-hr/gal of oil)	0.14	0.72
11	6.15	(3)	Internal-combustion electricity-generating plants	Fuel-consumption efficiency (kw-hr/cu. ft. of gas)	0.10	0.87

Table 6.18: Coefficients of Long-Term Scaling and Disequilibration in the Process of Technological Innovation

Case	Source Table	Source Equation	Technology	Measure of technology	Coefficient of scaling (β_2)	Coefficient of disequilibration (λ)
1	6.4	(1)	Farm tractor	Fuel-consumption efficiency (hp-hr/gal of fuel)	0.36	0.86
2	6.4	(2)	Farm tractor	Mechanical efficiency	0.27	0.82
3	6.4	(3)	Farm tractor	Horsepower-to-weight ratio	0.59	0.37
4	6.10	(1)	Locomotive	Tractive effort	0.30	0.97
5	6.10	(3)	Aircraft	Air speed	0.50	0.96
6	6.16	(1)	Steam-powered electricity-generating plants	Fuel-consumption efficiency (kw-hr/lb of coal)	0.133	0.85
7	6.16	(2)	Internal-combustion plants	Fuel-consumption efficiency (kw-hr/gal of oil)	0.187	0.84
8	6.16	(3)	Internal-combustion plants	Fuel-consumption efficiency (kw-hr/cu. ft. of gas)	0.214	0.86

The results are presented in Table 6.19. At the $\alpha = 0.05$ level, there exist highly significant differences between different groups of learning coefficients. Thus, the role of learning in innovations in computer technology is significantly different from that of learning in innovations in air and marine transportation technology. The role of learning in the latter case is in turn significantly different from learning in innovations in electricity generation, and so forth. There is one exception to this pattern: The difference between the learning coefficients in group 1 (concerning innovations in tractors and locomotives) and group 4 (concerning innovations in electricity generation) is insignificant. In general, however, the efficacy of learning in innovative activity differs from case to case.

The central conclusion of the foregoing analysis is that the process of learning in technological innovation is object specific. Its potency in any given case primarily depends on the form and complexity of the system (Chapters 4, 9). Thus the development of every technology is governed by a process of learning that is at least partly unique. This confirms the previously proposed principle of technological insularity from a different premise (Chapter 3). Evidently, one main reason for the observed lack of interindustry transmission of technical know-how is that technological learning is context dependent. There is comparatively little information transfer in industrial innovation simply because there is a large element of exclusivity in the process of technological learning.

Next, we have compared the significance of the learning versus scaling processes in technological innovation on an *intraindustry* basis, with the results presented in Table 6.20. At the $\alpha = 0.05$ level, the estimated F ratios are insignificant in all three cases. Hence we have the important result that within any given industry the role of long-term learning tends to be comparable to the role of long-term scaling in the process of technological innovation.

We can also indirectly assess the role of the learning and scaling processes in technological innovation by comparing the coefficients of disequilibration (λ) in various cases. There are two advantages in doing so. First, the estimates of λ are highly significant in literally every case under consideration. Frequently, the estimated values of λ exceed the corresponding standard errors by several orders of magnitude. Second, λ is a universally valid measure for comparing the rates of technological evolution because it is appropriately normalized. Thus, recall that its values must lie between 0 and 1 ($0 < \lambda < 1$).

The results from the analysis of variance of disequilibration coefficients in learning and scaling processes are presented in Table 6.21. At the $\alpha = 0.05$ level, there exist highly significant differences between the disequilibration coefficients in various cases of learning. Moreover, there is no exception to the significance of observed differences (Table 6.21A).* This confirms the earlier

---

*see p. 196

Table 6.19: Significance of Differences in the Long-Term Coefficients of Learning in Various Cases of Technological Innovation

Technology	Mean value of long-term coefficient of learning (β_i)	F matrix ($v_1 = 1$, $v_2 = 7$)		
		Y_1	Y_2	Y_3
Tractor and locomotive (Y_1)	0.15			
Tank ship and aircraft (Y_2)	0.50	8.45	—	—
Computing systems (Y_3)	2.80	484.55	273.75	—
Electric power plants (Y_4)	0.11	0.17	9.61	450.47

Table 6.20:   An Intraindustry Comparison of the Relative Significance of
Long-Term Learning and Scaling Processes in Technological Innovation

	Agriculture	Transportation	Electricity generation
Mean value of long-term coefficient of learning (β_1)	0.186	0.346	0.106
Mean value of long-term coefficient of scaling (β_2)	0.406	0.40	0.178
F ratio	5.75	0.04	5.79
Degrees of freedom (ν_1, ν_2)	(1, 4)	(1, 3)	(1, 4)

conclusion that the role of learning in the process of innovation tends to be of a differential nature. With regard to the disequilibration coefficients in various cases of scaling, however, the differences are consistently insignificant (Table 6.21B). Thus, it seems that the role of scaling in the process of innovation tends to be of a partially invariant nature.[6] Finally, we have compared the significance of the differences in the disequilibration coefficients associated with the learning and scaling processes in technological innovation on an *inter-industry* basis (Table 6.21C). At the $\alpha = 0.05$ level, the estimated F ratio is insignificant, indicating that learning seems to play as important a role in the process of innovation as does scaling. This confirms the earlier finding that the significance of learning and scaling in innovation tends to be comparable.

7.   THE PUTTY–CLAY PRINCIPLE OF TECHNICAL PROGRESS

One main finding of this chapter deserves further comment. It is that advances in technology generally depend on the gradual acquisition of skills through actual participation in the production activity. Technical progress is largely a matter of learning by *direct* experience. The implication is that there are built-in obstacles to the transfer of technology, since innovations depend not so much on knowledge imported from without as they do on experience

---

[6]Note that this discussion of learning and scaling processes concerns only the *macrolevel* of innovative activity. A treatment of the very same processes at the *microlevel* of innovative activity will be found in Chapter 9.

Table 6.21:   An Interindustry Comparison of the Coefficients of Disequilibration in Technological Innovation

A.   Industry

Industry	Mean value of coefficient of disequilibration in the learning process (λ)		F matrix ($\nu_1 = 1$, $\nu_2 = 7$)	
			Y_1	Y_2	Y_3
Agriculture (Y_1)	0.626				
Transportation (Y_2)	0.980	Y_2	39.56	—	
Computer (Y_3)	0.120	Y_3	65.08	187.50	—
Electricity generation (Y_4)	0.833	Y_4	13.53	6.82	129.00

B.   Industry

Industry	Mean value of coefficient of disequilibration in the scaling process (λ)		F matrix ($\nu_1 = 1$, $\nu_2 = 5$)	
			Y_1	Y_2
Agriculture (Y_1)	0.683			
Transportation (Y_2)	0.965	Y_2	3.20	—
Electricity generation (Y_3)	0.816	Y_3	1.40	0.53

C.   Coefficient of disequilibration in

Coefficient of disequilibration in	Overall mean value	F ratio ($\nu_1 = 1$, $\nu_2 = 7$)
Learning	0.687	
Scaling	0.817	1.041

from within. To put it differently, technical know-how lacks the ubiquitous relevance of pure scientific knowledge. It is to a considerable extent both product and plant specific. This is not to say that it cannot be effectively transferred from one organization to the other. Rather, success in technology transfer hinges upon meticulous alterations in the design of the chosen technique to suit the requirements of differing production systems. The transfusion of technical know-how across system boundaries is therefore invariably a costly and time-consuming process.

It is interesting to note that the inference above is further supported by the results from a wide variety of other investigations in this area. First, studies of the communication of information in R&D organizations have led to two diametrically opposed conclusions. On the one hand, the evidence indicates that there is a highly significant, positive relationship between the performance of *scientists* in universities as well as in industrial laboratories and the extent of their communication with the outside world (Hagstrom, 1965; Farris, 1972). On the other hand, the evidence also suggests that there is a very consistent, negative relationship between the performance of *engineers* in industrial organizations and the degree of external communication (Allen, 1964; Baker et al., 1967). According to our analysis, the observed dichotomy is to be expected insofar as the relevance of technical know-how, unlike that of pure scientific knowledge, is limited to the system of its origin. As stated so well by Allen et al. (1979, p. 695),

Science may be said to be universal. . . . A scientist . . . is fully capable of understanding the nature of the problems and approaches employed by other scientists in his specialty anywhere in the world. The universal nature of the problems and the existence of shared language and methods permit effective communication across organizational and even national boundaries.

Technology, on the other hand, is not universal. Technology is highly localized in that problems are defined in terms of interests, goals, and local culture of the organization in which they are being attacked. Similar technological problems may become defined in very dissimilar ways by organizations working on them because these organizations have often different objectives and value systems . . . .

A consequence of organizational differences is that technological problems are defined to fit the particular strengths, professional orientation, and objectives of the organization. Certain types of solutions, which may be perfectly acceptable in one organization, simply will not work when applied to the same problem in another organization. . . . As a result, the externally defined solutions perform less well, and an inverse relationship is observed between external consultation and technical performance.

Second, a number of in-depth studies of international trade reveal that technology transfer generally occurs in the form of installments. It is seldom possible (or even useful) to ferry technical know-how from one country to the other in its entirety. This is true even in the case of countries with established industries. The Japanese coproduction of the U.S. military aircraft, the Lockheed F-104 Starfighter, during the 1960s is a case in point (Hall and Johnson, 1970). Whereas the technology for the airframe could be substantially transferred within a relatively short period of time, the transfer of technology for the jet engine took nearly four years of substantial effort. In contrast to these two items, there was little or no transfer of technology for manufacturing electronic gears. More generally, it is very rarely feasible to transfer technology lock, stock, and barrel, even if this entails only the embodied form of technical know-how.

Third, and finally, studies of the choice of technology in developing countries reveal that frequently a few techniques dominate the entire spectrum of those available in a given industry in that they require less of both capital and labor per unit of output. In certain cases, such as the steel and textile industries, technical progress has been so strongly localized that at present there is no viable alternative to a handful of established techniques. The failure of the Chinese experiment with small blast furnaces is a case in point (Eckaus, 1960), as is the less than wholly successful Indian attempt to utilize the Ambar Charkha (an improved version of the hand spinning wheel) in the production of textiles (Sen, 1962). The finding that technical progress tends to be at least partly localized does not of course imply that the range in the choice of technology is inherently restricted. It does, however, mean that innovative activity within any given field does not enhance the capability of the different production processes (e.g., capital-intensive versus labor-intensive techniques) in an equitable manner. The state of art does not advance simultaneously on all fronts because there is comparatively little cross-fertilization of technical know-how.

The central conclusion that emerges from these considerations is that technical know-how is rarely as broad based as it is commonly believed to be. It is often true to its origin rather than applicable across the board. Much technical progress occurs in driblets. In consequence, the resulting know-how is often fragmentary and context dependent. The implication is obvious: Although it is possible to direct R&D activity toward any objective, the relevance of the resulting know-how is largely limited to the chosen frame of reference. This suggests what may be called a putty–clay principle of technical progress: While a priori the development of a technology may be likened to the molding of putty, which can be shaped in any way, *ex post facto*, it involves an element of unchangeability, in the manner of clay.[7] That is, the initial choice of a technology (e.g., between biogas plants and nuclear reactors) is of supreme importance because it circumscribes the entire course of subsequent innovative activity. Very simply, technological development is a process of irreversible transformations.

Obviously, the putty–clay principle of technical progress is distinct from, but closely related to, the previously proposed principle of technological insularity. The former refers to the context-dependent nature of technical know-how, the latter to the lack of its transmissibility. The two are therefore complementary.

8.   PRINCIPAL CONCLUSIONS

This chapter has presented a general theory of long-term technological development. There are two central propositions of the theory. One, the learning by doing hypothesis, is that technological innovations originate in accumulated experience of a practical nature. The other, the specialization via scale hypothesis, is that technological innovations depend on the scale of the larger system designed to secure their effective utilization. It is found that a wide variety of innovations in farm machinery, transportation equipment, computing devices, and electric power systems can be adequately explained by these hypotheses. Thus an explanation of the otherwise baffling "residual factor" in the temporal variations in productivity turns out to be relatively simple.

In the investigations reported here, certain other widely held views of technological innovation are also examined, including the "demand-pull, technology-push" hypothesis and the theory of induced innovations. Considered at face value, none of these hypotheses is found to hold except in isolated instances. In general, they simply do not seem to stand up to the facts. As discussed earlier, the reasons for their apparently unsatisfactory performance are not difficult to see. Briefly, the notion of demand for technology errs in its parochial assumption that the potential for increased sales of a product by itself guarantees advances in its technological capabilities. Indeed, demand may well be a necessary but not a sufficient condition for technological innovation to materialize. At the other extreme, the notion of technology-push errs in its equally one-sided assumption that innovations arise of their own accord. The theory of induced innovations suffers from much the same defect as the

---

[7]In the economic literature it is customary to regard capital as being pure putty or putty–clay. In the former case it is always possible to vary the amount of labor per unit of capital. In the latter case, substitution of labor for capital is possible on an *ex ante* but not on an *ex post* basis. The results of the present investigation confirm what has long been suspected but never adequately proved (see, e.g., Atkinson and Stiglitz, 1969): Technology shares the characteristics of putty–clay capital.

exaggerated notions of demand-pull and technology-push. Further, it makes the rather strong assumption that technological change is inherently charac-terized by a labor-saving bias. However, it fails to explain adequately why firms have a preference to save one factor of production relative to another rather than to minimize total costs. In summary, notwithstanding the popu-larity of some of these theories of technological innovation, their conceptual basis leaves much to be desired.

This is not to reject the above hypotheses as fallacious. Rather, they are untenable in the (extreme) form in which they have generally been put forth. According to the results of our investigations, variables such as demand, tech-nological possibility, and relative price ratio do have a bearing on the course of innovative activity. However, their influence on observed advances in tech-nology is often mediated through other variables. Thus our analysis shows that the ratio of price of capital to that of labor is an important factor in governing the farm size structure, which in turn is a key determinant of innovations in farm mechanization. More important, the postulated role of learning and scal-ing processes in technological development in turn implies a substantial role of both market size and technological possibility in innovative activity. Con-ceivably, the *rates* of learning and scaling in innovation themselves depend on, *inter alia,* the demand for the product as well as purely technological variables insofar as the latter are limiting factors in production. In summary, the hypotheses under consideration are obviously relevant to the investigation of technical change process. The important point, however, is that we must ac-knowledge a certain fine structure of interactions between the factors on the supply side and the demand side in order to understand observed patterns of innovative activity. It is for this reason that the popular theories of innovation may be justifiably criticized in their failure to see the forest for the trees.

The results of this chapter then point to a number of noteworthy features of the learning process in technological development. They may be sum-marized as follows.

First, there exist significant time lags between the acquisition of produc-tion experience and the development of new techniques. This is exemplified by the frequent shifts in the slope of the log-linear relationship between innova-tion performance and cumulated production quantity variables (see Figures 6.2, 6.7–6.9, 6.12 and 6.13 illustrating the learning by doing hypothesis). From a theoretical point of view, the implication is that there are important differ-ences in the dynamical aspects of the two types of learning involved in the *use* and the *development* of a technology. From a substantive point of view, the observed phenomenon confirms a point repeatedly noted earlier: Innovations depend on gradual modifications of a technique. Technical advances do not materialize at once; rather, they take place in stages.

Second, the *extent* of learning in different cases of technological development is systematically related to certain peculiarities of the different institutional backgrounds. According to the results of the empirical analysis, innovations in the farm equipment and electric power industries are attributable to learning in both the machine-producing and the machine-using sector of the economy (see also Chapter 8). In contrast to this, innovations in the transportation sector have been made possible by learning solely in the machine-producing sector. These results are consistent with a priori information. As regards the farm equipment industry, there is considerable evidence that agriculture extension services have played an important role in innovation processes not only as a source for disseminating relevant know-how but also as an intermediary between farmers and manufacturers. In contrast, the initiative for innovations in transportation has invariably come from the suppliers rather than the users of the technology.

Third, while important differences characterize the various cases of technological development under consideration, they share a certain commonality: The crucial factor in technical progress is learning in the capital-producing rather than in the capital-using sector.[8] Thus one uniform inference from the results of the analyses of different cases is that (cumulated) gross investment is an essential variable in the process of innovation. This confirms what has long been suspected but never adequately proved; to a certain extent, *all* investment has the character of investment in R&D.

Fourth, there exist wide variations in the returns from learning in different cases of innovative activity. The potency of learning in any given case primarily depends on the nature of technology. Thus different technologies vary greatly in their susceptibility to advances that result from learning in each case.

The fifth point is closely related to the fourth. Learning plays a far more prominent role in the development of high-technology items (such as digital computers and aircraft) than in the development of traditional technologies (such as farm tractors and locomotives). Thus the long-term coefficients of learning associated with advances in high-technology products are invariably higher than those in the traditional technologies. One is reminded of the "Leontief paradox" here: With capital relatively plentiful and labor relatively scarce, we would expect the export industries of a country such as the United States to be more capital intensive than those industries competing against imports. Instead, Leontief (1953) found that the capital stock per employee in import-competing industries in the United States was some 30% higher than that in export industries. That is, the production of high-technology items for

---

[8]For a similar conclusion see Raj (1969), Rosenberg (1963), and Vietorisz (1969).

export was less capital intensive in comparison with traditional technology items intended for the domestic market. In the light of our findings this is to be expected inasmuch as development of high-technology items is founded on a greater, not lesser, accumulation of *human* capital.

Sixth, and finally, learning in innovation processes occurs in many different patches rather than in the form of a seamless whole. It is generally based on the acquisition of firsthand experience in the production activity. There are definite bounds to its expedience. Technical know-how lacks the generality of pure scientific knowledge. Rather, its significance is often limited to a given production setup. There is a large element of sectarianism in the development of new techniques. Equivalently, there is comparatively little fallout in the course of technical progress.

In the light of the foregoing conclusions, it is easy to provide an explanation of observed interindustry differences in the growth of productivity. To take the bull by its horns, the differential productivity performance of various industries is attributable to the differential nature of learning in the innovative activity. The differences in learning are in turn due primarily to differences in the nature of the technology employed in different fields. Thus our findings indicate that there exists a remarkably systematic relationship between the technical progressivity of an industry and its observed capability for learning in the development of new techniques. Moreover, interindustry disparities in the growth of productivity often tend to persist over long periods of time because there exist a great many obstacles to the cross-fertilization of technical know-how.

While the role of learning in innovative activity tends to be of a differential nature, the scale of technological operation introduces an element of invariance into it. Learning is an important factor in the adaptation of technology to the system of its production; scaling is an important variable in the adaptation of technology to the system of its utilization.

A close examination of the phenomenon under consideration reveals that the process of adaptation between technology and the conditions of its use tends to be of a mutual causal nature. (As discussed earlier, this is likewise true of adaptation between technology and the systems of its production.) It is very generally the case that a technology, during the initial stages of its evolution, branches off in multiple directions so as to meet the requirements imposed by its task environment. This is evidenced by the emergence of many different versions of the same technique. Once the evolution of a technology is well under way, many of its different versions rejoin to form an integrated system capable of performing many different tasks. Thus there is what may be called a "general-purpose design effect" leading to the construction of a single machine capable of serving many hitherto diverse functions. This single machine may be likened to a universal aunt. It is indeed striking that this effect is observable in a wide variety of technologies ranging from locomotives to computing devices. The whole phenomenon is illustrative of a very general thesis: While an adaptive system must have the capability to evolve, an evolutionary system need not have the capability to adapt.[9]

There are a number of policy implications of these findings. First, it is common to conceive of technological progress in terms of a series of segmented steps ranging from research and development to innovation and diffusion. According to this popular scheme, each such step depends only on its precursor. If the considerations advanced here are any guide, various steps in this process are appropriately regarded as being mutually interlocked. Each step depends on all others. Specifically, our results show that the organization designed to secure the use of a device plays as important a role in technical progress as does the system of its production. Thus it is suggested that the research and development activity in the production of a new technology is best pursued in close cooperation with its end users.

Second, while formal research and development activity is obviously important, it is only one among many factors in the successful development of new techniques. The most significant of all variables in long-term technical progress is learning by experience, which depends on all investment, not merely the one in the formal research and development activity. Moreover, harvesting the fruits of learning requires that the capital-producing sector of the economy be adequately protected through tariffs and investment subsidies during the infant stage of its growth. At the same time, however, it is important to facilitate the industry-wide diffusion of new technical know-how so as to avoid subsidizing the repetitive component of learning.

Furthermore, there is little or no justification for protecting the capital-using sector since learning is found to occur primarily in the capital-producing sector of the economy. Indeed, any protection offered to the capital-using sector may even be tantamount to penalizing the capital-producing sector because this is likely to eliminate the motivation of the former to seek improvements in the equipment supplied by the latter. Instead, an appropriate policy is to subsidize vocational training of the work force through the establishment of pilot plants.

Third, learning plays as important a role as does scaling in the process of innovation. Thus technological development is not merely a matter of installing a certain infrastructure of industry. It depends as much on the acquisition of production skills. The conclusion to be drawn is that a large-scale system of plant and equipment, by itself, is largely superfluous. Any such facility must

[9]For a similar viewpoint, see, e.g., Weinberg (1975).

necessarily be supplemented by adequate training of the work force, a point so obvious that it is often overlooked in developmental planning.

One final implication of our findings is for the subject of technology transfer. It is that the development and transfer of technology ought to be regarded as an integral part of innovative activity from the very outset. This is also evidenced by the fact that attempts to transfer ready-made technology from industrialized to the developing countries have frequently proved to be failures. By the same token, systematic efforts to develop technology with due regard to the production possibilities and conditions of use in third world countries have been highly successful.[10] The lesson to be drawn is that technology transfer can never be a wholly adequate substitute for independent research and development activity. However, the two can most fruitfully supplement each other.

---

[10]Included in this category are such diverse examples as development of solar power facilities for pumping water in desert areas (Girardier and Vergnet, 1976), use of bamboo wells for irrigation in conditions where it is difficult to hold back the soil around shallow wells (Dommen, 1975), and development of portable power weeders and axial flow threshers for tropical agriculture (Khan, 1976). Indeed, this list of successful examples of technological development can be substantially extended.

Chapter 7

TECHNOLOGICAL CYCLES

1. THE ROLE OF INNOVATIONS IN BUSINESS CYCLES

Whenever an economic recession occurs, or an inflationary period sets in, the call for a surge in technological innovations can be heard. Implicit in this clamor for innovation is a view of technology as the source of long-term economic growth. There is of course considerable evidence in support of this view (Chapter 1). However, a close examination of the case for an innovation-oriented public policy reveals that there is much less to it than meets the eye (Carey, 1980). Upon separation of the wheat from the chaff, what is left out is one basic idea that does deserve further scrutiny. It was originally put forth by Joseph Schumpeter (1934, 1939), and it boils down to the statement that observed fluctuations in economic activity are somehow related to the development of new techniques.

As noted in Chapters 2 and 3, the starting point of Schumpeter's work is the distinction between the initial act of innovation by a few heroic entrepreneurs and its subsequent imitation by a host of followers. Only a very few individuals are capable of launching a new enterprise. However, once someone has crossed the Rubicon, others can, and readily do, emulate. The introduction of a major innovation is therefore often accompanied by a swarm of secondary innovations and imitations. Thus, innovations are seldom uniformly distributed. Rather, they appear "discontinuously in groups or swarms" (Schumpeter, 1934, p. 233). This discontinuity in the advent of innovations in turn gives rise to severe fluctuations in the rate of investment. Specifically, the "swarm-like appearance of new enterprises" following a major innovation inevitably results in a period of boom in capital investment. The whole process is accentuated by the underlying cumulative mechanism leading to the "secondary waves" of business expansion across interrelated industries. "Many things float on this 'secondary wave', without any new or direct impulse from the real driving force" (Schumpeter, 1934, p. 226). In short, an explanation of economic boom is to be found in the clustering of innovations.

At the same time, however, innovations disturb the existing pattern of economic relationships among production possibilities, availability of raw materials, prices, consumer income, etc. The system is driven away from its earlier state of equilibrium. The changes caused by the innovation can be absorbed only gradually. Thus, economic boom is necessarily followed by a period of recession involving readjustment of prices, costs, and production quantities. Frequently, the recessionary forces acquire a dynamics of their own, thereby plunging the system into a state of economic depression. As Schumpeter himself puts it (1934, p. 224), "Depression is nothing more than the economic system's reaction to the boom, or the adaptation to the situation into which the boom brings the system." The recovery of the system in turn depends on another gradual process of adjustment. This adjustment process may eventually succeed in propelling the system toward a state of equilibrium. The whole process is repeated, starting from a new burst of innovative activity.

It is of considerable significance that the rejuvenation of the system depends on radical innovations as distinguished from changes in existing techniques. Economic revival takes the form of long cycles of about 50 years in duration, a concept generally associated with Kondratieff. By way of evidence for this Schumpeter points out that the upturn in the first Kondratieff cycle (1787–1800) was triggered by the advent of steam power, the second (1843–1857) by the spread of the railroads, and the third (1898–1911) by the combined influence of electricity and the automobile. In summary, techno-logical innovation is the primum mobile of observed fluctuations in economic growth.

Whatever be its other merits, the formal theoretical basis of the Schumpeterian viewpoint remains somewhat shaky. It has never been adequately corroborated. Rather, it has continued to rest upon essentially conjectural reasoning. As Kuznets has put it, in his characteristically thoughtful manner (1954b, pp. 122–123):

> The critical evaluation . . . of what appear to be important elements in Professor Schumpeter's conclusions, viewed as a systematic and tested exposition of business cycles, yields disturbingly destructive results. The association between the distribution of entrepreneurial ability and the cyclical character of economic activity needs further proof. The theoretical model of the four-phase cycle about the equilibrium level does not yield a serviceable statistical approach. . . . The core of the difficulty seems to lie in the failure to forge the necessary links between the primary factors and concepts (entrepreneurs, innovation, equilibrium line) and the observable fluctuations in economic activity.

Others hold similar reservations. In the statement of Heertje (1973, p. 101):

> Schumpeter's work in this area must be regarded as really only a very ramified
> analysis and description of the phenomenon of innovation and its consequences.

In summary, although the Schumpeterian viewpoint is eminently plausible, it needs further verification. This chapter presents a systematic examination of fluctuations in technological activity. The objective is to determine what if any relationship exists between technical change and business cycles.

2.   THE ORIGIN OF TECHNOLOGICAL CYCLES

The following view of technological cycles is simply an outgrowth of one of our earlier findings: The process of innovation is best regarded as a non-deterministic system governed by causes of a cumulative nature. A detailed discussion of this point can be found in the preceding investigations (see especially Chapters 3 and 6). To recapitulate briefly, innovative activity is characterized by a great deal of experimentation. New techniques do not originate all at once; rather, they are developed through time. In particular, one central characteristic of the process of technological change is that its response to various stimuli tends to be of a cumulative nature. For example, many of the findings of R&D activity at a given point in time may prove to be important years later rather than at once. Indeed, very often the technical developments of today have their origin in experience accumulated in the past. Further, chance plays an important role in the process of technological innovation. Thus, one innovation often has several independent origins. In sum, the process of technological innovation is governed by the interplay of chance and cause on a cumulative basis.

This brings us to the next point. The temporal evolution of any system is likely to exhibit persistent fluctuations if there are time discrepancies in the reactions of its constituent parts to uncorrelated perturbations (Frisch, 1933). That is, a phenomenon subject to a cumulation of random changes is generally characterized by prolonged oscillations.[1] The crux of this thesis can be very simply explained as follows (Kuznets, 1954a). Suppose that the process under consideration involves both small and large changes. These two types of changes may well differ in the frequency with which they occur. The large changes, unless they are nullified by a sequence of small changes, are likely to

---

[1] Formally, as shown by Slutzky (1937), if an intercorrelated series is formed from a purely random series by m iterated summations of two successive terms followed by n first differences, then provided that the ratio m/n is a constant, the resulting series tends to a sine curve as $n \to \infty$. The Slutzky effect is exemplified by the apparent cycles that sometimes result from the application of the moving averages to the data.

give rise to a cyclical phenomenon insofar as their influence is extended far
beyond the time of their occurrence because of the process of cumulation. Further, random changes may well come at times in clusters. These clusters will
tend to combine with each other into a high or low level of saturation because
of the process of cumulation. In consequence, each of these clusters will tend
to exert its influence beyond the time of its occurrence. Thus it is demonstrably
true that cumulation or averaging of a frequency distribution of a random
character will not result in a smooth trend. Rather, it is expected to yield
oscillations arising from the prolonged influence of large single changes or of
large clusters of changes pursued by a sequence of smaller changes or of
smaller clusters.

As remarked earlier, the process of technological change is not entirely
random in character. It is indeed characterized by a secular trend. We can
therefore conceptualize it in terms of a skewed distribution.[2] The skewness of
the distribution will then account for the existence of a secular trend in technological activity. In addition, however, the process will be subject to (possibly
recurrent) oscillations arising from the cumulative operation of the underlying
causes.

A priori, therefore, the existence of recurring fluctuations in innovative
activity is not unexpected. The implication is that the process of technological
development involves two phenomena of change. First, as noted in Chapter 6,
there exists the phenomenon of persistent long-term growth characterized by
the existence of a certain trend of development. Second, as noted here, a
priori, there also exists the phenomenon of recurring fluctuations. These two
types of phenomena are distinct but related. For example, it is not implausible
that an understanding of oscillations in the innovative activity is essential in
order to account for the continuity of the process as a whole. It may also provide a framework wherein the growth and stagnation of technology can be
systematically analyzed.

3. SPECTRAL AND CROSS-SPECTRAL ANALYSIS

The following investigation into fluctuations in innovative activity utilizes
the technique of spectral and cross-spectral analysis (Fishman, 1969; Granger
and Hatanaka, 1964; Jenkins, 1965; Jenkins and Watts, 1968; Parzen, 1961;
Morgenstern, 1961). The technique is best understood as an analysis of the
variance of a time series in the frequency domain. It is based on the idea that a
stationary time series can be decomposed into a number of uncorrelated components, each associated with a period or frequency. From the resulting spectral density function or the autospectrum, we can measure the relative impor-

---

[2]For an analogous view of economic development see Kuznets (1954a).

tance of each of the frequency bands in terms of its contribution to the overall variance of the series.

The motivation behind the use of spectral analysis in the present investigation arises from its many advantages. First, unlike other methods of determining the rhythmic behavior in time series, it does not require that the phenomenon under consideration be strictly periodic. It is, first and foremost, an approach to analyzing the mixture of regularity and nonregularity that is invariably present in most time series. Its basic objective is to estimate the average power in a frequency band centered around the frequency under consideration rather than to test for a cycle of given frequency. If a given frequency band contributes an important proportion of the total variance, it may be concluded that there exists a "cycle" of the corresponding average frequency. Thus spectral analysis provides an approach whose reliability does not tack on prior information about the cyclicity of the phenomenon in question. Second, the spectral analytic approach makes possible the simultaneous determination of cycles of various durations: "It does not require the elimination of shorter cycles from the series before cycles of longer duration can be studied" (Adelman, 1956, p. 446). Finally, spectral analysis provides a sampling framework in which probabilities of the component outcomes in a time series depend on the previous outcomes. It is tailor-made for samples in which autocorrelation is known to be present (see, e.g., Naylor et al., 1969). In all, the technique is particularly well suited to analyzing the problem in hand because, as noted earlier, very little is known about the nature of fluctuations in technological activity.

Suppose we have a stationary time series $[x_t, t = 1, \ldots, n]$. Let m be the number of components into which we wish to decompose the series. The autocovariance of the series can be estimated as

$$C(i-1) = \frac{1}{n} \left[\sum_{t=1}^{n-(i-1)} x_t x_{t-(i-1)} - \frac{1}{n} \left(\sum_{t=1}^{n} x_t \right) \left(\sum_{t=1}^{n} x_t \right) \right] \qquad (7.1)$$

where $i = 1, 2, \ldots, (m+1)$.

The autocovariance function (7.1) can be converted to estimates of the power spectrum by means of a weighted autocovariance

$$C'(i-1) = [\lambda(i-1)][C(i-1)] \qquad (7.2)$$

where $[\lambda(i-1)]$ is a weighting function. The use of a window or weighting function is necessary so as to give equal and zero weights, respectively, to the power at all frequencies that lie within and outside the frequency band centered around the frequency under consideration.

The Parzen "lag window" is used in the present investigation. The weights depend on m (the number of frequency bands to be estimated) and are of the form

$$\lambda(i-1) = 1 - 6\left(\frac{i-1}{m}\right)^2 + 6\left(\frac{i-1}{m}\right)^3$$

for $1 \leqslant i \leqslant m/2$, and

$$\lambda(i-1) = 2\left(1 - \frac{i-1}{m}\right)^3 \tag{7.3}$$

for $m/2 + 1 \leqslant i \leqslant m+1$.

The weighted autocovariance is then subjected to a Fourier cosine transformation to yield the estimates of the power spectrum

$$\hat{F}(i-1) = \frac{1}{2\pi} \sum_{j=1}^{m+1} \beta C'(j-1) \cos \frac{(j-1)(i-1)\pi}{m} \tag{7.4}$$

for $i,j = 1,2, \ldots , m+1$ where

$$\beta = \begin{cases} 1 & \text{for } j = 1; j = m+1, \\ 2 & \text{for } j = 2,3, \ldots , m. \end{cases}$$

The power spectrum provides a decomposition of the variance of a time series in the frequency domain since

$$\sigma_x^2 = C(0) = \frac{1}{m} \int_{\lambda=0}^{m} f(\lambda) \, d\lambda. \tag{7.5}$$

Thus a natural interpretation of the power spectrum is that if a frequency component contributes a large proportion of the total variance, then it may be regarded as more important than a component whose contribution is less.

For a pair of time series samples $[x_t, t=1, \ldots , n]$ and $[y_t, t=1, \ldots , n]$, the estimates of the cross-spectral density function can be obtained by first computing the pairs of cross-covariance functions

$$C_{xy}(i-1) = \frac{1}{n}\left[\sum_{t=1}^{n-(i-1)} x_t y_{t-(i-1)} - \frac{1}{n}\left(\sum_{t=1}^{n} y_t\right)\left(\sum_{t=1}^{n} x_t\right)\right], \tag{7.6}$$

$$C_{yx}(i-1) = \frac{1}{n} \left[\sum_{t=1}^{n-(i-1)} y_t x_{t-(i-1)} - \frac{1}{n} \left(\sum_{t=1}^{n} y_t \right) \left(\sum_{t=1}^{n} x_t \right) \right] \quad (7.7)$$

where $i = 1,2, \ldots, m+1$. These functions are then weighted to form the weighted cross-covariance functions

$$C'_{xy}(i-1) = [\lambda(i-1)][C_{xy}(i-1)],$$

$$C'_{yx}(i-1) = [\lambda(i-1)][C_{yx}(i-1)] \quad (7.8)$$

where $i = 1,2, \ldots, m+1$. The Parzen weighted cross-covariance functions are then subjected to Fourier transformations to yield cross-spectral estimates of the cospectrum (the "real" part of the cross-spectrum) and the quadrature spectrum (the "imaginary" part of the cross-spectrum):

$$c(i-1) = \frac{1}{2\pi} \left\{ \sum_{j=1}^{m+1} \left[C'_{xy}(j-1) + C'_{yx}(j-1) \right] \cos \frac{(j-1)(i-1)\pi}{m} \right\}, \quad (7.9)$$

$$q(i-1) = \frac{1}{2\pi} \left\{ \sum_{j=1}^{m+1} \left[C'_{xy}(j-1) - C'_{yx}(j-1) \right] \sin \frac{(j-1)(i-1)\pi}{m} \right\} \quad (7.10)$$

for $i,j = 1,2, \ldots, m+1$.

It is customary to characterize the cross-spectrum by three functions: either the coherence, phase, and cross-amplitude spectra; or the coherence, phase, and gain spectra. The cross-amplitude spectrum measures the degree of association between the amplitudes of the periodic components of the two series over a common band of frequencies and it is calculated as

$$A_{xy}(i-1) = [c^2(i-1) + q^2(i-1)]^{1/2} \quad (7.11)$$

for $i = 1,2, \ldots, m+1$. Coherence is measured as the squared amplitude divided by the product of the estimated autospectral functions at a given frequency component. It may be estimated as

$$C^2(i-1) = \frac{c^2(i-1) + q^2(i-1)}{[\hat{F}_{xx}(i-1)][\hat{F}_{yy}(i-1)]} \quad (7.12)$$

for $i = 1,2, \ldots, m+1$. This statistic is the rough analogue of the coefficient of determination in correlation analysis and lies in the range $0 \leqslant C^2(i-1) \leqslant 1$.

Gain is interpreted in the same way as the regression coefficient of $[x_t]$ on $[y_t]$. It is a measure of the scalar by which the amplitude of the base series must be multiplied to produce values in series crossed on the base series, and it may be estimated as

$$G(i-1) = \frac{[c^2(i-1) + q^2(i-1)]^{1/2}}{\hat{F}_{xx}(i-1)} \tag{7.13}$$

for $i = 1,2, \ldots , m+1$.

Phase is the angular measure of the shift on the time axis of the crossed series relative to the base series at which coherence is maximized at that frequency. The phase is measured in radians and it is calculated as

$$\phi(i-1) = \tan^{-1}\left[\frac{q(i-1)}{c(i-1)}\right] \tag{7.14}$$

for $i = 1,2, \ldots , m+1$.

The phase angle can be converted into a time unit by computing $\tau(i-1)$:

$$\tau(i-1) = \frac{\phi(i-1)}{2\pi\,\omega(j-1)} \tag{7.15}$$

for $i, j = 1,2, \ldots , m+1$ and
$$\omega(j-1) = \frac{(j-1)\pi}{m}\,, \qquad j = 1,2, \ldots , m+1,$$

which is the measure of the frequency in terms of cycles per unit time. As a matter of interpretation, when $\tau(i-1)$ is positive, the phase shift of the series relative to the base series has been such that the base series leads the crossed series.

One crucial problem in the application of spectral analysis is that it is based on the assumption that the time series under investigation is stationary, that is, its mean and variance do not change over time. Since time series data on developmental processes are generally characterized by a secular trend in both mean and variance, they cannot be used as such. One way out of the problem is to employ the first differences of the original time series as input data in the spectral analysis. Such a procedure is, however, completely arbitrary and devoid of any a priori rationale. An alternative and a more commonly employed approach is to fit a certain trend to the original data on the basis of qualitative considerations and use the deviations from this fitted trend as input data in the spectral analysis. In either case, the snag is that if the trend is not adequately removed, the spectral estimates will be biased due to violation of the stationarity assumption. Further, the remaining trend will produce a peak at the zero frequency band along with the so-called leakage effect: the estimated spectrum will be smoothly decreasing for the next few frequency bands (Granger, 1966, p. 153). In consequence, the results would be biased against long cycles (i.e., cycles in the low-frequency range). On the other side of the ledger, the potential pitfall is to eliminate not only the trend but also some of the other low frequencies.

Obviously the whole problem boils down to an adequate conceptualization of what constitutes a trend. Note that the issue cannot be settled in purely statistical terms. Suppose that a trend is chosen solely on the basis of its fit to the data and that the resulting spectrum of residuals from the fitted trend exhibits certain peaks. But this implies that the original form of the trend was misspecified inasmuch as its residuals are found to be autocorrelated. According to this type of logic, an adequate specification of the trend, by its very definitions, precludes the existence of cycles in the data.[3] Thus, a purely statistical approach to trend specification is inherently unsatisfactory. Rather, a trend has to be specified on qualitative grounds. In particular, neither trend nor cycles can be adequately defined in an unconditional manner. The existence of a cyclical phenomenon is meaningful only *in relation to* the simplest form of trend that is plausible on a priori grounds.

In the following investigation, a log-linear progress function incorporating a chronological time variable is chosen as an appropriate form of the trend. The general significance of this type of model has already been discussed at length elsewhere (see especially Chapter 6). The choice of the functional form and the chronological time variable (rather than cumulated production) will be defended here on the grounds of simplicity. Thus, in all the case studies reported in this chapter, a linear least squares trend to the logarithms of the original data is fitted and deviations from this fitted trend are used as input to the spectral analysis.

The spectral methods have generally been applied to cases for which large amounts of data are available. Although it is desirable that the number of observations be greater than about 100, there is no agreed minimum number of observations required. A number of studies indicate that it is possible to obtain useful results based on fewer observations (Gudmundsson, 1971; Harkness, 1968). Indeed, the reliability of the results of spectral analysis depends on the ratio n/m, not on n alone.

---

[3]This inference may sound somewhat exaggerated. However, it is not so. For example, see Granger (1966), who concludes, on the basis of a purely numerical approach to trend elimination, that business cycles do not exist. If other independent evidence is any guide, this conclusion must be rejected as a patently untenable empirical finding of an otherwise excellent theoretical work.

The choice of the number of estimated components m is inversely related to the degrees of freedom as well as to the bandwidth of the lag window. Clearly, the smaller the bandwidth, the more maxima and interpeak intervals that can be observed. Choosing a large value of m facilitates localization of the period of the cycles involved. However, it can be shown that, irrespective of the sample size, the variance of the power estimates increases as the bandwidth decreases. We must therefore choose between the accuracy of the power estimates and the accuracy of the frequency intervals. In this study two choices of number of lags were made, equal to approximately one third and one fifth of the number of observations. Since the majority of the qualitative conclusions were found to have remained unchanged, results only of the latter choice are presented. These results therefore reflect greater accuracy in power estimates than in frequency intervals.

In the following analysis of data, we have chosen not to apply any tests of significance on the estimated spectra. This is because the small sample properties of the existing tests are yet to be determined. Instead, we will follow the practice of many researchers in this area and regard as "significant" any peak whose value is two times the previous low value (Harkness, 1968). Further, the ideas of long run and short run have been operationalized in terms of certain low frequencies (e.g., from 0 to 0.12 cycles per year) and high frequencies (e.g., from 0.25 to 0.50 cycles per year), respectively (Granger and Hatanaka, 1964). Admittedly, there is an arbitrary element involved in the cutoff point. Finally, like many other investigations in this area, we have relied on visual examination of the spectra to interpret the results of the study.

4. CASE STUDIES OF CYCLICAL PHENOMENA IN TECHNOLOGICAL CHANGE

In all, the following seven cases of technological change are examined. The specific variables and the types of data employed will be defended on the grounds of empirical necessity. The data in Case 1 were obtained from Schmookler (1972, Series 122, pp. 138–139) and Cases 2 and 3 from Schmookler (1966, Table A-2, pp. 223–226). The data in the remaining cases will be found in the Appendix to this book.

Case 1: Annual number of patented inventions in the construction industry, 1855–1952.

Case 2: Annual number of patented inventions in the farm equipment industry, 1837–1957.

Case 3: Annual number of patented inventions in the railroad industry, 1837–1957.

Case 4:   Innovations in farm tractor technology, 1922–1970. The specific measure of technology employed is the average of the maximum belt horsepower of tractor purchased.

Case 5:   Innovations in locomotive technology, 1904–1967. The specific measure of technology employed is the average tractive effort (in pounds) of the stock of locomotives in use.

Case 6:   Innovations in tank ship technology, 1900–1973. The specific measure of technology employed is the average service speed (in knots) of the stock of tankers in use.

Case 7:   Innovations in electricity generation, 1920–1970. The specific measures of technology employed are fuel-consumption efficiency of plants in terms of (a) kilowatt-hours per pound of coal and (b) kilowatt-hours per gallon of oil used.

In the interpretation of the results of the following investigation, it will be useful to keep in mind that certain important differences exist between the different case studies. Briefly, the first three case studies are devoted to the process of invention, the last four to the process of innovation (consisting of the commercial application of invention). All four case studies of innovation conceive of technical progress in terms of changes in the functional capability of the system. One main advantage of functional measures over other variables is that they enable us to obtain an objectively measurable index of technical progress in which various major and minor innovations are automatically weighted according to their contribution to a certain common dimension of the phenomenon (Chapter 2). Thus, for example, the data regarding the service speed of tank ships over the course of time constitute an appropriately weighted index of a wide variety of innovations ranging from the screw propeller and turbine engine to the geared turbine and welded construction.

The specific measures of functional capability chosen in the four studies of innovative activity differ from case to case. The data for Case 4 have been derived from the available information on gross investment in technology. In contrast, the data for Cases 5 and 6 have been obtained from information on the existing stock of plant and equipment. Insofar as a distinction should be made between product and process innovation, Cases 4, 5, and 6 are concerned with the former while Case 7 is largely (but not wholly) concerned with the latter.

The case studies proposed here also differ as regards the time period covered. Thus the range of the eight time series under consideration varies from a maximum of 123 years to a minimum of 49 years. There are nevertheless important commonalities among different case studies. Thus one group of cases (1–3) is devoted to a study of change in the number of techniques, the other group of cases (4–7) to a study of the resulting magnitude of change in the technology itself. Together the various cases thus provide a fairly broad coverage of the variety of circumstances in which technical change in products and processes takes place.

The estimated power spectra for the various cases are presented here in Figures 7.1–7.5. Table 7.1 provides the estimates of the periods where signifi-

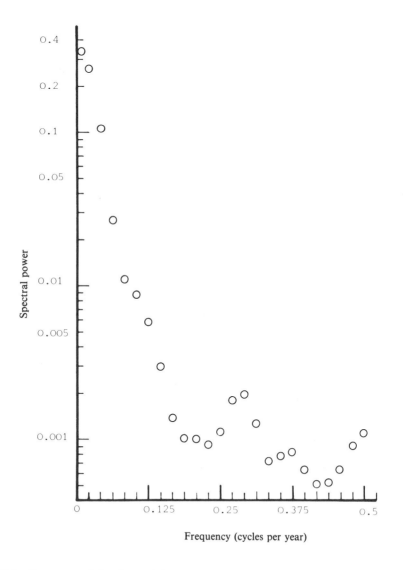

Fig. 7.1 Spectrum of the detrended series on inventions in the farm equipment industry, 1837–1957.

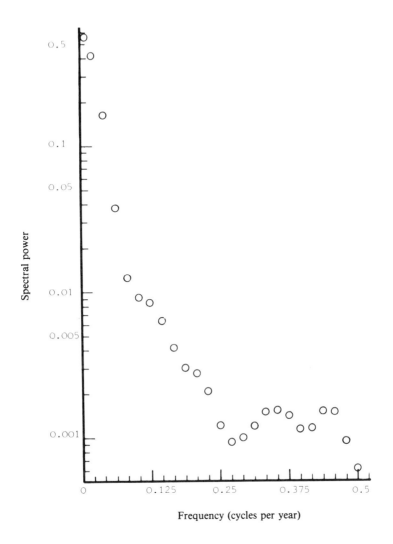

Fig. 7.2 Spectrum of the detrended series on inventions in the railroad industry, 1837–1957.

cant peaks occur in various cases. There is no indication of significant power associated with the zero frequency in any of the cases under consideration. Granger and Hatanaka (1964, p. 132) define a trend as all frequency components with periods greater than $n/2$ where n is the number of observations in the series. The results presented in Table 7.1 then indicate that the trend component has been effectively eliminated in all the cases under consideration. Yet, it can be seen that considerable power remains in low frequencies. Specifically, in all cases, the lower frequencies contribute much more to the variance

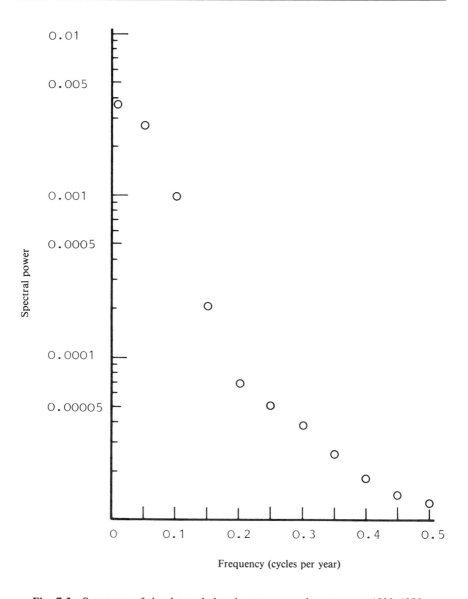

Fig. 7.3 Spectrum of the detrended series on tractor horsepower, 1922–1970.

of the process than do the higher frequencies. It may be concluded, therefore, that in the long run innovations in any given field tend to occur in clusters. This finding is consistent with the results from our earlier investigation into the origin of new techniques (Chapter 3).

The evidence clearly indicates that all seven cases of technological activity are characterized by cycles of varying duration. The spectrum of the inven-

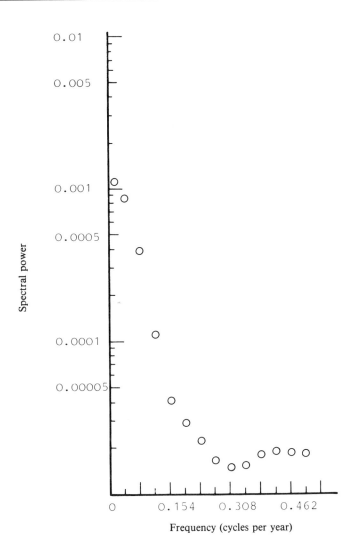

Fig. 7.4 Spectrum of the detrended series on the tractive effort of locomotives, 1904–1967.

tions in the construction industry shows that peaks occur at frequencies corresponding to periods of 40, 20, and 13.3 years. The last two peaks correspond to the first and second harmonics of the 40-year fluctuation. The spectra of inventions in both the farm equipment and railroad industries exhibit a long cycle of 48 years and its various harmonics. The spectrum of the tractor-horsepower series shows that peaks occur at frequencies corresponding to periods of 20, 10, and 6.66 years. The last two peaks correspond to the first

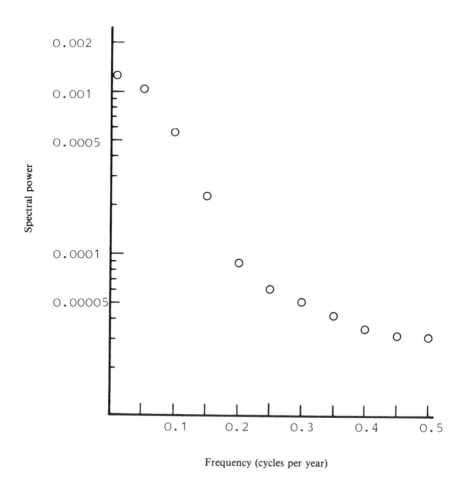

Frequency (cycles per year)

Fig. 7.5   Spectrum of the detrended series on the efficiency of electric power plants in KW-hr/lb. of coal, 1920–1970.

and the second harmonics of the 20-year fluctuation. In the case of the series on the tractive effort of locomotives, peaks occur at frequencies corresponding to periods of 26, 13, and 8.66 years. These correspond to the 26-year cycle and its first two harmonics. In the case of the series on the speed of tank ships, peaks occur at frequencies corresponding to periods of 30, 15, 10, and 7.5 years. The last three peaks are clearly the three harmonics of the 30-year fluctuation. Finally, in the case of the series on the fuel-consumption efficiency of electric power plants, peaks occur at frequencies corresponding to 10 and 6.6 years. The presence of harmonics in these series may be regarded as further evidence substantiating the existence of cycles in the phenomenon under con-

Table 7.1:   Duration of Identified Technological Cycles (in Years)

Case	Periods where peaks occur
1. Inventions in construction industry, 1855–1952	40, 20, 13.3
2. Inventions in farm equipment industry, 1837–1957	48, 24, 16, 6.85
3. Inventions in railroad industry, 1837–1957	48, 24, 16
4. Innovations in farm tractor technology, 1922–1970	20, 10, 6.6
5. Innovations in locomotive technology, 1904–1967	26, 13, 8.66
6. Innovations in tank ship technology, 1900–1973	30, 15, 10, 7.5
7. Innovations in electricity generation, 1920–1970	(a) 10, 6.66 (b) 10, 6.66

sideration (Granger and Hatanaka, 1964, p. 63). In conclusion, the process of technological change is characterized by the persistence of oscillations.

More important, the periods of the cycles found in technological activity are in numerical agreement with the periods of the well-known Kondratieff cycle (about 50 years in duration), the Kuznets cycle (about 15–25 years in duration), and the Juglar cycle (about 7–11 years in duration) found in economic time series. The resemblance among them is indeed striking. It indicates that the origin of business cycles may well be in activities of a technological nature. This is not to say that the results of our analysis prove that innovations are in fact the prime cause of economic fluctuations. They do, however, demonstrate that the relation of business cycles to the "exogenous" factors of technological change is hardly as remote as is sometimes believed.[4]

The results of this investigation therefore do not refute the essential aspects of Schumpeter's thesis. However, the reasoning underlying that thesis must be qualified in several respects. According to Schumpeter, fluctuations in economic growth stem from a highly discontinuous occurrence of *major* innovations in restricted industrial areas. The results of our investigation indicate that innovations tend to cluster not only spatially, but also temporally

---

[4]For a discussion of the contemporary view of business cycles see, e.g., Gordon and Klein (1965) and Volcker (1978).

within any given industry, presumably because the development of every technology is governed by causes of a cumulative nature. Moreover, both major and minor innovations tend to cluster, inasmuch as there exist not only long cycles but also cycles of relatively short duration in the process of technological development. Above all, given a causal link between technological and economic activities, an explanation of observed economic fluctuations is to be found in the cyclicity of technological development rather than in the bunching of major innovations.

In conclusion, Schumpeter's emphasis on radical innovations as the only source of the cyclical phenomenon in economic growth seems to be unwarranted. This is not to deny that there may be an element of discontinuity in certain technological breakthroughs. Rather, it is difficult to see how some major innovation can spark off a long cycle of economic growth given that both the development and the diffusion of a new technique are usually gradual processes.[5] If the results of the present investigation are any guide, the observed oscillations in long-term economic growth stem from recurrent fluctuations in day-to-day technological activity rather than from the advent of a few radical innovations.

The results of the present investigation also indicate that there may well be certain differences in the characteristics of product and process innovations. Thus the periods of the cycles in product innovation (Cases 4,5,6) are considerably greater than the periods of the cycles in process innovation (Case 7). This is a plausible result in view of the findings of a number of surveys indicating that process development efforts consume less than a quarter of the U.S. industrial R&D expenditure while the remaining three quarters are allocated to product development and improvement (see Chapter 2). The implication is that the impact of product innovation on economic activity may be considerably greater than that of process innovation.

Nevertheless, there is a striking similarity between the general shape of the estimated spectra in various cases. Thus, all the spectra are roughly L-shaped while exhibiting several relative peaks. That is, the power of the spectra generally decreases as frequency increases. Thus it may be concluded that events that affect the process of technological development in the long run are more important than those that alter it in the short run.[6] It is reassuring to find that we are led to a similar conclusion from a variety of different starting points (see, e.g., Chapter 9).

It is commonly recognized that there is a certain amount of "spillover" in the course of innovative activity. That is, experience accumulated in one field may turn out to be valuable in providing solutions to technical problems in

---

[5]For a similar viewpoint see Ray (1980).

[6]For a partially analogous view of economic development see Granger (1966).

some other field. It is therefore of considerable interest to examine what if any relationship exists between the cyclical aspects of technological change in different fields.

The first case investigated here concerns the possible relationship between inventions in the farm equipment and railroad industries. A priori, it is not implausible that a relationship does in fact exist between the two because technological development in both industries owes a great deal to advances in mechanical engineering. The results of the investigation into the possible relationship between the two are presented in Figures 7.6–7.8. As indicated by the coherence diagram (Figure 7.6), interrelatedness between the two series is generally poor except in periods corresponding to very low frequencies. The cross-amplitude spectrum of the two series indicates that significant peaks occur at frequencies corresponding to periods of 48, 24, and 16 years (Figure 7.7). The salient features of the cross-amplitude spectrum therefore do not differ from those of the individual spectrum. The gain plot is represented in Figure 7.8. Together with the phase diagram (not presented here) it indicates that the series on inventions in the farm equipment industry lags behind the series on inventions in the railroad industry by a very small but fixed amount.

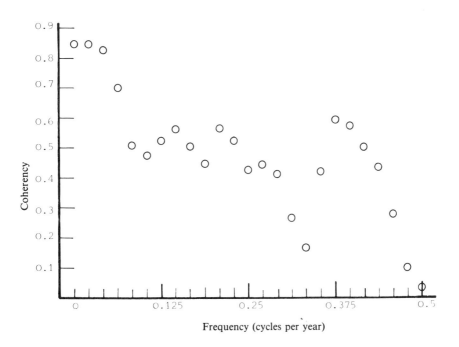

Fig. 7.6   Estimates of coherency between the two series concerning inventions in the farm equipment and railroad industries, 1837–1957.

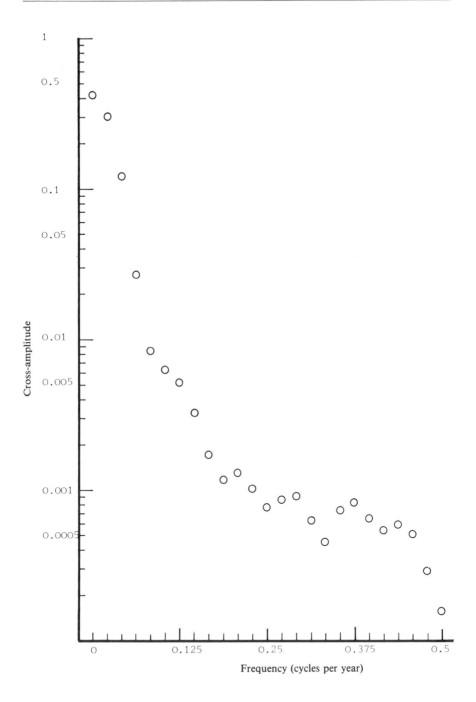

Fig. 7.7 Cross-amplitude spectrum of the two series concerning inventions in the railroad and farm equipment industries, 1837–1957.

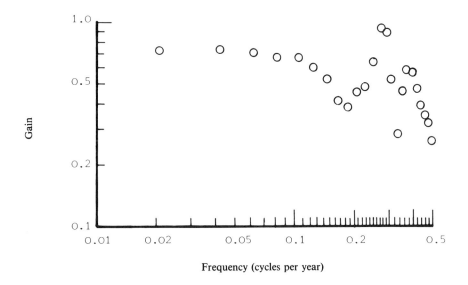

Fig. 7.8 Estimates of the gain between the two series concerning inventions in the railroad and farm equipment industries, 1837–1957.

In general there is little or no phase shift between the two series. In summary, there seems to have been relatively little regular transfer of technical know-how between the farm equipment and railroad industries. Rather, any fallout resulting from the inventive activity in these two fields appears to have occurred on a restricted, *long-term* basis.

The results of the cross-spectral analysis of the two series on technological change in locomotives and farm tractors are presented in Figures 7.9–7.11. As indicated by the coherence diagram (Figure 7.9), the degree of interrelatedness between the two series is very low indeed. The cross-amplitude spectrum of the two series indicates that significant peaks occur at frequencies corresponding to periods of 9 and 3 years (Figure 7.10). Thus it is suggested that there exists a short-term association between the periodic components of the two series. The gain plot is presented in Figure 7.11. Together with the phase diagram (not presented here) it indicates that the relationship between the two series is such that the series on locomotive tractive efforts generally leads the series on tractor horsepower and higher frequencies are amplified more than the lower frequencies. That is, the relationship between the two series behaves like a high-pass filter. All in all, the results provide virtually no support for the existence of fallout in the development of the two technologies under consideration.

The results of cross-spectral analysis of the two series on the fuel-consumption efficiency of electric power plants are presented in Figures

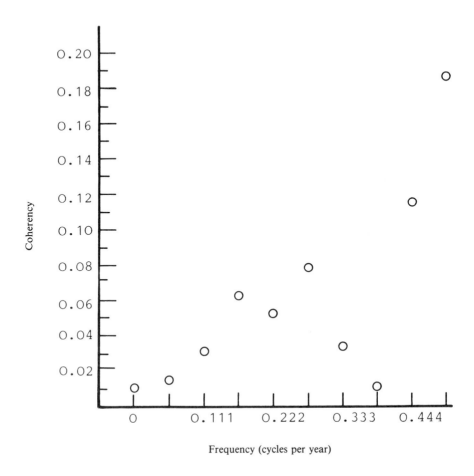

Fig. 7.9 Estimates of coherency between the two series concerning the tractor horsepower and the locomotive tractive effort, 1922–1967.

7.12–7.14. As indicated by the coherence diagram (Figure 7.12), interrelatedness between the two series is generally very high. This is somewhat expected in that the two series concern highly interrelated aspects of the same technology. The cross-amplitude spectrum of the two series indicates that significant peaks occur at frequencies corresponding to periods of 10 and 6.66 years (Figure 7.13). The essential aspects of the cross-amplitude spectrum therefore do not differ from those of the individual spectrum. The gain plot is presented in Figure 7.14. Together with the phase diagram (not presented here) it indicates that the relationship between the two series is such that the series on the efficiency of the use of coal in electric plants leads the series on the efficiency of

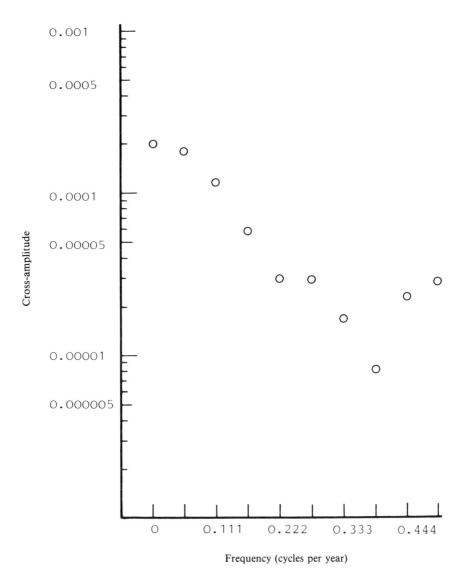

Fig. 7.10 Cross-amplitude spectrum of the two series concerning the tractor horsepower and the locomotive tractive effort, 1922–1967.

the use of oil in electric plants by a small but fixed amount. In general, there is very little phase shift between the two series. By and large, these results reveal little new beyond the findings of the foregoing two investigations of technology transfer in industrial innovation.

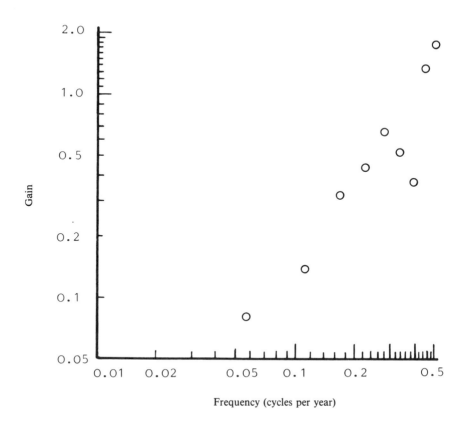

Fig. 7.11 Estimates of gain between the two series concerning the tractor horsepower and the locomotive tractive effort, 1922–1967.

5. PRINCIPAL CONCLUSIONS

This chapter has presented an attempt to examine the nature and significance of fluctuations in technological activity. Our findings indicate that a cyclical phenomenon is an integral part of the process of technical change. The oscillations tend to be quite pervasive and not anomalies in the course of technological development.

The results of the present investigation further indicate that there exists a striking similarity between the cyclical aspects of technological and economic activities. The similarity between the two transcends mere qualitative considerations. The periods of the technological cycles found here are in quantitative agreement with the periods of certain well-known economic cycles. In conclusion, the origin of business cycles may well be in activities of a technological nature. This possibility has the important implication that a long-term eco-

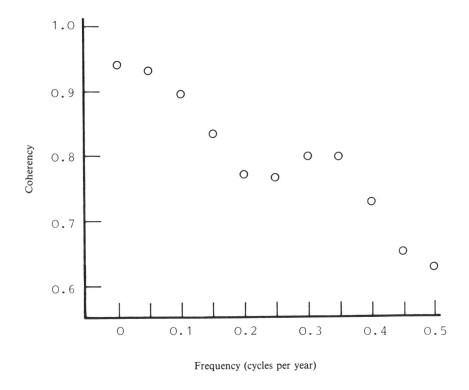

Fig. 7.12 Estimates of coherency between series on the fuel consumption efficiency of electric power plants in KW-hr/lb. of coal and KW-hr/gal. of oil, 1920–1970.

nomic stabilization policy must be at least partially based on scrutiny of research and development activity.

Moreover, if the findings of our analysis are any guide, the observed cycles of long-term economic growth are caused by a multitude of changes in industrial technology rather than by the advent of any single technical breakthrough. In particular, the notion of a *major* innovation as the sole source of an economic miracle seems largely unfounded. Rather, it is seemingly incremental innovations that hold considerable promise for the future. In conclusion, the objective of a sustained long-term economic growth is best served by a policy that seeks to promote gradual, step-by-step advances in certain selected techniques rather than only the so-called radical advances in technology.

The results of the present investigation also provide further support to some of our earlier findings (see Chapters 3 and 6). First, advances in technology seldom tend to be uniformly distributed. There is a preponderance of

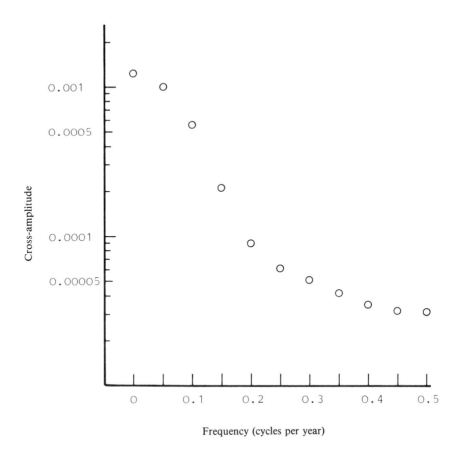

Fig. 7.13   Cross amplitude spectrum of series on the fuel consumption efficiency of electric power plants in KW-hr/lb. of coal and KW-hr/gal. of oil, 1920–1970.

evidence to indicate that innovations, like Shakespearean sorrows, occur in clusters: "They come not single spies, but in battalions!"

Second, it is apparent that the know-how acquired in the development of a technology cannot be wholly transferred from one industry to the other even if the two are closely related. A large part of technical know-how tends to be industry specific. Moreover, its transmission across different fields is typically held up over protracted periods of time. There is of course some fallout in the course of research and development activity. The important point, however, is that transferral is sporadic. There is considerable sluggishness in the inter-industry flow of technical know-how. This confirms the previously proposed principle of technological insularity.

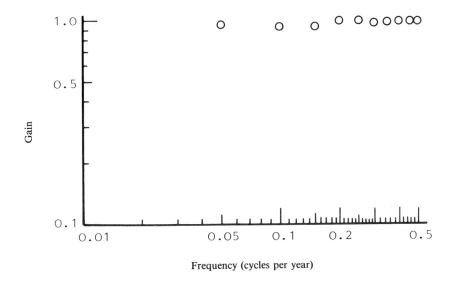

Fig. 7.14   Estimates of gain between series on the fuel consumption efficiency of electric power plants in KW-hr/lb. of coal and KW-hr/gal. of oil, 1920–1970.

Finally, the events that affect the technology over extended periods of time are found to be far more significant than those that influence it over brief spans. The implication is that technological planning should be regarded as a long-term process. It cannot be effectively conducted on a stop-gap basis. At the risk of laboring the obvious, let it be said that the objective of innovation-oriented public policy is not to create a nine days' wonder; rather, it is to develop an enduring system.

Chapter 8

THE MAXIMUM CAPABILITY OF TECHNOLOGY

1. THE STEPWISE GROWTH OF PLANT AND EQUIPMENT SIZES

It is apparent that there exist significant differences in the size and vintage of any given technology employed across different firms within a given industry and among industries themselves. Invariably, technical capabilities tend to be distributed in a nonuniform manner. What is more, the distributive inequality often feeds on itself. For example, it may initially lead to significant differences in the productivity of the various enterprises, thereby giving rise to further inequality in their innovative performance. Evidently, an analysis of change in the *average* level of existing technical capabilities does not cover all the vital aspects of innovation and productivity growth. An investigation into the *maximum* or minimum capability of technology is equally important. Thus a distinction is sometimes drawn between the *general* and the *best-practice* technology of the industry. The former refers to the technology in common use, the latter to the most advanced technology available at any given time. It has been observed that the average productivity of an industry is usually only about half that of the best practice (Salter, 1969). Moreover, there is some indication that the observed disparity becomes a further cause of industry-wide stagnation of innovative activity, as in the case of the steel and textile industries. However, very little is known about the origins of best-practice technology.

In one of the pioneering studies in this area, Simmonds (1969a,b, 1972) observed that the growth of certain types of technology occurs in a stepwise fashion. That is, the observed pattern of growth is characterized by instantaneous jumps separated from each other by limited periods of constancy. Moreover, increase in the capacity of technology keeps pace with the growth

Devendra Sahal, Patterns of Technological Innovation   ISBN 0-201-06630-0

of production. As an example, the capacity (size) of the largest U.S. ethylene plant has increased in a stepwise fashion *while remaining a constant fraction* of the total U.S. production during the period 1945–1970. Likewise, the size of the largest Canadian ethylene plant has increased in a fixed proportion to the total Canadian production over the same period of time. This is shown in Figure 8.1. The growth of the largest plants producing ethylene, benzene, synthetic ammonia, and vinyl chloride monomer has taken place in the same way.

As Simmonds pointed out, there are serious economic consequences of the observed pattern of growth. It can lead to periodic overcapacity because increase in plant size generally far exceeds the growth of the market. Often,

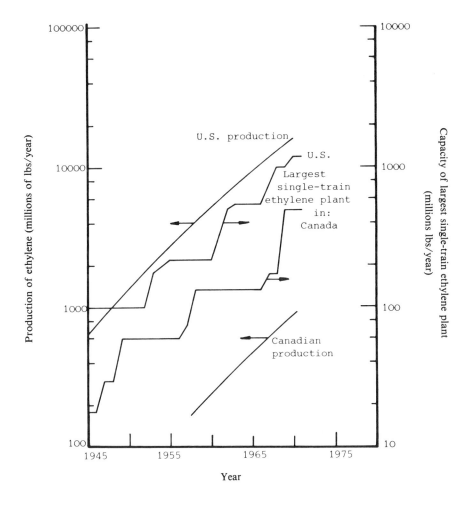

Fig. 8.1   Growth of the largest ethylene plant in relation to total ethylene production in Canada and the United States. After Simmonds (1969).

this triggers a price-cutting mechanism, thereby nullifying some of the economies of scale derived from large-scale plants. The buyers of the product do not remain oblivious to this situation. Frequently, they make forward offers to purchase the product at lower prices in anticipation of the price decline. This in turn leads to further price cutting until a break-even point is reached. The whole situation becomes one of a cyclical relationship between growth–overcapacity and underutilization–price decline–profit loss–market growth–full capacity utilization–profit gain–growth.

The phenomenon discussed above is not peculiar to the petrochemical industry. Martino and Conver (1972) show that the capacity of the largest turbines in the United States has also increased in a stepwise manner while remaining a fixed proportion of the total installed capacity over the course of time. Evidence further indicates that this type of growth applies to many other variables. For example, the increase of the maximum temperature, pressure, and unit capacity of steam-powered electricity-generating plants during the years 1900–1960 is best described in terms of a stepwise pattern of growth (Ling, 1964, pp. 25–27). The maximum capability of the transmission circuits in the United States has increased in a similar way during the period 1950–1970 (cf. Federal Power Commission's 1970 power survey).

In summary, there is considerable evidence to indicate that the growth of maximum capability of many types of technology occurs in a stepwise manner. Nevertheless, the theoretical basis of the observed pattern requires further clarification. To begin with, it should be noted that step functions occur ubiquitously. Further, they depend on scale of measurement. Ashby (1952, pp. 88–90) gives a number of examples of this. The conductivity across the contacts of a postoffice relay exhibits a highly continuous trajectory of slowly accelerating, then decelerating, and finally declining values if observed over microseconds. If the conductivity is observed from second to second, however, it varies almost exactly in a step-function form. In general, any smooth function plotted over a certain interval of time will result in a step function if the time interval of the basic observations is made sufficiently large. This should not be taken to imply that a step function is merely an artifact of measurement. A step function does exhibit striking simplicity of behavior in contrast with a full function. The fact that certain phenomena can be conceived in such simple terms deserves an explanation.

An even more puzzling aspect of the phenomenon has to do with the observed constancy of the ratio of the maximum unit capacity to the total installed capacity in both the electricity generation and petrochemical industries. What, if any, significance is there to the existence of this ratio? How can it remain constant over long periods of time, despite the fact that the environment surrounding it is changing all the time? Granted its validity, the existence of such a simple invariant factor is illustrative of a point noted earlier: The law of a complex system need not itself be complex. However, further investigation is required to determine why the observed pattern is what it is.

The objective of this chapter is to examine the mechanism underlying the advance of technical frontiers. The specific case chosen for this purpose is the growth of the maximum capacity of thermal and hydroelectric generating units in Canada during the years 1917–1972 (Figure 8.2). The background of the investigation is, in part, provided by the information contained in the Federal Power Commission's 1970 national power survey.[1]

2.   A CASE STUDY OF TECHNOLOGICAL INNOVATIONS IN ELECTRIC TURBOGENERATORS (1917–1972)

As shown in Figure 8.2, the maximum capacity of thermoelectric generators has increased from barely 0.8 MW in 1917 to 540 MW in 1972. The corresponding growth of hydroelectric units has been less spectacular, although still sustantial; it has increased from 18 MW to 475 MW during the same period of time. Several advances in design, metallurgy, and manufacturing techniques have contributed to the observed growth of technology (Beckwith, 1952; Krick, 1969; Rosenberg, 1974; Schroeder and Wilson, 1958; Wiedemann, 1958). While it may be obvious, it needs to be emphasized that today's large turbogenerators did not originate overnight. Rather, their development was made possible by a series of improvements in various aspects of technology. The earliest improvements came in stator ventilation in an attempt to improve cooling and to reduce core losses. By 1924 more efficient fans and new methods of cooling were developed. At the same time, core losses were reduced by the adoption of high-silicon steel. The resulting problem of brittleness was resolved first by the use of special fillets in the slot bottoms and wedge notches, and later by new methods of punching high-silicon steel laminates with improved ductility. The rotor capability was similarly enhanced by the use of improved material. Initially, the rotor coil retaining rings were made of high-strength alloy steel. With the use of nonmagnetic steel in 1932 it became possible to reduce both rotor leakage fluxes and local heating at the ends of the stator teeth. The rotor performance was further enhanced by the use of narrow ventilating slots along the teeth and pole faces. In this way much longer rotors became feasible.

---

[1]Some of the technical terms employed are as follows. A thermoelectric generator unit refers to a boiler–turbine–generator complex. The data on thermal units analyzed in this study include both fossil-fuel and nuclear power units. A plant is composed of one or more thermal or hydroelectric units. An electric power generating system consists of one or more plants and is commonly referred to as a power system. Finally, an interconnected or integrated power system refers to a system of electricity generation and transmission. It is common to express the size of a unit, plant, or system in terms of its output in megawatts, which is abbreviated MW.

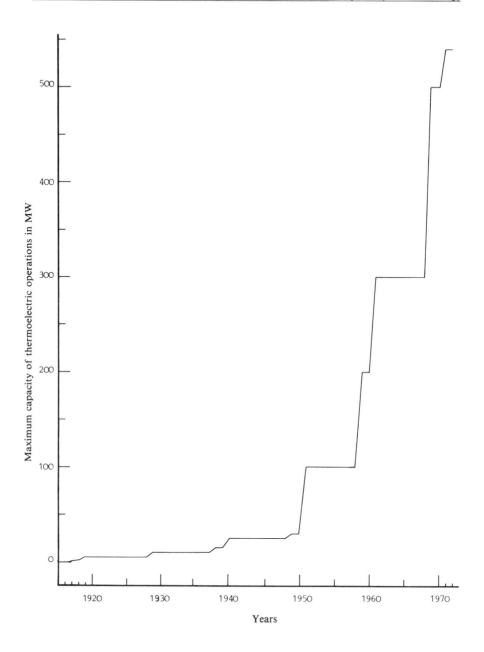

Fig. 8.2   Growth of maximum capacity of thermoelectric generators, 1917–1972.

Air was the universal cooling medium for all turbogenerators of this time. By the mid 1940s it became possible to employ hydrogen instead of air. This was essentially an outcome of quite some experimental effort dating back to

1928. Hydrogen cooling made possible as much as a 55% increase in the rating of any given generator. The resulting problem of wide temperature difference between the copper and the stator teeth due to an increase in the rate of heat transfer through the stator coil was ultimately solved by the development of new means of insulation. Thus asphalt was replaced by thermosetting epoxy or polyester resin as a binder. Insulation nevertheless continued to be a major thermal obstacle in machines of higher rating. An important breakthrough in this area was the development of direct cooled rotors beginning in 1951. Later, the application of direct cooling was extended to the stator windings. By 1954, about 150% increase in capacity was reportedly achieved due to direct cooling at high gas pressure. Simultaneously a number of new methods of more uniform cooling of the rotor were developed. They made it possible for the gas to enter at several different locations of the rotor conductor space without undue loss of forging strength. Liquid stator cooling with oil was introduced in 1956. Beginning in the early 1960s a number of rotors with direct water-cooled windings were placed in operation. In this way, the reliability of the machine has further improved.

In retrospect, two main features of the growth of turbogenerator capacity stand out. First, advances have occurred in several different aspects of technology, such as methods of ventilation, construction and insulation materials, and the medium of cooling, rather than in any one single area. Typically, these advances were of an incremental nature, although collectively their impact has been substantial. The new ideas had first to be patiently tried out before they became feasible. Virtually every innovation in this area has had a long history of experimentation. The earliest hydrogen-cooled generators date back to 1928. Nevertheless, the use of hydrogen became fully feasible only after the war. Similarly, a number of attempts were made to design a rotor with direct water-cooled windings as early as 1914 and 1947. However, liquid-cooled rotor windings became a real possibility only in the 1960s. That is, the fruition of many of the important advances had to await the accumulation of relevant experience.

Of significance is the fact that much of the relevant experience was acquired not only in the design and production of technology as discussed above, but also in its utilization. Thus it should be emphasized that often the initiative for change in turbogenerators came from the electric utility industry as well as from the builders of turbogenerators. For instance, when the limit to the growth of scale of the reciprocating steam engine was reached by the turn of the century, the utility industry compelled the manufacturer to introduce steam turbines despite widespread doubts about the feasibility of the implied increase in the scale of technology (F. MacDonald, cited by Hughes, 1971). According to one classic example, the Allis Chalmers Company had orders for seven turbines even before a blueprint of the design existed (G. Orrok, cited by Rosenberg, 1974). Over the years, the electric utility industry has retained its initiative in seeking improved economies of scale. It has actively promoted and even demanded changes in technology. The manufacturers' position has often been one of acting on the proposals of the electric utility industry. Further, the sustained growth in the scale of turbogenerators has been made possible through experience acquired in the manufacturing process as well as through the identification and elimination of defects in the actual performance of the technology. Some very important changes in technology have come about in an attempt to remedy defects that became evident only after several years of operating experience. Thus the installation of each new electricity-generating unit has significantly added to the existing stock of experience through feedback. In turn, the accumulated experience in the use of technology has contributed to its further growth. All in all, advances in the scale of technology have been made possible by the accumulation of both operating and manufacturing experience.

A second prominent feature of the phenomenon under consideration is that changes in the capacity of individual units appear to have been governed by the size of the electricity-generating system. For example, the available data for the United States during the period 1933–1965 clearly point to the existence of an association between the unit size and the system size. In general, larger systems have been at the forefront in the adoption of new and larger units, while small systems have been laggards in this respect due to the limitations imposed by the system size per se (Hughes, 1971). It has been observed that the relationship between the unit size and the system size has weakened somewhat as a consequence of increasing numbers of pooling agreements between power systems. It nevertheless continues to persist. There are two important reasons for this.

First, the larger the electricity-generating unit at a given speed, the larger is the ratio of kilowatt output to the weight of the rotating parts, and the lower its inherent stability during major system disturbances (as measured in terms of its constant of inertia, WR^2, where W is the rotating weight in pounds and R is the radius of gyration in feet). *Ceteris paribus*, for a system designed with a specific stability criterion, changes in the unit size are therefore governed by the system size. Second, as unit sizes increase, reserve capacity requirements necessarily increase. The reserve requirements can be only partly reduced by pooling agreements and coordination between utilities. In consequence, growth of the unit capacity has generally been dictated by the system size.

In summary, the history of innovations in turbogenerators provides another admirable illustration of the two general propositions of long-term technological development put forward in Chapter 6. Thus, for the case in hand, two main determinants of innovation emerge in the light of available evidence. One is the operating and the manufacturing experience; the other is the scale of the larger system embodying the technology.

3.   RESULTS OF EMPIRICAL ANALYSIS

The test of the theoretical propositions under consideration requires suitable measures of experience and scale. They are chosen here as total installed capacity and cumulated number of years in the manufacture and use of the technology. Total installed capacity measures both the size of the power system and the operating experience. The cumulated number of years variable is approximated in terms of calendar years on the grounds of empirical necessity. Clearly, it is a measure not only of operating but also of manufacturing experience.

The results from the test of a simple log-linear relationship between maximum capacity of turbogenerators and the chosen measures of scale and experience are presented in Table 8.1. As indicated by the values of the coefficient of determination, the relationship performs very well in its explanation of observed changes in the maximum capacity of turbogenerators. The coefficients of both explanatory variables are highly significant. However, according to the results from the Durbin–Watson test, the residuals in these relationships are autocorrelated. Consequently, both the explanatory power and the significance of the regression coefficients have been overestimated. Nevertheless, the performance of the relationships is still very good and it may be concluded that the results do not refute the role of learning and scale factor in the process of technological change under consideration. The postulated relationships are further illustrated in a diagrammatic form in Figures 8.3–8.5. It is apparent that both the slope and the form of the growth functions depicted in Figures 8.4 and 8.5 tend to differ from those of the initial growth functions in Figures 8.2 and 8.3. It should be noted, moreover, that the observed differences are only partly, not wholly, attributable to the logarithmic transformations of data. Evidently, the sizes of the instantaneous jumps and periods of constancy in the original stepwise pattern of growth are greatly reduced when the chosen measure of technology is plotted against certain internal, system-specific measures of time such as scale and accumulated experience. The conclusion to be drawn is that a step function is essentially a progress function or a scaling law in disguise. This is not very surprising. As noted earlier in this chapter, step functions are conditional upon the scale of measurement. The patterns based on theoretical considerations in these diagrams then indicate that existence of stepwise growth is essentially illustrative of certain situations where the time of observation is long *in comparison with* the average time of the relevant intrinsic activities in the evolution of technology.

Table 8.1   Relationships Among Maximum Capacity of Turbogenerators,
Total Installed Capacity, and Elapsed time, (1917–1972)[a]

Hydroelectric Units

Eq. (1) $\log Y = -$ 1.06 + 0.75 $\log X$
                  (0.14)   (0.04)

$R^2 = 0.89$, $S = 0.11$, $F = 418.03$, $d^* = 0.38$, $N = 54$

Eq. (2) $\log Y = -$ 0.82 + 1.64 $\log T$
                  (0.18)   (0.11)

$R^2 = 0.80$, $S = 0.15$, $F = 215.10$, $d^* = 0.24$, $N = 54$

Thermal units

Eq. (1) $\log Y = -$ 1.82 + 1.12 $\log X$
                  (0.13)   (0.04)

$R^2 = 0.93$, $S = 0.19$, $F = 714.04$, $d^* = 0.37$, $N = 54$

Eq. (2) $\log Y = -$ 4.84 + 3.94 $\log T$
                  (0.28)   (0.17)

$R^2 = 0.91$, $S = 0.23$, $F = 510.66$, $d^* = 0.29$, $N = 54$

[a]Definitions: The variable Y is the maximum capacity of turbogenerators, X the total installed capacity where both X and Y are measured in megawatts and T is the cumulated number of years in the manufacture and use of the technology. The origin of this time scale was arbitrarily placed at 17, the last two digits of the earliest year for which the data are available. R^2 is the coefficient of determination, S the standard error of the estimate, F the ratio of the variance explained by the model to the unexplained variance, d^* the Durbin–Watson test statistic, and N the total number of observations. The standard errors of the coefficients are indicated in parentheses.

Seen in this light, an explanation of the observed regularity turns out to be relatively simple. In particular, the significance of the ratio of maximum unit capacity to total installed capacity lies in the role played by the phenomena of long-term learning (i.e., the accumulation of relevant experience) and scaling in the process of technological innovation.

The results also indicate that the coefficients of total installed capacity in relation to maximum capacity are significantly different from unity in the case of both hydroelectric and thermal units at the usual significance level of $\alpha = 5\%$. Roughly speaking, however, the values of the estimated coefficients suggest that the ratio of the maximum capacity of thermal units to the total installed thermal capacity has remained fairly constant during the last 55 years,

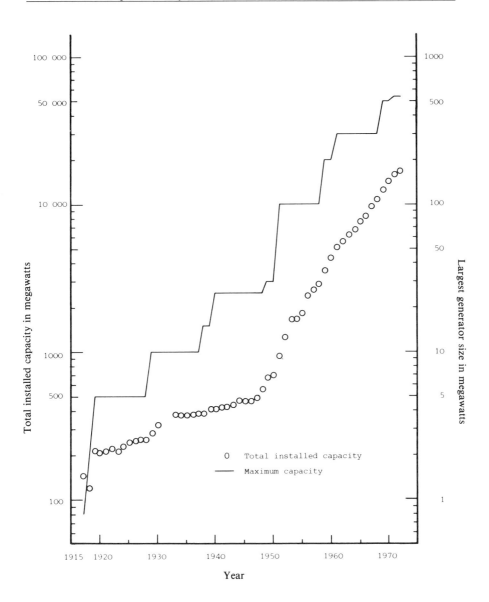

Fig. 8.3   Growth in maximum capacity of thermoelectric generators and total installed capacity, 1917–1972.

while the corresponding ratio in the case of hydroelectric units has declined, presumably because of the limitation imposed by the size of the water resources at any given location. Another indication of the relative constancy of the ratio of maximum unit capacity to total installed capacity is given by the

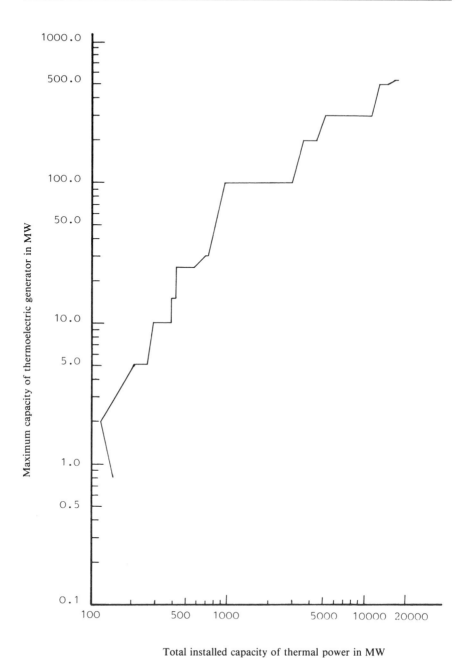

Total installed capacity of thermal power in MW

Fig. 8.4   Relationship between maximum capacity of thermoelectric generator and total installed capacity of thermal power, 1917–1972.

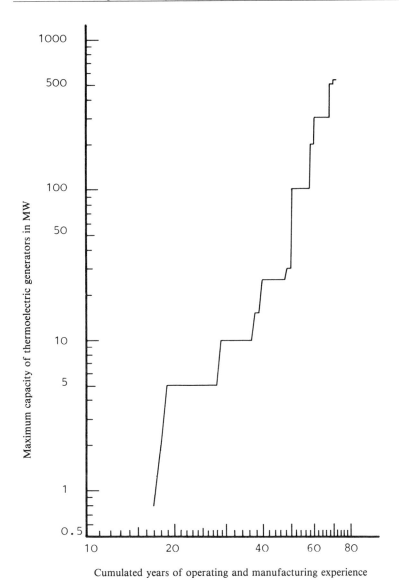

Fig. 8.5   Relationship between maximum capacity of thermoelectric generators and cumulated years of manufacturing and operating experience, 1917–1972.

computed values of the coefficient of variation. In the case of thermal and hydroelectric units they are 0.44 and 0.39, respectively, during 1917–1972 and 0.35 and 0.22, respectively, during 1930–1972. Judging by these values there is relatively little variation in the parameter under consideration. Specifically, the size of the largest turbogenerator of either type has remained fixed at approximately one hundredth of the total installed capacity during the latter time period.

Despite the good fit of the simple log-linear relationship to the data, its formal basis remains somewhat shaky. In particular, it makes no provision for any eventual slowdown in the growth of technology. For the case at hand, four limiting factors in the growth of technology deserve particular emphasis. To begin with, there is recent evidence pointing to diseconomies due to a decrease in reliability as a result of scaling-up in electricity generation. There is some indication that larger units are not always more economical (Abdulkarim and Lucas, 1977; Fisher, 1978a,b; Lee, 1978). Second, the size of the energy resources available locally or at nearby sites may eventually impose a constraint on the maximum capacity of an individual electricity-generating unit. This is, of course, most evident in the case of hydroelectric turbines. However, notwithstanding the development of new transportation devices, such as unit trains, it is an inhibiting factor in the growth of thermal units as well. This is because beyond a certain point, economies of scale in production are necessarily outweighed by the cost of transportation of fuel from distant reserves. Third, and a potentially more important constraint in the growth of both hydroelectric and thermal units, is an economical source of condenser cooling water. Hitherto, it was common to utilize water from, say, a river in a "once-through" system. However, the quantity of circulating water required for a large plant utilizing this arrangement is quite high despite numerous improvements in the design. Moreover, recently enacted government regulations restrict the amount of heat that can be discharged into a river or lake. In the face of the declining number of natural water facilities, the use of cooling towers is expected to grow. In this way, the existing limit to the capacity of the individual units may be pushed further. However, new constraints to the growth of technology arise. Besides adding to the investment (in the form of extra land required) and operating costs, the use of cooling towers is bound to alter the character of problems from one of thermal pollution to one, eventually, of environmental changes as a consequence of discharging large amounts of waste heat into the atmosphere. Last but not least, it is being increasingly recognized that building big plants may no longer be the best strategy. Rather, it may be economical to construct small plants in which the waste heat can be reused. Against the background of diminishing energy resources, the trend toward the use of small-scale technology is expected to grow in the future.

These considerations suggest that it might be worthwhile to consider an alternative specification of the theoretical propositions such as the one proposed in Chapter 6. In symbols, let y be the maximum capacity of technology and x its determinant. It may be recalled that the alternative form of the model is given by the equation

$$\log Y_t = \alpha(1-\lambda) + \beta(1-\lambda) \log x_t + \lambda \log Y_{t-1} + v_{1t} + \gamma v_{2t} \quad (8.1)$$

where

$$\lambda = 1/(1+B), \quad \gamma = (1-\lambda)/\lambda, \quad \text{and} \quad 0 < \lambda < 1. \quad (8.2)$$

In the formulation above, B is a measure of growth rate and $(v_{1t} + \gamma v_{2t})$ is a disturbance term whose successive values can be shown to be serially uncorrelated.

The results from the test of the model are presented in Table 8.2. The parametric estimates of the relationships are generally well determined except in the case of hydroelectric technology, where the explanatory variables themselves are highly correlated. At the $\alpha = 0.01$ level, the results of the Dur-

Table 8.2 Test of the Proposed Hypotheses of Learning, Scaling, and Technological Innovation in the Case of Long-Term Growth in the Maximum Capacity of Turbogenerators, 1917–1972.[a]

Hydroelectric units

Eq. (1) $\log Y_t = -0.13 + 0.087 \log X_t + 0.90 \log Y_{t-1}$
          $\qquad\quad (0.14) \quad (0.080) \qquad\qquad (0.10)$
          $\qquad R^2 = 0.95, S = 0.07, F = 502.94, d^* = 2.07, N = 53$

Eq. (2) $\log Y_t = -0.04 + 0.067 \log T_t + 0.97 \log Y_{t-1}$
          $\qquad\quad (0.11) \quad (0.14) \qquad\qquad (0.08)$
          $\qquad R^2 = 0.95, S = 0.07, F = 493.6, d^* = 2.19, N = 53$

Thermal units

Eq. (1) $\log Y_t = -0.46 - 0.33 \log X_t + 0.68 \log Y_{t-1}$
          $\qquad\quad (0.16) \quad (0.09) \qquad\qquad (0.08)$
          $\qquad R^2 = 0.98, S = 0.11, F = 1078.3, d^* = 1.63, N = 53$

Eq. (2) $\log Y_t = -0.40 + 0.40 \log T_t + 0.87 \log Y_{t-1}$
          $\qquad\quad (0.40) \quad (0.32) \qquad\qquad (0.07)$
          $\qquad R^2 = 0.97, S = 0.12, F = 886.30, d^* = 1.68, N = 53$

[a]Definitions are as in footnote a, Table 8.1.

bin–Watson test are satisfactory in all cases. Finally, the performance of the model is excellent throughout. Thus the estimated relationships leave a mere 2-5% variance unexplained in the data. In conclusion, the origin of the stepwise pattern of growth of technology demonstrably lies in the process of accumulation of relevant experience and changes in the overall scale of the system.

According to the results presented in Table 8.2, the long-term coefficients of the total installed capacity in the case of hydroelectric and thermal units are 0.85 and 1.06, respectively. Thus it is indicated that the growth of the maximum capacity of thermal units is likely to remain in proportion to the growth of the total installed capacity even on a long-term basis. This is indeed a remarkable result for it means that the relative constancy of the ratio of maximum capacity to total installed capacity is not to be dismissed as a transient pattern. Its practical implications are clear: Contrary to the prevailing opinion of experts in the industry, growth in the maximum capacity of turbogenerators is likely to continue in the same way as in the past. This is, of course, not implausible if superconducting generators become feasible in the future: the conductors with negligible losses due to resistance when frozen to a temperature range near absolute zero (Snowden, 1972). The findings have equally important theoretical implications. They are discussed below.

4. STEP FUNCTIONS AND STRUCTURAL STABILITY

Step functions have been a subject of considerable discussion in the theory of cybernetics (Ashby, 1952, Chapter 9). According to this viewpoint, a step mechanism constitutes an important instrument of regulation and control. The reasoning is somewhat involved; however, the following illustrative examples may help make this clear. Consider the electric resistance of a fuse. It changes in a stepwise fashion, for it remains constant except when it increases to a very high value by a sudden jump. The example illustrates that change in the value of a step function is conditional upon the occurrence of certain critical states. In the context of this example, the critical state of the fuse is the value of the electric current, that is, number of amperes that will cause it to blow. Next, consider a system that must retain control of its essential variables against a set of disturbances. Suppose that it is equipped with a step mechanism. We can conceive of two broad types of disturbances: minor, slowly varying ones that occur on a more or less continuous basis, and particularly severe ones whose occurrence may well be unexpected. By definition, the latter impact upon the critical states of the system. However, past the critical states, the step mechanism of the system is activated, thereby ensuring that the disturbances do not remain unchecked. In this way the particularly severe disturbances prove to be self-destructive and the system is able to retain control of its essential variables. For example, the success of the human brain as a system of controlling a variety of functions may be attributable to the possible fact that its constituent elements, the nerve cells, form not only a chain but a circuit whose activity closely follows a stepwise pattern of change. Such a circuit tends to maintain two types of activity: the inactive and the maximal excitation, depending on whether the amplification factor is less than or greater than unity. Its critical states are the smallest excitation capable of raising it to full activity and the smallest inhibition capable of making it inert. There are, of course, many other possible causes of stepwise patterns of neuronal activity. But the essential point of this example is that the capability of certain mechanisms to behave in a stepwise fashion may well be the key to their success in regulation and control. The limitation of the underlying reasoning should be equally evident: it is largely of a metaphorical nature. That is, although it is plausible that step mechanisms are essential in regulation and control, the evidence is far from conclusive.

The results of this study shed new light on the implications of step functions. First, a stepwise pattern of growth is not merely a predetermined regularity. Rather, as discussed earlier, it is systematically related to the process of learning by experience. It is therefore to be expected that a step mechanism is an important means of self-organization.

Second, our findings show that a stepwise pattern of growth is a manifestation of an underlying nondimensional variable that is representative of the system as a whole. Thus, for the case at hand, the nondimensional ratio of maximum capacity to total installed capacity is shown to be a meaningful variable. Further, the relevant nondimensional variable may remain constant (as is the case of thermoelectric generators) or vary only within certain limits (as in the case of hydroelectric generators). The constancy or limited variation of a characteristic nondimensional variable of a system is a necessary condition for its structural stability. The nature and significance of nondimensional variables is discussed at length in Chapter 9. As an example, the effect of variation in the Reynolds number on a system of fluid flow may be readily cited. A low Reynolds number is characteristic of a laminar flow. An increase in the Reynolds number beyond a certain point, however, alters the character of the flow completely by abruptly giving rise to turbulence. In essence, the stability of a system depends on a limit to the variation in its relevant nondimensional characteristics. Thus, a step function may well be an important instrument of self-organization insofar as it implies the existence of a relatively constant nondimensional variable characterizing the system.

In the light of foregoing considerations it is easy to grasp the significance of the observed pattern of growth in the maximum plant capacity. It is prima facie a manifestation of the self-organizing nature of technical progress. Equally important, however, is that it is not an inexorable pattern. Rather, its origin lies in what is a largely if not a wholly controllable factor: the goodness of fit between innovation and its context. The meaning of the observed pattern is simply that success in innovative activity hinges upon how well a new technique is adapted to its production and utilization possibilities.

5.   **PRINCIPAL CONCLUSIONS**

This chapter has presented an attempt to examine the factors governing the maximum capability of technology. The results of the present investigation confirm that the growth of certain types of technology occurs in a stepwise fashion. Moreover, they provide considerable further support to the theory of technological innovation advanced in Chapter 6. Thus, the observed pattern of stepwise growth can be readily explicated in terms of the learning and scaling processes in innovative activity as outlined earlier. Equally important, the proposed theoretical framework accounts for a remarkably simple law of an otherwise complex system of technological innovation in the petrochemical and electric power industries: The ratio of maximum unit capacity to total installed capacity remains relatively constant over long periods of time.

The results of this investigation shed further light on two findings of Chapter 6. First, as shown here, increase in unit size is made possible by the accumulation of relevant experience. Thus scaling is primarily a matter of learning. Moreover, one central feature of this process is that the know-how acquired from operating on a large scale is not foregone if the scale is subsequently reduced. That is, in contrast with the traditional viewpoint, not all economies of scale are of a *reversible* type. In a fundamental sense, at least certain improvements in the performance of a system due to change in its scale are of an *irreversible* nature. To sum it up, large-scale technology is ineffectual without the acquisition of production skills.

Second, it is commonly recognized that one important source of economies of scale lies in the indivisibility of the inputs employed in the production process. The specific case study presented in this chapter suggests that in some cases the role of indivisibilities may be far more encompassing than has been recognized hitherto. As regards the petrochemical and electric power industries, there seem to exist what may be called "industry-wide indivisibilities" reflected in the existence of a stable relationship between unit size and total installed capacity. Further, such indivisibilities might operate on a long-term basis. The obvious implication is that countries with relatively small domestic markets for the products of these industries cannot afford to employ the best available technology because the scope for the use of the most efficient techniques depends on the total installed capacity. Moreover, the learning processes underlying innovation in these industries occur not only in the production but also in the utilization of technology. In conclusion, it seems that there is a clear case for protection of the petrochemical and electrical machine industries by tariffs and other means during the infant stage of their growth.

Chapter 9

THE INNER DYNAMICS OF
TECHNOLOGICAL INNOVATION

1.   TOWARD A MICROVIEW OF INNOVATIVE ACTIVITY

The results from the analyses presented in the preceding chapters point to
two striking features of technical progress. One is that it tends to be object
specific. The other is that it lacks transmissibility. To begin with, our investigation reveals the existence of technological guideposts implying that
much of the innovative activity in any given area tends to be centered upon a
specific pattern of design (Chapter 2). Moreover, there is a large element of
tardiness in the process of technology transfer (Chapters 3, 7). Finally, different technologies vary considerably as regards the extent of the advances that
result from the relevant learning processes in each case (Chapter 6). Not only
are these findings recurrently brought out, they are also mutually augmenting.
Thus the object specific character of technical know-how may be justifiably regarded as both a cause and a consequence of the lack of its transmissibility.

In turn, these findings raise a number of important issues. Among other
things, they imply that the observed interindustry differences in the growth of
productivity are caused by differences in the nature of the technology employed in different fields. The question that remains is, what, if anything, can
be done to bring about a willful change in the intersectoral productivity differential? Granted that the nature of an individual technology plays an important role in its own right, what are the relevant policy instruments? In its
essence, the question becomes, just what is implied by the nature of the technology? To put it differently: What constitutes the technological innovation
potential of a product or a process? Is it amenable to quantitative measurement? If so, how?

The issue is also one of central importance in the management of R&D activity (Sahal, 1976a,b,c, 1977, 1978b). Numerous examples may be cited. What is the extent of possible further improvement in the thermal efficiency of electric power plants at the current level of R&D expenditure? Is further saving in the material employed in the construction of tankers still feasible? What would be a reasonable extent of further improvement in the efficiency with which various energy resources are utilized?

In the formulation of a national policy for technological development, we are confronted with a series of analogous questions. If technical progress is indeed object specific, a national strategy ought to be based on a careful consideration of the growth and stagnation characteristics of alternative technologies. Moreover, in accordance with our putty–clay principle of innovative activity, there is an element of irreversibility in technical progress (Chapter 6). Thus it is critically important that the initial choice of a technology be based on an explicit assessment of available alternatives.

From a policy point of view, other important questions deserve attention. If technical progress is indeed characterized by a certain amount of insularity, as indicated by our results, it becomes all the more important to gain a detailed knowledge of the mechanism underlying the transfer of technical know-how across different industries and economies. We most emphatically do not wish to imply that learning in the innovative activity is wholly localized. Rather, to the extent that technical know-how is indeed transferred across firms and industries, we would like to know whether it is acquired via blueprints, trained personnel, or plant and equipment. In essence, who all are involved in the relevant learning processes? Moreover, just what is it that they are learning?

In an attempt to examine these questions, this chapter provides a closer look at two constant themes of this book. One concerns the process of learning and scaling in technological innovation (cf. especially Chapter 6). This is discussed in Section 2. The other concerns what may be called a principle of self-resemblance: A system during the course of its evolution sacrifices its morphological characteristics so as to remain invariant in its functional properties (cf. especially Chapter 4). This is discussed in Section 3. When these two themes are brought together in Section 4, there emerge a number of extremely interesting relationships about the lawlike aspects of innovation processes. These relationships in turn lead to a new finding: The distribution of technical capabilities in any given area follows the Pareto law. This paves the way for a formalization of the concept of technological innovation potential in Section 5. Section 6 presents a number of case studies of technical progress. The main findings of the investigation are reported in Section 7.

2.   A RECONSIDERATION OF LEARNING AND SCALING
IN TECHNOLOGICAL INNOVATION

One of the most striking aspects of engineering practice is that it is very rare to design a technique wholly on paper. The development of novel design concepts is almost always based on extensive experimentation carried out on a small scale. The use of small-scale modeling has a long history, dating back to the eighteenth century. One of the earliest attempts to make models of water wheels is found in the work of John Smeaton in 1759. William Froude conducted pioneering investigations into the performance of model ships in 1866. In the same year Otto Lilienthal initiated experiments on small-scale lifting vanes. Fargue began his work in the building of river models in 1875. About a decade later, Osborne Reynolds undertook small-scale experiments in connection with the improvement of ship canals. Reynolds' path-breaking work was continued by F. Vernon-Harcourt in 1886. By the turn of the century, the practice of modeling had become well established. Today it is generally recognized that models are an indispensable bridge to laboratory experiments and large-scale production. Their role in the development of new techniques is of fundamental importance. The reasons for this are plain: Inevitably there exists the risk that a new design may not in fact work in a satisfactory manner, and mistakes made on a large scale can be very costly.

In recent years, the introduction of digital computers has made possible analytic solutions to some engineering problems. However, conventional modeling based on trial-and-error principles still remains the only established technique for the resolution of more difficult problems. In many branches of engineering, such as hydraulics, reliable procedures based on physical laws are generally lacking. Some 300 years ago, Galileo is reported to have observed: "I have met with fewer difficulties in observations relating to the movements of heavenly bodies, notwithstanding their stupendous distances away, than in investigating the movement of flowing water which takes place before our very eyes." In the opinion of informed engineers, this is, by and large, true today as well.

It has been observed that the rise of science-related industries such as aerospace, chemicals, and electronics in modern times heralds a significant change in the system of technical progress (Freeman, 1974). It is of course undeniable that the days of craftsman-type organization belong to the past. Technical innovation today depends in no small measure on organized research and development activity. However, one fundamental aspect of innovation has remained unchanged: it is a process of learning from trial experiments. Wind-tunnel tests still underlie a great many advances in aircraft technology, as do pilot plant experiments in the field of chemicals. It is easy to see why this is so. In many instances of design, the relevant scientific laws may be well known. However, they hold only under a set of limiting conditions. Frequently, it is impossible to know if these conditions are in fact being fulfilled. As an example, the otherwise valuable principle of conservation of energy is of little use in problems involving the conversion of mass into energy. Consequently a designer must often resort to the semiempirical methods of modeling and simulation.

To put it somewhat differently, the process of design is always an exploration into the unknown. Obviously, its success hinges upon the existence of a certain discriminatory mechanism. There are of course many sources of selection in engineering problem solving and pure scientific knowledge is one of them. However, the most important source of selection in such an endeavor is previously accumulated experience of a practical nature.[1] This is evidenced by the fact that new designs are seldom attempted without reference to engineering handbooks containing data from past experience or to computer programs incorporating previously established rules. But despite the best possible effort made in the design, a pre-production prototype is often beset by various flaws. Identification and remedy of these defects are, in turn, conditional upon the accumulation of operating and test experience. In all, technological innovation depends on the acquisition of relevant knowledge via a *systematically organized* process of trial and error. As an example, the following view of innovations in aircraft technology is worth quoting (Miller and Sawers, 1968, pp. 247–248):

> The aircraft industry is thus more directly based on science now than it was before the problems of supersonic flight began to complicate the calculations of designers. The solutions that the designer finds to the problems set by the performance he wants to achieve still have to be his creation, but limits within which he can choose are narrower. Yet the degree of empiricism and so the margin for error remains a large one, for so elementary a problem as the inaccuracy of wind-tunnel tests continues to plague the industry, some 60 years after it was first recognized.

Advances in nuclear reactors provide another admirable illustration of the important role played by learning from trial experiments in technological innovation. As one expert in this area has put it (White, 1964, p. 147),

> The evolution of the boiling-water reactor has occurred by simplifying the design as a result of operating experience and test performance. Important technical

---

[1] Indeed, there is considerable evidence to indicate that success in nearly *all* types of problem solving depends on selectivity based on experience (Simon, 1969).

features contributing to equipment specification, to improved safety, and to improved operating characteristics, include the use of a single cycle with steam separation inside the reactor vessel and a pressure-suppression type of containment. Core and fuel performance have been improved by applying feedback information from operating and test experience.

As discussed earlier (Chapters 2, 3), science plays the role of a facilitating rather than an enabling factor in technical progress. According to the evidence put forward here, this is to be expected because development of new techniques is often based on empiricism in lieu of purely analytic method.

These considerations suggest what may be called a "trial, error, and selection" hypothesis of technological innovation. In essence, the development of new techniques involves many efforts of a hit-or-miss nature. Accumulated experience of a practical type is essential for the convergence of such a process. Obviously, this hypothesis is analogous to the previously stated hypothesis of learning by doing (Chapter 6). The former is relevant to the microlevel, the latter to the macrolevel of technical change.

The second hypothesis of this chapter is that the development of a technology is a function of its size. It is commonly recognized that the availability of new techniques opens new avenues for change in the size of equipment. According to this viewpoint, technological innovation is a cause of scale changes. For example, it is often said that the development of the jet engine made bigger airliners possible. According to the viewpoint advanced here, however, the converse is at least equally true. That is, changes in the scale of existing equipment constitute one of the most important causes of technological innovation. For example, the fluid bed, an innovation of considerable significance in the case of turbines, was necessitated by an increase in their size beyond the stress limits of the material. As will become apparent, such examples of technological innovation may be indefinitely multiplied.

It is a truism in engineering that substantial improvement in the capability of a technique must ultimately require a change in the scale of its construction. It has already been noted that experiments carried out on small-scale models play an important role in the development of new techniques. As will be discussed in Section 3, the values of certain nondimensional variables must be the same for both the model and the prototype in order to ensure complete similarity between them. However, this is very rarely the case because in general an object cannot preserve both functional and geometric similarity with changes in its size (Chapter 4). For example, fluid flow on a small scale tends to be laminar and on a large scale turbulent. Consequently, it is often impossible to maintain the same value of the nondimensional Reynolds criterion on widely different scales unless velocity and fluid properties are radically changed. Sometimes Reynolds similarity may require fluid velocities on a small scale to approach sonic velocities. Consequently, the nondimensional Mach criterion becomes significant for the model, but not for the prototype, thereby making it impossible to preserve similarity between the two. The general conclusion to be drawn is that it is seldom possible to maintain the same physical conditions on different scales. Much experimental work in the development of new techniques is therefore concerned with the design and construction of distorted models that are similar to prototypes in some but not in all respects. The systematic deviations between the two may occur in the configuration, loading characteristics, linear dimensions, boundary conditions, materials of construction, etc. The essence of innovative activity lies in the identification of, and adjustment for, these scale effects. Thus, the development of new techniques depends on successful change in the scale characteristic of the system under consideration.

All this is in marked contrast with the generally accepted viewpoint that change in scale is a figment of technical progress. The conventional belief, for example, is that supertankers and large furnaces were developed because increase in scale had been successful at the lower level. So designers simply tried to make the ships and furnaces bigger until the economies of scale were exhausted.

Despite its intuitive appeal, it is apparent that the conventional viewpoint does not hold up. In part, this is due to the reluctance of economists and other social scientists to examine innovative activity at the microlevel of technical change. For the remainder, it is a reflection of a general preoccupation with the immediate efficient cause at the expense of other relevant causes of the phenomenon.[2] Indeed, as our discussion clearly shows, scale is invariably a formal cause of technological innovation in determining the properties of the design.

We must further distinguish between two broad types of changes in technology as a consequence of change in its scale. One type of change may be such that it is *facilitated* by the change in its scale. As an example, the use of pneumatic tires on farm tractors—an innovation of major significance—was made possible by a reduction in the size of tractor drive wheels over the course of time in an attempt to improve its tractive capability on weak soils. The other type of change in technology may be such that it is *required* by change in its scale. Thus the use of dual rear wheels, auxiliary front-wheel drive, and four-

---

[2]This, of course, follows from the Aristotelian typology of four causes: the efficient cause (that by which change is brought about), the material cause (that in which the change is brought about), the formal cause (the resulting form of change), and the final cause (the purpose behind the change). As an example, we have an architecture as an efficient cause of a house, concrete and other materials of construction being its material cause. The cubical form and the other distinctive properties of the house constitute its formal cause, while the desire to possess a beautiful residence is its final cause.

wheel drive in the farm tractor were necessitated by an increase in the vehicle's absolute size in terms of overall weight. Yet other types of change in technology are often a result of a combination of the two types of change just noted.

These examples also indicate that certain changes in technology may be due to either change in the scale of one of its components (as in the case of pneumatic tires) or change in its absolute scale (as in the case of four-wheel drive). Yet other types of changes in technology may be due to change in the scale of some larger system of its use (see especially Chapter 6). Many technical improvements also depend on a change in the relative size or the shape of the system, although some of these changes of shape may not be measurable in any direct way. For example, improvement in the overall efficiency of the blast furnace (i.e., combined coke and fuel inputs per ton of pig iron) has been made possible by, *inter alia*, the use of a flexible throat (consisting of flexible steel armor plates along the inside wall of the top of the furnace) so that the distribution of the charge can be altered at will. All in all, it is hardly an exaggeration to say that the scope for change in the technology of a system depends on the choice of scale.

In summary, considerations of scale play a central role in the development of new techniques. The solutions to a wide variety of problems in heat, mass, and momentum transfer require the determination of relevant nondimensional criteria that are based on analysis of scale changes. Further, the success of a technique on a certain scale does not guarantee its success on a substantially different scale. Consequently much design and development work is concerned with the determination of the scale that is appropriate to the specified task of the new technology. From beginning to end, the problems of design and development of new techniques are problems of scale. It is therefore hardly surprising that a wide variety of important advances in technology, including both major and minor innovations, are demonstrably attributable to change in scale. These considerations suggest what may be called a "scale of trial" hypothesis of technological innovation. Evidently, this hypothesis is analogous to the previously proposed hypothesis of specialization via scale (Chapter 6). The former pertains to the microlevel, the latter to the macrolevel of technical change.

3.   A FORMALIZATION OF THE PRINCIPLE OF SELF-RESEMBLANCE

The origin of the proposed principle of self-resemblance lies in the considerations advanced in Chapter 4. In its essence, an evolutionary system is first and foremost a system capable of preserving its identity as a whole by orderly foregoing the identity of its parts. The substantive basis of the principle has already been discussed in detail. Its meaning is simply that the nature of evolutionary process constitutes a self-resembling system.

The proposed principle is formalized here by means of the modern theory of dimensions. It is based on treatises by a number of authors (Bridgman, 1931; Buckingham, 1914; Moran, 1971; Sahal, 1978c,d) and it especially draws on the pioneering work of Causey (1969) in this area. However, detailed knowledge of these earlier studies is not required. The following treatment of the theory is intended to be as self-contained as possible within the limitations of the available space.

Let $\{X_i : i = 1, \ldots, n\}$ be a set of variables characterizing the system under consideration. The variables X_i involved have positive numerical magnitudes or measures x_i which can be expressed in terms of a set of reference or primary scales S_j, $j = 1, \ldots, k$. A change to a new set of scales $\bar{S}_j$, $j = 1, \ldots, k$, where

$$\bar{S}_j = S_j / s_j \tag{9.1}$$

and s_j denotes a positive real number, also changes the positive measure x_i. The new value $\bar{x}_i$ will be related to the old by equations of the form

$$\bar{x}_i = s_1^{\delta_{i1}} s_2^{\delta_{i2}} \ldots s_k^{\delta_{ik}} x_i, \tag{9.2}$$

in which case we say that the *dimensions* of the variable denoted by x_i are given by the k-tuple

$$(\delta_{i1}, \delta_{i2}, \ldots, \delta_{ik}). \tag{9.3}$$

Accordingly, the matrix $\Delta = [\delta_{ij}]$ associated with expression (9.3) will be termed a dimensional matrix.

A simple illustration might help to clarify these concepts. For example, let X_i be the variable "mass density" with positive measure x_i, measured in terms of two ($k = 2$) primary scales of mass, M, and length, L. Suppose that the density measure x_1 is equal to 62.4 lb per cubic foot, as in the case of water when the primary scales are chosen to be 1 lb and 1 ft, respectively. A change in the scales to 1 kilogram (kg) and 1 meter (m), respectively, changes the density measure x_1 to $\bar{x}_1$, where

$$\bar{x}_1 = s_1^{\delta_{11}} s_2^{\delta_{12}} x_1 = \left(\frac{1}{2.2045} \right) \left(\frac{1}{3.2808} \right)^{-3} 62.4 \tag{9.4}$$

$$= 1000 \text{ kg per cubic meter,}$$

the dimension of the density variable x_1 being $(\delta_{11}, \delta_{12}) = (1, -3)$ in terms of the two primary scales of mass (M) and length (L).

One further important notion is that of dimensional homogeneity. Let the measure x of a variable X be expressed as a function of the measures $x_1, x_2, \ldots, x_n$ of the n variables $X_1, X_2, \ldots, X_n$:

$$x = \phi(x_1, x_2, \ldots, x_n). \tag{9.5}$$

To say that (9.5) is dimensionally homogeneous means that the function satisfies the following identity in $s_1, s_2, \ldots, s_k$:

$$\phi(s_1^{\delta_{11}} s_2^{\delta_{12}} \ldots s_k^{\delta_{1k}} x_1, \, s_1^{\delta_{21}} s_2^{\delta_{22}} \ldots s_k^{\delta_{2k}} x_2, \ldots,$$

$$s_1^{\delta_{n1}} s_2^{\delta_{n2}} \ldots s_k^{\delta_{nk}} x_n) = s_1^{\delta_1} s_2^{\delta_2} \ldots s_k^{\delta_k} \phi(x_1, x_2, \ldots, x_n) \tag{9.6}$$

where the δ_j, $j = 1, \ldots, k$, are the dimensions of X. In particular, consider a law of the form

$$\psi(x_1, x_2, \ldots, x_n) = 0. \tag{9.7}$$

To say that function (9.7) is dimensionally homogeneous means that, for every positive real $x_1, x_2, \ldots, x_n$ and $s_1, s_2, \ldots, s_k$,

$$\psi(x_1, x_2, \ldots, x_n) = 0$$

$$\Leftrightarrow \Psi(x_1^{\delta_{11}} s_2^{\delta_{12}} \cdots s_k^{\delta_{1k}} x_1, \, s_1^{\delta_{21}} s_2^{\delta_{22}} \cdots s_k^{\delta_{2k}} x_2, \ldots,$$

$$s_1^{\delta_{n1}} s_2^{\delta_{n2}} \cdots s_k^{\delta_{nk}} x_n) = 0. \tag{9.8}$$

It can be shown that changes of scale according to Eq. (9.2) constitute a k-parameter continuous group of transformations (Moran, 1971). Further, there exist certain absolute invariants associated with the group, that is, functions I such that, under the transformation given by Eq. (9.2),

$$I(\bar{x}_1, \bar{x}_2, \ldots, \bar{x}_n) = I(x_1, x_2, \ldots, x_n). \tag{9.9}$$

Of particular interest are absolute invariants of the form

$$\Pi = x_1^{\varepsilon_1} x_2^{\varepsilon_2} \cdots x_n^{\varepsilon_n} \tag{9.10}$$

where

$$\sum_{i=1}^{n} \varepsilon_i \delta_{ij} = 0, \qquad j = 1, \ldots, k. \tag{9.11}$$

This form of absolute invariant is termed a nondimensional variable. As will be seen, the importance of absolute invariants lies in the fact that they can be regarded as criteria of similarity.

A dimensionally homogeneous equation can be reformulated in terms of absolute invariants. This assertion is based on the so-called Π theorem, which can be stated as follows (Bridgman, 1931; Buckingham, 1914; Moran, 1971).

THEOREM 1.   Let $\phi(x_1, x_2, \ldots, x_n)$ be dimensionally homogeneous with respect to k primary scales $S_1, S_2, \ldots, S_k$. Then the equation

$$\phi(x_1, x_2, \ldots, x_n) = 0 \tag{9.12}$$

is equivalent to

$$\Phi(\Pi_1, \Pi_2, \ldots, \Pi_{n-k}) = 0 \tag{9.13}$$

for some function Φ, where k is the rank of the dimensional matrix and

$$\Pi_p = x_1^{\varepsilon_{p1}} x_2^{\varepsilon_{p2}} \cdots x_n^{\varepsilon_{pn}}, \qquad p = 1, \ldots, q, \tag{9.14}$$

are $q = n - k$ independent and nondimensional variables with exponents ε_{pi} subject to the condition that

$$\sum_{i=1}^{n} \varepsilon_{pi} \delta_{ij} = 0, \qquad j = 1, \ldots, k; \tag{9.15}$$

and, for $k < n$, nontrivial absolute invariants of the form (9.14) always exist. Further, a set of $n - k$ functionally independent absolute invariants of form (9.14) correspond to any $n - k$ linearly independent solutions to Eqs. (9.15).

It will be useful to have a concrete example as an aid to further development of our proposed terminology. Consider the flow of a viscous fluid through a tube governed by the well-known Poiseuille law,

$$J = \frac{\pi}{8} \frac{\Delta P}{L} \frac{r4}{\eta} \tag{9.16}$$

where J is the volume flow rate of the fluid per unit length, P the pressure gradient, L the length, r the radius, and η the viscosity. The dimensions of J, P, r, and η are L^2T^{-1}, $ML^{-2}T^{-2}$, L, and $ML^{-1}T^{-1}$, respectively, where M, L, and T denote the dimensions of mass, length, and time, respectively. The parameter η is illustrative of *dimensional constants* that appear in many physical laws. There are two types of dimensional constants: First, the universal constants, which are always observed to have one particular value for a fixed set of scales of measurements (e.g., the velocity of light, Avogadro's constant, and Boltzmann's constant); second, system-dependent or material constants, whose values are generally different for different systems, even under a fixed set of scales of measurement (e.g., Hooke's constant and the viscosity of a fluid). A system-dependent constant will be referred to as a *parameter*.

In the case of Poiseuille's law, there are three variables, namely, the volume flow rate of fluid per unit length, the pressure gradient, and the radius. If these quantities can be measured with respect to some fixed units, then a *state* of the fluid can be described by an ordered triple (J, P, r). Now suppose we have a class of fluids satisfying Poiseuille's law for various values of η. If a fluid is considered to constitute a system, then such a class is an example of a class of similar systems, since each fluid obeys a law of the same general form as the law obeyed by any other fluid of the class.

In general, let $E = \{E_h : h \in H\}$ be a set of systems, where H is a nonempty index set, such that for each h in H, E_h is a nonempty set. Each E_h is a system, and the elements of each E_h are its states. Clearly $\cup E_h$ is the set of all states of every system in E. Now let R_+ be the set of positive real numbers and let R_+^n be the set of ordered n-tuples of positive reals (where n is the number of variables). Further, let the state measurements be denoted by μ, a vector-valued function from $\cup E_h$ into R_+^n such that μ uniquely determines the states of each system E_h. An n-dimensional class of systems is then defined by an ordered pair $C = (E, \mu)$; $\mu(C) = \{\mu(E_h) : h \in H\}$ is called the image of C under μ. Thus $\mu(C)$ is a set of laws: Clearly every $\mu(E_h)$ in $\mu(C)$ is a law governing the system E_h of C.

The important properties of R_+^n are as follows. Let the vector $V = (v_1, \ldots, v_n)$ be an element of R_n^+. Multiplication of two such vectors is defined by $VV' = (v_1 v_1', \ldots, v_n v_n')$. Clearly, under this operation R_n^+ forms an Abelian group in which $V^{-1} = (v_1^{-1}, \ldots, v_n^{-1})$ is the inverse of V in which the identity element is $1 = (1, 1, \ldots, 1)$. R_+^n is also a vector space over the field of real numbers. The scalar multiplication is defined as an exponentiation operation; that is, if $V \in R_+^n$ and ζ is a real number, then $V^\zeta = (v_1^\zeta, \ldots, v_n^\zeta)$. If ζ and ξ are reals, then $(V^\zeta)^\xi = V^{\zeta\xi}$. The operation of a matrix on a vector is defined in a similar way. Thus if $U \in R_+^m$ and $D = [d_{ij}]$ is a real $n \times m$ matrix, U^D is defined to be the vector $V \in R_+^n$ with components

$$
\begin{aligned}
v_1 &= u_1^{d_{11}} u_1^{d_{12}} \cdots u_m^{d_{1m}}, \\
v_2 &= u_1^{d_{21}} u_2^{d_{22}} \cdots u_m^{d_{2m}}, \\
&\vdots \\
v_n &= u_1^{d_{n1}} u_2^{d_{n2}} \cdots u_m^{d_{nm}}.
\end{aligned}
\tag{9.17}
$$

Certain subgroups of R_+^n are of particular interest. Let D be a real $n \times m$ matrix and $G^D = \{U^D: U \in R_+^m\}$. That is, G^D is the image of R_+^m under the linear transformation given by D. Clearly, G^D is a subspace and subgroup of R_+^n. Also, if $R \subseteq R_+^n$, we can define $U^D R = \{U^D V: V \in R\}$. Further, if R_1 and R_2 are subsets of R_+^n, R_1 is defined to be G^D-similar to R_2 if and only if there exists a U^D in G^D such that $R_1 = U^D R_2$. Accordingly we can write $[R]_G D = \{U^D R: U^D \in G^D\}$; that is, $[R]_G D$ is the G^D-equivalence class of R. It can then be readily verified that an equivalence class $\Re$ is a complete class of similar subsets of R_+^n if and only if there is a real $n \times m$ matrix D and a subset R of R_+^n such that $\Re = [R]_G D$. An n-dimensional class of systems $C = (E, \mu)$ is therefore called an (n, m) class of similar systems with matrix D if and only if there is a real $n \times m$ matrix D and a subset R of R_+^n such that $\mu(C) = [R]_G D$. G^D is called the group of C.

Another important notion is that of a continuous dimensional function, λ, of a measure v. Let $C = (E, \mu)$ be a class of similar systems. Clearly C possesses a measure $\mu: C \to R_+^n$. If C also possessed a measure $v: C \to R_+^l$, then a *continuous dimensional function* is a function $\lambda: R_+^m \to R_+^l$ such that for each U in R_+^m

$$\lambda(U) = U^{D_\lambda} \tag{9.18}$$

where $D_\lambda = [d_{\lambda gj}]$, a real $l \times m$ matrix, is a given dimensional matrix. Thus, for each $g = 1, 2, \ldots, l$, the gth component of $\lambda(U)$ is the positive real number

$$\lambda_g(U) = u_1^{d_{\lambda g1}} u_2^{d_{\lambda g2}} \cdots u_m^{d_{\lambda gm}}. \tag{9.19}$$

One final notion that will be needed is that of a law $\mathcal{L}(V, W)$ governing a class of similar systems, C. Let $\mu(C) = [R]_G D$; then define $\mathcal{L}(V, W)$ as follows: For every V in R_+^n and every W in R_+^l

$$\mathcal{L}(V, W) \Leftrightarrow [\exists R \varepsilon \mu(C)] \{ [v(R_1) = W] \wedge [V \varepsilon R_1] \}. \tag{9.20}$$

That is, the $(n + l)$-place relation $\mathcal{L}(V, W)$ is called the law of C with respect to v if the values of the parameters in this law are values of the measure v. As an example, since the values of the parameter η in Poiseuille's law depend on the fluid, and since η has dimensions, it is natural to consider η a measure of fluid.

The utility of these concepts lies in the fact that they facilitate an important operationalization of the notion of similarity between two or more systems. The crux of the operationalization is given by the following theorem, which is stated here without proof (Causey, 1969).

THEOREM 2.   Let C be a class of similar systems with matrix D. Suppose that C obeys the law $\mathcal{L}(V, W)$ with respect to the measure v. Then, for every V in R_+^n, every W in R_+^l, and every U in R_+^m,

$$\mathcal{L}(V, W) \Leftrightarrow \mathcal{L}(U^D V,\ U^{D_\lambda} W). \tag{9.21}$$

Comparison of condition (9.21) with condition (9.8) shows that $\mathcal{L}(V, W)$ is dimensionally homogeneous (note that x_i in Eq. (9.7) represents variables and dimensional constants). In conclusion, a continuously measurable class of similar systems obeys a dimensionally homogeneous law.

Recall, however, that according to Theorem 1, a dimensionally homogeneous law is equivalent to a relationship between the nondimensional variables of the system under consideration. Thus Theorems 1 and 2 together lead to the important proposition that the nondimensional variables of a system can be regarded as criteria of similarity in comparison with other systems.

In summary, two systems are *similar*, despite absolute differences in their sizes and other characteristics, if the values of the relevant nondimensional variables are the same for both. Likewise, if this is true of a system during its various stages of evolution, it can be regarded as a *self-resembling* system.

In what follows, it will be shown that certain simple laws of technological innovation can be derived by assuming that some of the absolute invariants of a system remain constant during its evolution.

4. THE PARETO DISTRIBUTION OF TECHNICAL CAPABILITIES

The following attempt at the operationalization of the theoretical propositions advanced in Section 2 is best explained by means of an example. Suppose that the evolution of a system in time is given by the equations

$$dy/dt = \sigma y(t) \quad \text{and} \quad dx/dt = \chi x(t) \tag{9.22}$$

where t is time and x and y are two variables of the system under consideration. From the initial conditions $y(0) = 1$ and $x(0) = 1$, we obtain

$$y = \exp(\sigma t) \quad \text{and} \quad x = \exp(\chi t). \tag{9.23}$$

The scale changes for the variables may be expressed via the transformations

$$\bar{y} = \tau^{\delta_1} y \quad \text{and} \quad \bar{x} = \tau^{\delta_2} x \tag{9.24}$$

where $\tau = \exp(t)$, $\delta_1 = \sigma$, and $\delta_2 = \chi$. Thus we have two variables, x and y, and a one-parameter group of transformations involving only time. Therefore, according to the Π theorem stated earlier, there exists only one absolute invariant of the nondimensional form. It is given by

$$\pi(x,y) = y^{d_2}x^{-d_1}, \tag{9.25}$$

for then $\pi(x,y) = \pi(\bar{x},\bar{y})$. If the evolution of the system exhibits the property of self-resemblance, this invariant must be a constant, say c_1 (cf. Section 3); that is,

$$y^{d_2}x^{-d_1} = c_1 \tag{9.26}$$

or equivalently,

$$y = c_2x^p \tag{9.27}$$

where $p = \sigma/\chi$ and $c_2 = c_1^{1/\chi}$. Clearly, the exponent p of this relationship is determined such that the conditions of self-resemblance are fulfilled.

Utilizing the general approach of the illustrative example, we can formalize the two hypotheses presented in Section 2 very simply as follows. In symbols, let y be a functional property of the system under consideration, z the size, and x a record of the cumulative number of events leading to the improved performance of the system. As an example, consider the evolution of nuclear technology. If y is a measure of its technical capability, such as thermal efficiency, z and x can be measured, for example, in terms of the size of the reactor core and cumulated number of reactors built, respectively. In analogy with the derivation of Eq. (9.27), the relationships between these variables are given by the equations

$$y_i = a(x_i)^b, \tag{9.28}$$

$$y_i = A(z_i)^B \tag{9.29}$$

where the exponents are coefficients of self-resemblance. Equations (9.28) and (9.29) are termed here the "learning function" and the "scale function" relationships. In what follows, their implications will be discussed.

The relationship given by Eq. (9.28) is essentially a general theoretical formulation of the well-known unit progress function concept that experience acquired in the production process plays a significant role in the improvement of productivity (Wright, 1936). In symbols, let $\mathscr{P}$ be an index of productivity and $\mathscr{Y}$ the cumulated production. The progress function (also sometimes called a learning curve) is commonly specified as

$$\mathscr{P}_i = \alpha(\mathscr{Y}_i)^{\beta} \tag{9.30}$$

where a and b are constants. For example, in the so-called manufacturing progress function, V is the cumulative total of units produced and P is the amount of direct labor employed (Conway and Schultz, 1959); b is the slope of the logarithmic progress function and it is frequently equal to -0.32. That is, a 10% increase in cumulated output is on average associated with a 3.2% decline in direct labor requirement. The specific definitions of the variables, of course, differ among different applications of the concept.

One central implication of the proposed formulation is that the process of technological innovation is governed by the Pareto distribution

$$F(q) = P(Q < q) = 1 - \alpha q^{-\beta} \tag{9.31}$$

where $P(Q < q)$ is the probability that a random variable Q assumes a value less than q, and α and β are constants. Specifically, it can be shown that a progress function is isomorphic to the cumulative density function of the Pareto distribution (Sahal, 1979b). To begin with, Eq. (9.28) can be rewritten as

$$x = gy^{-h} \tag{9.32}$$

where $g = a^{-1/b}$ and $h = -1/b$. The following two observations can be made about the cumulative production quantity variable x. First, by its definition, x increases monotonically with decreasing y. Second, x is also the *number* of observations, $N(y)$, greater than or equal to y; that is,

$$x = N(y) \geq y. \tag{9.33}$$

Further, since

$$N(y)/N = 1 - F(y) = P(Y \geq y), \tag{9.34}$$

the combination of Eqs. (9.32), (9.33), and (9.34) gives

$$F(y) = P(Y < y) = 1 - \gamma y^{-h} \tag{9.35}$$

where $\gamma = g/N$. Equation (9.35) is, however, identical with the expression for the Pareto distribution, Eq. (9.31). Further, the progress function coefficient, b, is interpretable in terms of the Pareto coefficient, β. Specifically

$$b = -1/\beta. \tag{9.36}$$

The essence of the proposed formulation is that the system preserves similarity of a holistic nature with change in its scale. As discussed earlier, however, an object cannot preserve similarity in all of its aspects with change in scale. In particular, it must change geometrically if it is to remain functional at widely different scales. Thus, one further implication of the proposed formulation is that the development of new techniques generally involves changes in the morphological characteristics of the existing techniques. This is best expressed in terms of the so-called allometric relationship first proposed by Huxley (1932). In symbols, if z is a measure of the size of the whole system and s the size of any of its parts,

$$\frac{ds/dt}{s} = u \frac{dz/dt}{z} \qquad (9.37)$$

where u is a constant. By substitution of dz/ds for $(dz/dt)(dt/ds)$, we obtain

$$\frac{1}{s} = \frac{u}{z} \frac{dz}{ds} \, . \qquad (9.38)$$

Integration of both sides gives

$$\log s = \log m + u \log z, \qquad (9.39)$$

or equivalently,

$$s = mz^u \qquad (9.40)$$

where $\log m$ is a constant of integration.

Two essential aspects of the allometric relationship should be noted. First, it implies that the ratio of the two specific growth rates $s'(t)/s(t)$ and $z'(t)/z(t)$ is constant. Obviously, it is not required that both specific growth rates be constant; the necessary condition is only that their ratio be a constant. Second, when geometric similarity is preserved with change in size, we have the case of isometric growth. The criterion of isometric growth is that Eq. (9.40) satisfies the condition of dimensional homogeneity. For example, if the variables s and z are of the same dimension, isometric growth implies that the exponent u of the relationship (9.40) is equal to 1. The case of allometric growth, on the other hand, implies a systematic departure from the condition of dimensional homogeneity. The meaning of allometric growth is therefore simply that subsystems must grow at a different rate from the system as a whole. Also, the rates of growth of different subsystems must also be generally different.

One final implication of the proposed formulation is that the distribution of variations in size obeys the Pareto law. By combination of Eqs. (9.28) and (9.29), we obtain

$$x = \zeta z^{-\theta} \qquad (9.41)$$

where $\zeta = (A/a)^{1/b}$ and $\theta = -B/b$. Further, since

$$x = N(z) \geqslant z \qquad (9.42)$$

and

$$N(Z)/N = 1 - F(z) = P(Z \geqslant z), \qquad (9.43)$$

the combination of Eqs. (9.41), (9.42), and (9.43) gives

$$F(z) = P(Z < z) = 1 - \eta z^{-\theta} \qquad (9.44)$$

where $\eta = \zeta/N$. Equation (9.44) is, of course, identical with the usual expression for the Pareto distribution. Further, the parameter θ in Eq. (9.44) is an estimate of the Pareto coefficient of the size distribution under consideration and it is interpretable in terms of the two coefficients of self-resemblance,

$$\theta = -B/b. \qquad (9.45)$$

Hence, the Pareto coefficient of size distribution is best regarded as a generalized coefficient of self-resemblance.

While our attempt has been to develop a theory of technical capabilities, it should be noted that the theoretical propositions presented here are very generally applicable.[3] Thus, according to the first proposition, experience plays a valuable role in improvement in the performance of a system. According to the second proposition, performance is also a function of system size. There are a number of important implications of the theory. Thus, it accounts for the existence of the progress function and allometric growth models. Equally important, it indicates that the size distribution of an evolutionary system is expected

---

[3]In passing it may be noted that some of the terms employed here are revisions of those used earlier (Sahal, 1978a). Thus the present use of "self-resembling system" seems preferable to the earlier use of "self-similar system" because it is seldom possible to preserve the constancy of *all* the relevant nondimensional variables during the actual course of evolution. Also, the use of the term "allometry" in its classical sense of morphological changes seems preferable to the earlier use in a much broader sense of all types of scale-related changes.

to obey the Pareto law. It remains to be seen whether these predictions hold in the development of new techniques. For the present, it should be noted that they are certainly borne out by available empirical evidence in other unrelated areas. Thus, a wide class of size distributions relating to biological, demographic, economic, and linguistic phenomena have been observed to obey the Pareto law (Ijiri and Simon, 1977; Steindl, 1965; Sahal, 1978a, Zipf, 1949). It can be said that the theory developed here accounts for such evidence.

5.   ASSESSMENT OF TECHNOLOGICAL INNOVATION POTENTIAL

The proposed framework for the assessment of technological innovation potential rests upon the learning function and the scale function relationships presented in the preceding section. Accordingly, we will first consider the problems in estimating the parameters of these relationships.

It will be useful to have a concrete example as an aid to discussion of the scale function relationship in technological innovation. Consider the evolution of a mechanical system in terms of changes in its size (z) and horsepower characteristic (H). We can employ a number of alternative measures of size, such as weight (W) or volume (V). The scale changes for the variables under consideration can be expressed via transformations

$$\bar{H} = ML^2T^{-3}H, \qquad \bar{W} = MLT^{-2}W, \qquad \bar{V} = L^3V \qquad (9.46)$$

where M, L, T denote the dimensions of mass, length, and time, respectively.

The following two simplifying assumptions are made: (1) Gravity (G) affects all objects in the same way, regardless of their size; that is, $\bar{G} = LT^{-2}G$ and $\bar{G} = G$. (2) Density (D) is also a constant, that is, $\bar{D} = ML^{-3}D$ and $\bar{D} = D$.

Thus, the example involves a three-parameter group and five variables, of which two are constants. According to the Π theorem stated earlier (Theorem 1), there exist only two invariants, given by

$$I(W,V) = WV^{-1} = GD \quad \text{and} \quad I(H,W,V) = H^2W^{-2}V^{-1/3} = G, \quad (9.47)$$

for then $I(W,V) = I(\bar{W},\bar{V})$, since $G = \bar{G}$ and $D = \bar{D}$, and $I(H,W,V) = I(\bar{H},\bar{W},\bar{V})$, since $G = \bar{G}$. If the system exhibits self-resemblance, the two invariants above must be constants, say, c_3 and c_4; that is,

$$WV^{-1} = c_3 \quad \text{and} \quad H^2W^{-2}V^{-1/3} = c_4 \qquad (9.48)$$

or

$$H = c_5 (W)^{7/6}. \tag{9.49}$$

Thus, we have the a priori theoretical relationship that the horsepower varies as the 7/6 power of weight. In general, let y be any functional property of the system. Suppose that the scale changes for the variable can be expressed via the transformation

$$y = M^{\delta_1} L^{\delta_2} T^{\delta_3} \tag{9.50}$$

where δ_1, δ_2, and δ_3 are real "numbers." The resulting theoretical relationship is then given by

$$y = C(W)^k, \qquad k = \delta_1 + \delta_2/3 + \delta_3/6. \tag{9.51}$$

The theoretical relationships for kinematic, hydrodynamic, and other types of systems can be derived in an analogous way, although the underlying assumptions necessarily depend on the type of system under consideration.[4]

Clearly, the derivation of relationship (9.51) is based on two simplifying assumptions. First, the material employed in the construction of technology does not change during the course of its evolution. Second, the form of the technology also remains unchanged. However, neither of these assumptions is generally valid. It is therefore necessary to introduce an adjustment factor in the estimation of the theoretical model so as to account for change in the material and morphological characteristics of the technology. Furthermore, it should be noted that data on technological development are inevitably subject to measurement errors. As an example, until the late fifties, data on the pull-speed characteristics of a farm tractor were based on the use of power as the controlled variable. However, recent use of various devices for causing travel speeds to vary inversely with changes in drawbar pull has led to the adoption of drawbar pull as the controlled variable. The essential point of this example is that the evolution of a technology generally necessitates changes in the ways and means of obtaining data on its characteristics. It is therefore necessary to introduce another adjustment factor in the estimation of the theoretical model so as to account for errors of measurement.

The proposed theoretical formulation will therefore be estimated in the form

---

[4]It is interesting to note that relationships like Eq. (9.51) are frequently used in actual engineering practice. Moreover, they have also been utilized to a limited degree in physiological analyses (Günther, 1966; Stahl, 1962). However, their theoretical explanation has been generally lacking. It remains to be seen whether these relationships are also relevant to *long-term* innovation processes. For the present it should be noted that the proposed principle of self-resemblance provides an a priori theoretical rationale of why these relationships are what they are.

$$y = c + d(z)^k \tag{9.52}$$

where the constant d is a measure of change in the morphological and material characteristics of the system, c is an adjustment factor to account for errors involved in the measurement of y, and k is a theoretical exponent. Note that the values of both c and d depend on the system of units employed.

The learning function relationship (9.28) or its equivalent Pareto distribution version given by Eq. (9.35) can be estimated in many different ways. Here we will consider only two very simple forms of the relationship. The combination of Eqs. (9.32) and (9.33) gives

$$N(y) = gy^{-h}. \tag{9.53}$$

But $N(y)$ is also the rank of y in the descending order of its values. Thus an alternative form of Eq. (9.53) is

$$y = c_6 R_1^{-l}, \qquad l = 1/h, \qquad c_6 = g^l \tag{9.54}$$

where R_1 is the rank of the y values such that the largest value of y is ranked number 1, the second largest number 2, etc.

If, however, y is ranked in the ascending order of its values, the theoretical relationship takes on the form (cf. Eq. (9.32))

$$y = c_7 R_2^{b}, \qquad b = -1/h, \qquad c_7 = g^{-b} \tag{9.55}$$

where R_2 is the rank of the y values such that the smallest value of y would rank number 1, the second smallest number 2, etc.

For purposes of illustration, both forms of the model will be employed in the following case studies. The relationship between the two forms can be easily ascertained. From Eqs. (9.36), (9.54), and (9.55), the relationship between the exponents of the two forms is given by

$$l = -b = 1/\beta. \tag{9.56}$$

Therefore, from Eqs. (9.35), (9.54), (9.55), and (9.56)

$$c_6 = c_7 = (\gamma N)^{1/\beta}. \tag{9.57}$$

Thus, the only difference between the two forms (9.54) and (9.55) is the difference in the signs of their exponents. Equations (9.54) and (9.55) will be referred to as "rank function" relationships. For the purpose of estimation, they leave much to be desired. Their chief attraction is simplicity. They will be defended on the grounds of computational necessity.

Within the proposed framework of lawlike relationships, the technological innovation potential of a system can be assessed as follows. First, consider the scale function relationship. Equation (9.52) reveals that the growth in the technological capability of a system due to change in its scale depends on the parameter d apart from the theoretically fixed exponent k. Specifically, the greater the value of d, the greater the implied advance in technology due to deliberate changes in the material and morphological characteristics of the system via changes in its scale. It is therefore natural to measure the scale-dependent technological innovation potential of the system in terms of an appropriately normalized measure such as

$$\mathscr{E}(\mathscr{S}) = \frac{d}{s(d)} \tag{9.58}$$

where $s(d)$ is the standard error of the parameter d. The proposed measure of innovation potential can also be tested as a t statistic at any given level of significance. Further, for any two comparable cases, if the sets of coefficients in the scale function relationship (9.52) differ from each other, it may be concluded that significant differences in evolutionary potential exist. It is said that "the great designer was not hampered by the knowledge of the theory of dimensions." The proposed measure of $\mathscr{E}(\mathscr{S})$ is consistent with this observation.

It should be recalled that certain innovations depend on a change in the scale of some larger system for the use of technology rather than a change in the scale of the physical equipment itself (Chapter 6). It is therefore useful to distinguish between two types of scale-dependent technological innovation potential. One corresponds to change in the scale of the technique itself. It will be denoted by $\mathscr{E}(\mathscr{S}_p)$. The other corresponds to change in the scale of organization for the utilization of technology. It will be denoted by $\mathscr{E}(\mathscr{S}_o)$. The form of Eq. (9.52) is applicable to measurement of both; only the relevant scale variables differ. Clearly, the distinction between $\mathscr{E}(\mathscr{S}_p)$ and $\mathscr{E}(\mathscr{S}_o)$ is relative rather than absolute.

Next consider the rank function relationship given by Eq. (9.55). The slope of this relationship, b, represents the ratio of an infinitely small relative change in the chosen capability of technology associated with a correspondingly small change in the accumulated know-how. Clearly, the greater the value of the parameter b, the greater the potential for evolution of a technique by means of learning processes. Thus, the relevant measure of technological innovation potential in this case is given by

$$\mathscr{E}(\mathscr{R}) = b. \tag{9.59}$$

It is furthermore nondimensional.

We can look at $\mathscr{E}(\mathfrak{R})$ from another point of view. As discussed earlier, the exponent b of the learning function relationship is related to the Pareto coefficient β ($b = -1/\beta$). Further, it is well known that the Pareto coefficient is a measure of inequality in the distribution of a variable. Specifically, the less the Pareto coefficient, the higher the inequality. A consideration of the rank function relationship makes this clear. Let i and j be the ranks of any two members in this relationship and let y_i and y_j be their respective sizes. If the largest size is ranked number 1, the second largest number 2, etc. (i.e., $y_i > y_j$, $j > i$), we have (cf. Eq. (9.54))

$$\frac{y_i}{y_j} = \left(\frac{j}{i}\right)^l. \tag{9.60}$$

Obviously, the greater the l, the larger the big members are and the greater the inequality is. Alternatively, if the smallest size is ranked number 1, the second smallest number 2, etc. (i.e., $y_i > y_j$, $i > j$), we have (cf. Eq. (9.55))

$$\frac{y_i}{y_j} = \left(\frac{i}{j}\right)^b. \tag{9.61}$$

Thus, the greater the b, the larger the big members and the greater the inequality. In turn, the higher the inequality, the higher the redundancy of a given distribution. This can be seen clearly upon considering the formal expressions for entropy and redundancy, given by

$$H(y) = -\sum_{i=1}^{N} y_i \log y_i, \qquad 0 \leqslant H(y) \leqslant \log N, \tag{9.62}$$

and

$$R = 1 - \frac{H}{H_{max}}, \tag{9.63}$$

respectively. Complete equality occurs when $y_i = 1/N$ for all i and $H(y) = \log N$. Complete inequality occurs when $y_i = 1$ and $y_j = 0$ for all $j \neq i$, so that $H(y) = 0$. Clearly, the measure of redundancy varies directly with the degree of inequality. In conclusion, a relatively high value of the exponent b in the rank function relationship implies that the redundancy of the evolutionary system is relatively high.

It should be noted that redundancy is one of the major factors in determining the self-organizing capability of a system (Sahal, 1979a). As an example, the instability exhibited by arctic and boreal fauna is at least partly due to the fact that these communities are not sufficiently redundant to damp out disturbances (Macfadyen, 1963). As another example, suppose that certain parts of the human brain are removed. It still abounds in variety but its redundancy is reduced. In consequence, the patient becomes "stimulus bound," that is, he acts only in the present (McCulloch, quoted by Arbib, 1969, pp. 204–206). His speech becomes less ordered, his sentences shorter. The essential point of these examples is that, *ceteris paribus*, the redundancy of a self-organizing system must be relatively high. It is therefore natural to regard the exponent *b* of the rank function relationship as a measure of technological innovation potential. Accordingly, $\mathscr{E}(\mathfrak{R})$ will be termed a redundancy-dependent technological innovation potential of the system.

It may well be asked whether the proposed measures are best interpreted as measures of evolutionary efficiency rather than of evolutionary potential. Do they not reflect what the state of affairs has been rather than what it will be in the future? The answer is that the proposed measures do provide an indication of what the evolutionary efficiency of a technology has been. However, insofar as they are essentially parameters of lawlike relationships (i.e., relationships that remain invariant over long periods of time), they are also measures of evolutionary potential.

6. CASE STUDIES: LAWLIKE ASPECTS OF TECHNOLOGICAL INNOVATION IN FARM MACHINERY, TRANSPORTATION SYSTEMS, DIGITAL COMPUTERS, AND NUCLEAR REACTORS

This section presents various applications of the theory to technological innovation in aircraft, digital computers, farm tractors, passenger ships, tankers, and nuclear power reactors. Evidently, the case studies cover both traditional and so-called high-technology items. The choice of the specific variables employed in the proposed relationships has been dictated by empirical necessity. It turns out that virtually every case of technological change can be treated at least approximately as a case of change in some dynamic (mechanical) property of the system.

According to our earlier discussion of the nature and significance of simple laws (Chapter 1), the proposed relationships are expected to hold only over a limited range of data. It remains to be seen whether they are in fact stable relationships. Indeed, they cannot be said to represent the lawlike aspects of the system unless they show a certain degree of permanence. However, it would be wrong to insist that they cannot change over time. Moreover, recall that the standard statistical tests of significance are not wholly appropriate for the test of simple lawlike relationships. Thus, in what follows, we will try to determine the plausibility of any given theoretical model on the basis of *several* criteria, such as its goodness of fit in any given case, the sensitivity of its fit to change in the underlying assumptions, and the variety of cases to which it applies.

By way of illustration, the evolution of certain chosen measures of technological innovation in tank ships and farm tractors over the course of time is shown in Figures 9.1 and 9.2. The pattern of evolution of tank ship technology corresponds to two successively smaller S-shaped curves. In the case of farm tractors, growth is very slow during an initial period, followed by an exponential trend throughout the remaining period. Clearly, the pattern is nonlinear in both cases.

Figures 9.3–9.7 illustrate the performance of the scale function relationship of technological innovation in a wide variety of cases. They depict the relationship between certain functional measures of technology and its scale, which is transformed according to the a priori theoretical relationship (9.51). Had the material and morphological characteristics of the technology remained unchanged with change in scale, each of these graphs would depict a

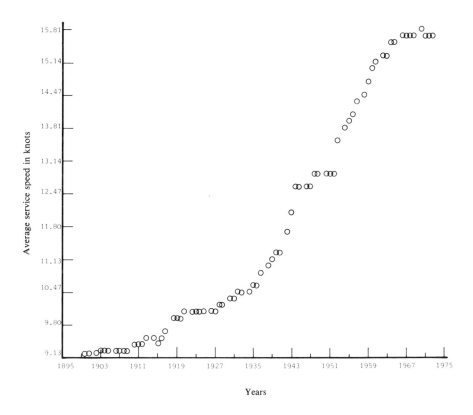

Fig. 9.1 Evolution of the service speed of tank ship technology, 1900–1973.

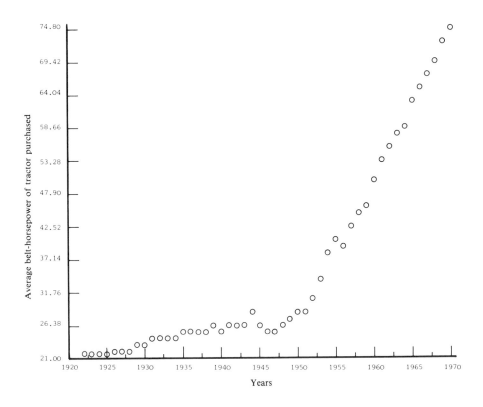

Fig. 9.2 Evolution of belt-horsepower of farm tractor technology, 1922–1970.

straight line with a slope equal to unity. In the majority of cases, the observed pattern corresponds to two straight lines of different slopes over two different ranges of the transformed scale variables. The slope of these straight lines is generally different from unity. There are a number of important implications of the pattern. First, the theoretically postulated transformation of the scale variables generally linearizes the pattern of change in the functional character-istics of the technology. This is entirely in keeping with the hypothesis of this study. Second, it is indicated that the process of innovation generally involves changes in the material and morphological characteristics of the technology. This is to be expected. Third, the parameters of the scale function relationship are evidently changed with change in the scale beyond a certain threshold. That is, simple lawlike relationships do in fact exist. However, their applicabil-ity tends to be limited to a certain range. Beyond this range, deviations from a *given* lawlike relationship tend to be generalizable. In summary, certain sys-tematic patterns of technological development can be identified and explained.

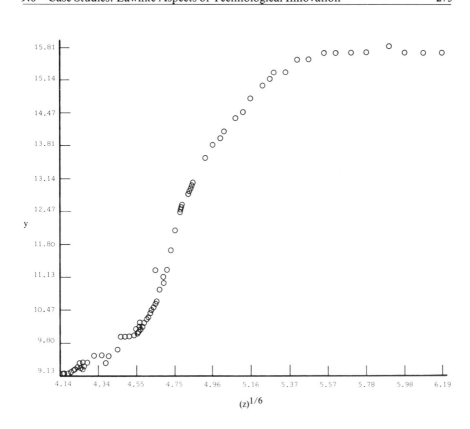

Fig. 9.3   Relationship between service speed (y) and deadweight tonnage of tank ships (z), 1900–1973.

Tables 9.1 and 9.2 present the parametric estimates of the scale function relationship for the various cases under consideration. Estimates for separate time periods are also provided in cases where the parameters of the relationship change from one period to the next. The relationships between change in technology as a consequence of change in the scale of equipment are presented in Table 9.1. Next, the case of change in the scale of the larger organization in the use of technology is considered, with the results as presented in Table 9.2. As indicated by the values of the coefficient of determination and the estimate of the standard error, the relationship is significant in 19 out of 22 cases at the $\alpha = 0.01$ level. It may therefore be concluded that the theoretical relationship holds over long periods of time. It should be noted that the development in (chronological) time is of comparatively little interest. Rather, the indicated divisions of the time period should properly be interpreted in terms of the corresponding range of the scale variable. The essential aspect of these results is

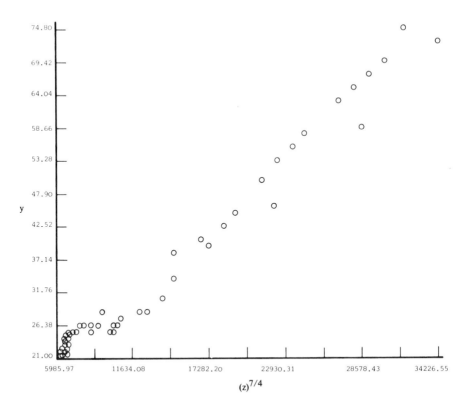

Fig. 9.4 Relationship between belt-horsepower of farm tractor (y) and average acreage per farm (z), 1922–1970.

that they disclose the existence of stable functional relationships over certain ranges of the scale variable.

Another point deserves attention. As discussed earlier (Chapter 6), the *long-term* advances in the speed of tankers and aircraft have occurred largely independently of increase in their absolute size. Indeed, there is evidence of nonlinearity in the relationship between the speed and the weight of tankers (Figure 9.3). According to the results presented in Table 9.1, the performance of the theoretically derived speed–weight relationship is nevertheless relatively good in both cases. It therefore seems safe to conclude that even when the observed advances in technology are seemingly independent of its scale, the dynamics of the underlying innovative activity is probably mediated through system size.

It would, however, be a mistake to push this inference too far. It should certainly not be taken to imply that *all* advances in technology originate from

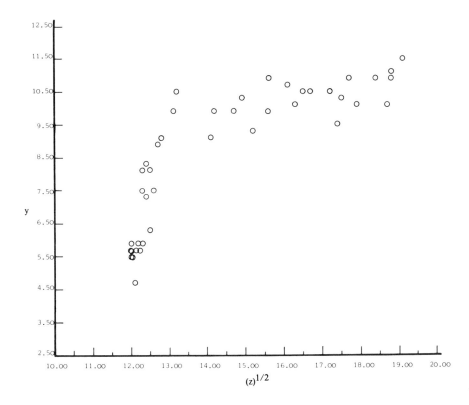

Fig. 9.5 Relationship between fuel-consumption efficiency of farm tractor in horsepower-hours per gallon (y) and average acreage per farm (z), 1921–1968.

changes in scale. Thus, note that certain cases of technological change cannot be adequately explained in terms of a change in the absolute scale of the system. Included in this category are change in the fuel-consumption efficiency of the farm tractor (case 3, Table 9.1) and in the fuel-consumption efficiency of the nuclear reactor (case 5c, Table 9.1, as well as case 7b, Table 9.1). These results are hardly unexpected. In all three cases, the pure effects of *absolute* size are relatively minor. We have already discussed the case of advances in the fuel-consumption efficiency of tractors in Chapter 6. As regards the case of nuclear reactors, improvement in fuel burnup has been achieved by various means, most of which are unrelated to absolute scale of technology. Thus the substitution of uranium dioxide for uranium metal has permitted higher temperatures, thereby making possible the production of high-quality steam. A second source of technical advances lies in the improvement of fuel-enrichment techniques and the use of improved alloys such as zirconium in-

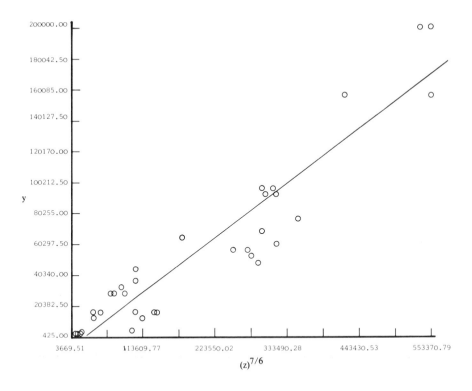

Fig. 9.6 Relationship between horsepower of passenger ships (y) and tonnage (z), 1840–1961.

stead of stainless steel. Thus it has become possible to prevent the loss of neutrons to the environment surrounding the reactor core. Further, a number of improvements in fuel fabrication have made possible high thermal conductivity and relatively uniform distribution of temperature throughout the thickness of the fuel rod. Finally, improvements in the pressure tube design have led to greater effectiveness in recovery of heat. None of these inovations appears to have been induced by changes in overall scale per se. In general, it is evident that some of the most important advances in technology may well be largely independent of its overall scale. Several alternative explanations of such advances may be noted.

First, as discussed earlier in this chapter, certain changes in technology may well be a function of the scale of one of its components, rather than of its overall size. It is clear that unless the relationships between change in the size of the relevant component and the overall size of the systems obey an isometric formulation (which is most unlikely according to the considerations advanced

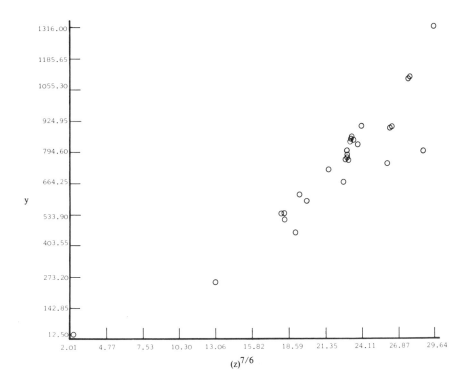

Fig. 9.7   Relationship between electric power output in megawatts (y) and volume of the active core (z) in boiling water reactors.

in this study), the empirical scale function relationships in these cases are bound to be inadequate because they have been estimated from data on overall size. Accordingly, while the relationship between fuel-consumption efficiency and plant size is insignificant (case 5c, Table 9.1), the relationship between thermal efficiency and core size performs rather well in its explanation of observed changes (Eq. (5b), Table 9.1)). Second, the observed advances may be due to change in the scale of some larger system of use rather than change in the scale of the technology itself. As a case in point, note that the explanatory power of the relationship between the fuel-consumption efficiency of tractors and the average acreage per farm is quite good (case 2, Table 9.2). That is, changes in certain characteristics of technology such as horsepower primarily take place via change in the overall scale of the system (cf. Table 9.1), while changes in certain other characteristics of technology such as fuel-consumption efficiency take place via change in the scale of some component of the system or change in the scale of some larger system of use. Other explanations of a general nature will be discussed below.

Table 9.1:   Parametric Estimates of the Scale Function Relationship at the Level of Physical Equipment[a]

Case	Technology	Period	Dependent variable	Independent variable	Estimated relationship	R^2	S	N
1a	Tank ship	1900–1973	Average service speed in knots $[y] = [LT^{-1}]$	Average dead-weight tonnage $[z] = [L^3]$	$y = -9.59 + 4.48\,(z)^{1/6}$ $\quad\;\;(0.92)\;\;(0.19)$	0.88	0.80	74
1b	Tank ship	1900–1962	Average service speed in knots $[y] = [LT^{-1}]$	Average dead-weight tonnage $[z] = [L^3]$	$y = -16.61 + 6.02\,(z)^{1/6}$ $\quad\;\;(1.09)\;\;(0.23)$	0.91	0.55	63
1c	Tank ship	1963–1973	Average service speed in knots $[y] = [LT^{-1}]$	Average dead-weight tonnage $[z] = [L^3]$	$y = 13.57 + 0.37\,(z)^{1/6}$ $\quad\;\;(0.65)\;\;(0.11)$	0.54	0.10	11
2a	Aircraft	1932–1965	Average airspeed in miles per hour $[y] = [LT^{-1}]$	Average seating capacity $[z] = [L^3]$	$y = -188.9 + 213.59\,(z)^{1/6}$ $\quad\;\;(25.07)\;\;(14.07)$	0.88	17.94	34
2b	Aircraft	1932–1953	Average airspeed in miles per hour $[y] = [LT^{-1}]$	Average seating capacity $[z] = [L^3]$	$y = -62.98 + 135.05\,(z)^{1/6}$ $\quad\;\;(14.54)\;\;(8.84)$	0.92	6.31	22
2c	Aircraft	1954–1965	Average airspeed in miles per hour $[y] = [LT^{-1}]$	Average seating capacity $[z] = [L^3]$	$y = -791.27 + 516.11\,(z)^{1/6}$ $\quad\;\;(59.71)\;\;(29.75)$	0.97	7.35	12

Table 9.1 (Continued):   Parametric Estimates of the Scale Function Relationship at the Level of Physical Equipment[a]

Case	Technology	Period	Dependent variable	Independent variable	Estimated relationship	R^2	S	N
3	Farm tractor	1921–1941, 1948–1968	Average fuel-consumption efficiency in horse-power-hours per gallon $[y] = [ML^{-1}T^{-2}]$	Total ballasted weight in pounds $[z] = [MLT^{-2}]$	$y = -\ 2.26\ +\ 0.54\ (z)^{1/3}$ $(4.22)\quad(0.20)$	0.15	1.94	42
4	Passenger ship	1841–1961	Horsepower $[y] = [ML^2T^{-3}]$	Tonnage $[z] = [L^3]$	$y = -\ 6022.57\ +\ 0.32\ (z)^{7/6}$ $(5212.34)\quad(0.02)$	0.94	18996.2	34
5a	Nuclear reactors[b]	Cross-sectional data	Electric power output in megawatts $[y] = [ML^2T^{-3}]$	Active core volume in cubic meters $[z] = [L^3]$	(i) $y = 224.40\ +\ 28.99\ (z)^{7/6}$ $(121.97)\quad(5.94)$ (ii) $\log y = 1.05\ +\ 1.65\ \log z$ $(0.16)\quad(0.15)$	0.34 0.73	216.17 0.15	48 48
5b	Nuclear reactors[b]	Cross-sectional data	Thermal efficiency $[y] = [I]$	Active core volume in cubic meters	(i) $\log y = 1.17\ +\ 0.30\ \log z$ $(0.03)\quad(0.03)$	0.69	0.03	48
5c	Nuclear reactors[b]	Cross-sectional data	Fuel-consumption efficiency in megawatts-day per tonne $[y] = [L^2T^{-2}]$	Plant size in megawatts $[z] = [MLT^{-3}]$	(i) $y = 10{,}373.45\ +\ 2073.54\ (z)^{2/7}$ $(6069.03)\quad(907.18)$ (ii) $\log y = 3.91\ +\ 0.16\ \log z$ $(0.14)\quad(0.05)$	0.10 0.19	5585.62 0.09	48 48

Table 9.1 (Continued):   Parametric Estimates of the Scale Function Relationship at the Level of Physical Equipment[a]

Case	Technology	Period	Dependent variable	Independent variable	Estimated relationship	R^2	S	N
6a	Boiling-water reactors	Cross-sectional data	Electric power output in megawatts $[y] = [ML^2T^{-3}]$	Active core volume in cubic meters $[z] = [L^3]$	$y = -226.89 + 43.94\,(z)^{7/6}$ $\quad\;\; (81.52) \quad (3.57)$	0.85	99.95	28
6b	Boiling-water reactors	Cross-sectional data	Thermal efficiency $[y] = [I]$	Active core volume in cubic meters $[z] = [L^3]$	$\log y = 1.07 + 0.38\,\log z$ $\qquad\;\;\; (0.02) \quad (0.02)$	0.91	0.02	28
7a	Pressurized-water reactors	Cross-sectional data	Electric power output in megawatts $[y] = [ML^2T^{-3}]$	Active core volume in cubic meters $[z] = [L^3]$	$y = -394.61 + 76.61\,(z)^{7/6}$ $\quad\;\;\; (203.94) \quad (12.19)$	0.68	151.51	20
7b	Pressurized-water reactors	Cross-sectional data	Thermal efficiency $[y] = [I]$	Active core volume in cubic meters $[z] = [L^3]$	$\log y = 1.48 + 0.03\,\log z$ $\qquad\;\;\; (0.05) \quad (0.04)$	0.02	0.014	20

[a]Definitions: R^2 is the coefficient of determination; S is the standard error of estimate. Standard errors of estimated coefficients are given in parentheses. The dimensions of the variables are in square brackets. M, L, T denote the dimensions of mass, length, and time, respectively. I indicates that the variable is nondimensional. N is the number of observations. At the $\alpha = 0.01$ level, the estimated relationship is significant in all cases except cases 3, 5c(i), and 7b. With the exception of 7b the relationship in the latter cases is significant at the $\alpha = 0.05$ level.
[b]Both pressurized-water and boiling-water reactors are included.

Table 9.2:   Parametric Estimates of the Scale Function Relationship at the Organizational Level[a]

Case	Technology	Period	Dependent variable	Independent variable	Estimated relationship	R^2	S	N
1a	Aircraft	1932–1965	Average airspeed in miles per hour $[y] = [LT^{-1}]$	Total route miles $[z] = [L]$	$y = -18.26 + 0.83\,(z)^{1/2}$ $(17.95)\quad(0.07)$	0.81	22.22	34
1b	Aircraft	1932–1949	Average airspeed in miles per hour	Total route miles	$y = 53.85 + 0.48\,(z)^{1/2}$ $(16.49)\quad(0.08)$	0.69	10.63	17
1c	Aircraft	1950–1965	Average airspeed in miles per hour	Total route miles	$y = -387.07 + 2.05\,(z)^{1/2}$ $(65.57)\quad(0.22)$	0.86	16.13	16
2a	Farm tractor	1922–1970	Average belt horsepower $[y] = [ML^2T^{-3}]$	Average acreage per farm $[z] = [L^2]$	$y = 10.12 + 0.0019\,(z)^{7/4}$ $(0.71)\quad(0.00004)$	0.97	2.58	42
2b	Farm tractor	1921–1941, 1948–1968	Average fuel-consumption efficiency in horsepower-hours per gallon $[y] = [ML^{-1}T^{-2}]$	Average acreage per farm	$y = -1.25 + 0.69\,(z)^{1/2}$ $(1.11)\quad(0.07)$	0.69	1.19	42
2c	Farm tractor	1921–1941	Average fuel-consumption efficiency in horsepower-hours per gallon	Average acreage per farm	$y = -45.85 + 4.28\,(z)^{1/2}$ $(5.26)\quad(0.42)$	0.84	0.67	21
2d	Farm tractor	1948–1968	Average fuel-consumption efficiency in horsepower-hours per gallon	Average acreage per farm	$y = 6.52 + 0.23\,(z)^{1/2}$ $(1.15)\quad(0.07)$	0.38	0.49	21

[a]Definitions: See footnote a Table 9.1. At the $\alpha = 0.01$ level, the estimated relationship is significant in all cases.

Figures 9.8–9.10 present the plots of allometric relationships in the development of nuclear reactors and passenger ships. The parametric estimates of these relationships are presented in Table 9.3. On the basis of the conventional form of the t test, it may be concluded that the estimated relationships differ significantly from the corresponding isometric relationships in seven out of eight cases at the $\alpha = 0.01$ level. Specifically, the development of nuclear reactors is found to be accompanied by a disproportionate change in the volume of the vessel in relation to the volume of the core. Further, an increase in core size is accompanied by a disproportionate increase in the complexity of the reactor measured in terms of the number of rods employed for control and safety. Likewise, there exists an allometric relationship between beam length and total length of passenger ships. However, changes in beam length have kept pace with the tonnage of the passenger ships. In conclusion, the process of technological development is generally characterized by changes in the complexity and the morphological characteristics of the system.

Tables 9.4 and 9.5 present estimates of the scale-dependent technological innovation potential for the various cases under consideration. According to

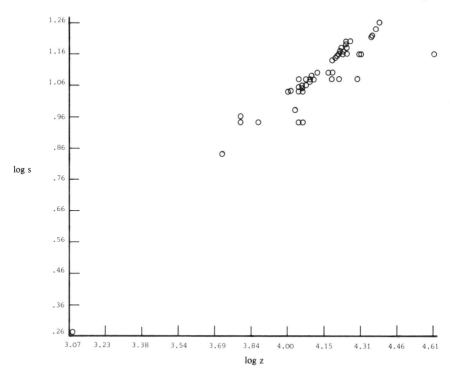

Fig. 9.8 Relationship between volume of active core (s) and volume of vessel (z) of nuclear reactors.

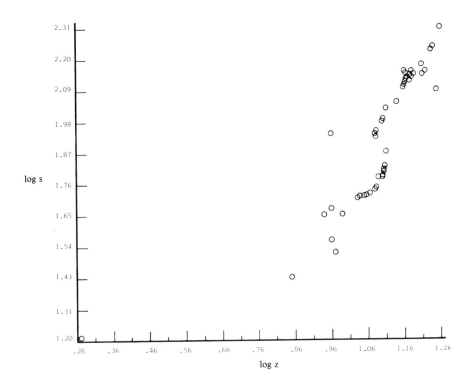

Fig. 9.9 Relationship between number of rods for control and safety (s) and volume of active core of nuclear reactors (z).

the results from the t test, the regression coefficients underlying these estimates are always significantly different from unity (generally at the $\alpha = 0.01$ level), thereby confirming the observation made earlier that innovations have been made possible by change in the material and morphological characteristics of the technology via change in its scale. Equally important, development of every type of technology is characterized by change in its scale-dependent innovation potential over the course of time. Moreover, there are marked variations in innovation potential across comparable types of technologies.

To what extent, if any, are the observed differences in the scale-dependent technological innovation potential significant? This is determined here for a number of comparable cases by means of a certain form of the F test involving a comparison of the sets of coefficients in two relationships (Johnston, 1963, pp. 136–138). The results are presented in Table 9.6. At the $\alpha = 0.01$ level, the differences are significant in all the cases considered here. Thus, the scale-dependent innovation potential of both tankers and farm tractors significantly

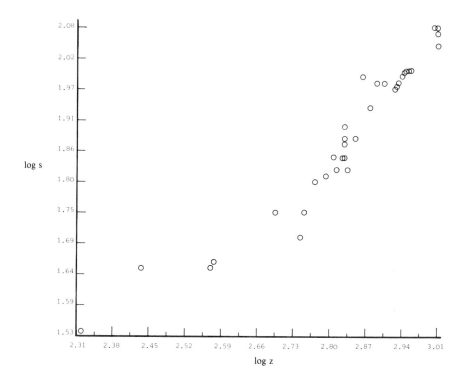

Fig. 9.10   Relationsihp between beam length (s) and total length (z) of passenger ships, 1840–1961.

declined during the later stages of their evolution. In contrast, the later stage of evolution of aircraft technology is marked by an increase in its innovation potential. Finally, the innovation potential of boiling-water reactors is significantly greater in comparison with that of pressurized-water reactors.[5] Summarizing, we have the important result that the process of technological development takes place in certain well-defined stages characterized by changes of innovation potential.

Table 9.7 presents the results of an investigation into the possible causes of differences in the evolutionary potential of different types of technology. These results indicate that there is no significant difference in the morphological aspects of technological change in boiling-water and pressurized-water reactors. However, the two reactor types significantly differ in terms of the

---

[5]This result is of independent interest in the context of the current debate on the role of nuclear power in meeting future energy requirements. However, its validity cannot be further ascertained. To the best of the author's knowledge, no other study has attempted such an assessment of the innovation potential of alternative techniques.

Table 9.3:   Parametric Estimates of Allometric Relationship[a]

Case	Technology	Period	Dependent variable	Independent variable	Estimated relationship	R^2	S	N
1a	Nuclear reactors[b]	Cross-sectional data	Volume of reactor vessel in square meters times millimeter $[s] = [L^3]$	Volume of active core in cubic meters $[z] = [L^3]$	$\log s = 2.64 + 1.36 \log z$ $\quad\quad\quad (0.09)\quad(0.08)$	0.86	0.08	48
1b	Nuclear reactors[b]	Cross-sectional data	Number of control rods $[s] = [I]$	Volume of active core in cubic meters $[z] = [L^3]$	$\log s = 0.46 + 1.34 \log z$ $\quad\quad\quad (0.14)\quad(0.13)$	0.69	0.13	48
2a	Boiling-water reactors	Cross-sectional data	Volume of reactor vessel	Volume of active core	$\log s = 2.69 + 1.32 \log z$ $\quad\quad\quad (0.10)\quad(0.09)$	0.89	0.08	28
2b	Boiling-water reactors	Cross-sectional data	Number of control rods	Volume of active core	$\log s = 1.05 + 0.89 \log z$ $\quad\quad\quad (0.06)\quad(0.05)$	0.93	0.05	28
3a	Pressurized-water reactors	Cross-sectional data	Volume of reactor vessel	Volume of active core	$\log s = 2.46 + 1.52 \log z$ $\quad\quad\quad (0.27)\quad(0.26)$	0.65	0.08	20
3b	Pressurized-water reactors	Cross-sectional data	Number of control rods	Volume of active core	$\log s = 0.31 + 1.35 \log z$ $\quad\quad\quad (0.13)\quad(0.13)$	0.86	0.04	20
4a	Passenger ships	1840–1961	Beam length in feet $[s] = [L]$	Total length in feet $[z] = [L]$	$\log s = -0.43 + 0.82 \log z$ $\quad\quad\quad\;\; (0.13)\quad\;(0.04)$	0.91	0.04	34
4b	Passenger ships	1840–1961	Beam length in feet	Tonnage $[z] = [L^3]$	$\log s = 0.59 + 0.29 \log z$ $\quad\quad\quad (0.03)\quad(0.07)$	0.98	0.02	34

[a]Definitions: See footnote a Table 9.1. At the $\alpha = 0.01$ level, the estimated relationship is significant in all cases.
[b]Both pressurized-water and boiling-water reactors are included.

Table 9.4: Estimates of the Scale-Dependent Technological Innovation Potential at the Level of Physical Equipment[a]

Number	Case (Table 9.1)	Technology	Dimension of technology	Technological innovation potential $\mathscr{E}(\mathscr{J}_p)$
1	1b	Tank ship, 1900–1962	Service speed	26.17
2	1c	Tank ship, 1963–1973	Service speed	3.36
3	2b	Aircraft, 1932–1953	Airspeed	15.28
4	2c	Aircraft, 1954–1965	Airspeed	17.34
5	3	Farm tractor, 1921–1941, 1948–1968	Fuel-consumption efficiency	2.70
6	4	Passenger ship, 1840–1961	Horsepower	16.00
7	6a	Boiling-water reactor	Electric power output	12.31
8	7a	Pressurized-water reactor	Electric power output	6.28
9	6b	Boiling-water reactor	Thermal efficiency	19.00
10	7b	Pressurized-water reactor	Thermal efficiency	0.75

[a]The regression coefficients underlying the estimates of $\mathscr{E}(\mathscr{J}_p)$ are significantly different from unity at the $\alpha = 0.01$ level in all cases except farm tractor technology. In the latter, the coefficient is significantly different from unity at the $\alpha = 0.05$ level.

Table 9.5:   Estimates of The Scale-Dependent Technological Innovation Potential at the Organizational Level[a]

Number	Case (Table 9.2)	Technology	Dimension of technology	Estimate of technological innovation potential $\mathcal{E}(\mathcal{T}_o)$
1	1b	Aircraft, 1932–1949	Airspeed	6.00
2	1c	Aircraft, 1950–1965	Airspeed	9.32
3	2c	Farm tractor, 1921–1941	Fuel-consumption efficiency	10.19
4	2d	Farm tractor, 1948–1968	Fuel-consumption efficiency	3.28
5	2a	Farm tractor, 1922–1970	Horsepower	47.50

[a]At the $\alpha = 0.01$ level, the regression coefficients underlying the estimates of $\mathcal{E}(\mathcal{T}_o)$ are significantly different from unity in all cases.

Table 9.6:   Significance of Differences in the Scale-Dependent Technological Innovation Potential

Number	Case	Table	Technology		Dimension of technology	Significant difference?
1	1b	9.1	Tank ship	1900–1962	Service speed	Yes
	1c		Tank ship	1963–1973		
2	2b	9.1	Aircraft	1932–1953	Airspeed	Yes
	2c		Aircraft	1954–1965		
3	1b	9.2	Aircraft	1932–1949	Airspeed	Yes
	1c		Aircraft	1950–1960		
4	2c	9.2	Farm tractor	1921–1941	Fuel-consumption efficiency	Yes
	2d		Farm tractor	1948–1968		
5	6a	9.1	Boiling-water reactors		Electric power output	Yes
	7a		Pressurized-water reactors			
6	6b	9.1	Boiling-water reactors		Thermal efficiency	Yes
	7b		Pressurized-water reactors			

Table 9.7:   Causes of Variation in the Technological Innovation Potential

Number	Case (Table 9.3)	Technology	Aspect of technological change	Significant difference?
1	2a	Boiling-water reactor	Relationship between volume of reactor vessel and volume of reactor core	No
	3a	Pressurized-water reactor		
2	2b	Boiling-water reactor	Relationship between number of control rods and volume of active core	Yes
	3b	Pressurized-water reactor		

change in their complexity due to a change in their size. Thus, one possible reason for the significant difference in the innovation potential of the two reactor types is simply that the technological change has been accompanied by different degrees of increase in their physical complexity.

Figures 9.11–9.14 illustrate the performance of rank function and rank size relationships in various cases of technological innovation. Theoretically, the relationship between functional measure and rank or between size and rank should result in a straight line on double logarithmic paper. Evidently, however, there are deviations from the theoretical relationship in virtually all the cases under consideration. It should be noted that the empirical distributions of the Pareto law invariably exhibit the observed anomalies. Typically,

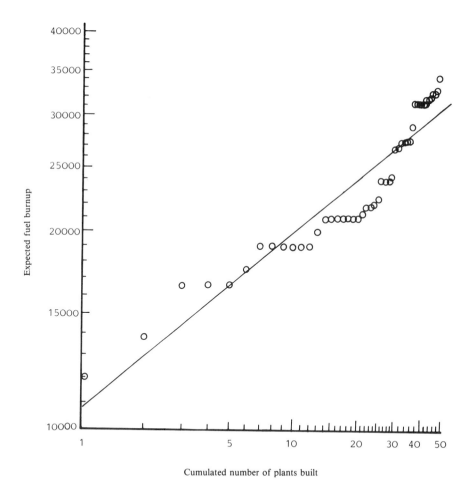

Fig. 9.11 Distribution of fuel-consumption efficiency of nuclear reactors.

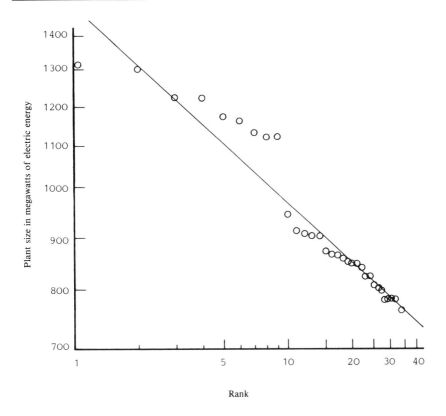

Fig. 9.12   Size distribution of nuclear power plants.

they tend to be concave upward and are linear or nearly so only over the lower range of data. Moreover, their tails tend to be highly irregular (Ijiri and Simon, 1977; Steindl, 1965; Zipf, 1949). All this is to be expected. From a conceptual viewpoint, as remarked earlier, it is a characteristic of the simple laws that they do not hold over the entire range of variables. Further, there is a certain uniformity in the deviations. Thus, in all the cases under consideration the deviations occur at either the upper or the lower range of distribution. The theoretical relationship generally holds well above a certain minimum value of the variable in any given case. From a substantive point of view, the implication is clear: Although a system's performance is indeed a function of its size, performance cannot continue to improve if the size of the system is indefinitely made to become either very large or very small. That is, there exists an optimal range of variations in the scale of technology.

   The parametric estimates of the rank function relationship are presented in Table 9.8. The relationship has been estimated in the log-linear form. One

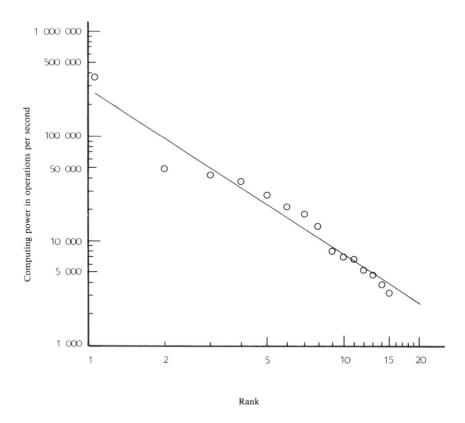

Fig. 9.13 Distribution of the technological capability of digital computers for scientific use produced in 1961.

major disadvantage of this form of the relationship is that the results are sensitive to the origin of the independent variable. Thus, the empirical estimates differ somewhat, even when theoretically they should not, depending on whether the dependent variable in the relationship is ranked in the increasing or decreasing order of its values. However, use of the log-linear forms of the relationship has been dictated by the computational facilities available. The fit of the relationship to the data is generally very good. It may therefore be concluded that the observed distribution of technological advances generally obeys the Pareto law. This finding has the important implication that the decisive factor in the evolution of technological systems is the differential rate of development of certain types of designs *within* a given field.

Recall that the exponents in the rank function relationships are inversely related to the Pareto coefficients. The resulting estimates of the redundancy-dependent technological innovation potential are presented in Table 9.9. At

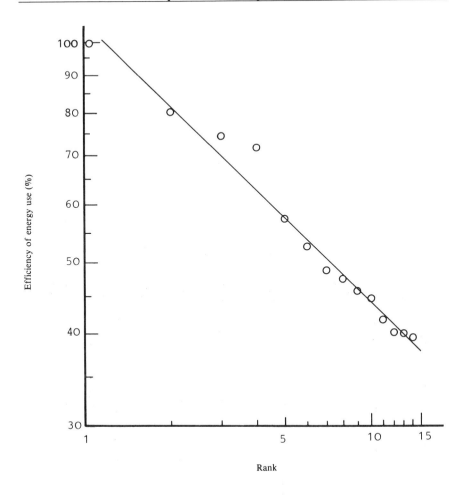

Fig. 9.14 Distribution of efficiencies in the use of various energy resources.

the two extremes, while the potential for innovation in digital computers is the highest, the potential for innovation in nuclear reactors is the lowest. The potential for innovation in the use of energy resources is between this range. Once again, the results point to significant intrinsic differences between the innovation potential of different techniques.

Table 9.10 presents the parametric estimates of the rank size relationship in the case of nuclear technology. The resulting estimates of the Pareto coefficient are provided in Table 9.11. Recall that the Pareto coefficient of the size distribution (θ) is theoretically a ratio of the exponents of the scale function and rank function relationships ($\theta = -B/b$). It can be readily verified that the empirically estimated Pareto coefficient in the case of core size distribution

Table 9.8: Parametric Estimates of the Rank Function Relationship

Case	Technology	Dimension of technology (y)	Range of y considered	Estimated relation	N	R^2	S
1	Computing devices for scientific use introduced in 1961	Speed in operations per second	Upper range	$\log y = 5.45 - 1.58 \log R_1$ $(0.08) \quad (0.09)$	15	0.96	0.11
2	Computing devices for commercial use introduced in 1961	Speed in operations per second	Upper range	$\log y = 5.46 - 1.69 \log R_1$ $(0.10) \quad (0.10)$	18	0.97	0.16
3	Energy resources	Efficiency of use	Upper range	$\log y = 2.02 - 0.38 \log R_1$ $(0.01) \quad (0.018)$	14	0.97	0.02
4	Nuclear plants in commercial operation	Thermal efficiency	(a) Entire range	$\log y = 1.38 + 0.09 \log R_2$ $(0.02) \quad (0.01)$	48	0.46	0.04
			(b) Upper range	$\log y = 1.46 + 0.04 \log R_2$ $(0.003) \quad (0.002)$	45	0.91	0.004
5	Nuclear plants in commercial operation	Expected fuel burnup	(a) Entire range	$\log y = 4.03 + 0.27 \log R_2$ $(0.01) \quad (0.01)$	48	0.89	0.04
			(b) Upper range	$\log y = 3.54 + 0.59 \log R_2$ $(0.05) \quad (0.03)$	23	0.93	0.01
			(c) Lower range	$\log y = 4.10 + 0.18 \log R_2$ $(0.008) \quad (0.007)$	25	0.96	0.01

Table 9.9:   Estimates of the Redundancy-Dependent Technological Innovation Potential[a]

Number	Case (Table 9.8)	Technology	Dimension of technology	Pareto coefficient	Comparative assessment of redundancy-dependent innovation potential
1	2	Digital computers for commercial use	Computing power in operations per second	0.59	1
2	1	Digital computers for scientific use	Computing power in operations per second	0.63	2
3	3b	Energy resources	Efficiency of use	2.63	3
4	5a	Nuclear plants	Fuel-consumption efficiency	3.70	4
5	4a	Nuclear plants	Thermal efficiency	11.11	5

[a]The highest level of innovation potential is assigned rank 1, the next highest rank 2, etc.

Table 9.10: Parametric Estimates of the Rank Size Relationship

Case	Technology	Measure of size (S)	Range of size considered	Estimated relation	N	R^2	S
1	Nuclear plants in commercial operation	Core size	(a) Entire	$\log S = 1.41 - 0.25 \log R_1$ $\qquad\quad (0.06)\quad(0.04)$	48	0.89	0.12
			(b) Upper	$\log S = 1.32 - 0.15 \log R_1$ $\qquad\quad (0.01)\quad(0.009)$	40	0.87	0.02
2	Nuclear plants in commercial operation	Plant size	(a) Entire	$\log S = 3.37 - 0.40 \log R_1$ $\qquad\quad (0.13)\quad(0.09)$	48	0.28	0.25
			(b) Upper	$\log S = 3.18 - 0.19 \log R_1$ $\qquad\quad (0.01)\quad(0.01)$	32	0.92	0.02

Table 9.11:   Pareto Coefficients of the Nuclear Reactor Size Distribution

Case Number (Table 9.10)		Dimension of technology	Range of size considered	Pareto coefficient
1	1a	Core size	(a) Entire	4.0
	1b		(b) Upper	6.6
2	2a	Plant size	(a) Entire	2.5
	2b		(b) Upper	5.2

(case 1a, Table 9.11) is approximately equal to the theoretically predicted value based on the relevant scale function and rank function relationships (case 5b, Table 9.1, and case 4a, Table 9.8). Its theoretically derived value is 3.33. Upon direct estimation, its value is obtained to be in the range 2.85 (when the smallest value of y is ranked 1) to 4.04 (when the largest values of y is ranked 1, as in Table 9.11). Thus there is good agreement between the theory and the data.

It should be noted, however, that the Pareto law does not adequately hold over the entire range of nuclear plant sizes. Following the theory advanced here, the size above which the law does apply is the minimum efficient size of technology. For the case in hand, this is estimated to be 762 MW. It is interesting to note that, starting from a very different premise, earlier investigations into economies of scale estimated the minimum efficient size of light-water reactors to be around 800 MW (Huettner, 1975, pp. 81–83). Thus there is excellent agreement between the results of this study and those of other unrelated studies.

Finally, recall that growth in the fuel-consumption efficiency of nuclear reactors turns out to be one of the few cases in which the scale function relationship does not hold. However, it can be readily explained in terms of the rank function relationship. There is an important lesson behind the observed exception. It is that learning is an essential complement to scaling in technical progress.

7.   PRINCIPAL CONCLUSIONS

This chapter has presented a microview of technological innovation processes. In its essence, the element of trial and error enters in a fundamental way in the development of new techniques. This should not be taken to imply that R&D activity is an *ad hoc* enterprise. Rather, there is seldom an escape from experimentation in the design of new techniques. Consequently, the trial-and-error phenomenon is itself *regulated* through various means such as wind tunnels and pilot plants. As discussed in the preceding chapters, technological innovation depends on the process of learning. According to the considerations advanced here, innovation also involves a great deal of "learning to learn."

The focal point of these learning processes is shown to be changes in the scale of the object system. The conventional belief has it that the scale of an object depends on the availability of relevant technology at any given point in time. Indeed, barriers to economies of scale often prove to be temporary with the advent of relevant innovations. According to the evidence presented in this chapter, this is a much too simplistic viewpoint. In reality, it is seldom possible to significantly change the scale of a system without affecting its form and structure. Thus the scaling of an object is an important design problem in its own right. Viewed from close quarters, one main objective of R&D activity is to determine a suitable scale for the new technology and to ensure that the scale alteration will in fact stick. Contrary to the popular view, changes of scale are not merely a by-product of successful innovation. Rather, innovations originate during the course of successful changes in scale.

The evidence futher indicates that the experience acquired in the design of a technique at one scale is not wholly transferable to its design at another widely different scale. It is not only that technical progress is, to a certain extent, object specific. It is, in part, scale specific as well. This clarifies the nature of learning in innovation processes: It is, *inter alia*, concerned with making the best possible use of experience acquired from one level of scale to another level. In sum, learning by doing in technological innovation is at least partly a process of learning by scaling.

This chapter has next outlined a framework for the identification of certain invariant characteristics of evolutionary systems. Within this framework, two fundamental relationships between learning, scaling, and technological innovation are shown to exist. They provide a hitherto lacking theoretical explanation of the well-known progress function and of allometric growth models. In the present context the implication of these relationships is simply that technical progress involves changes of both a *cumulative* and a *differential* nature. The two relationships are complementary. Together, they also account for the fact that a wide variety of empirical size distributions obey the Pareto law.

The proposed relationships between learning, scaling, and technological innovation are of a deterministic nature. In turn, however, they indicate that technical progress is a probabilistic process. Specifically, the distribution of technical capability within any given field obeys the Pareto law. The implication is that the process of innovation is governed by a plurality of factors inasmuch as the presence of a systematic influence is incompatible with the existence of a probabilistic phenomenon. In a sense, there is neither a unique "learner" nor a single "lesson" in the course of innovative activity. Rather, advances in technology depend on a wide variety of agents at work. Accordingly, the concept of learning in the development of new techniques is to be understood in a very broad sense. Similarly, the process of scaling is to be regarded in terms of the influence of a host of variables. Very simply, technical progress depends on myriad causes with none holding clear sway over others.

These results of quantitative analysis are supported by other qualitative evidence. As indicated by detailed case studies of the innovation process (cf. Chapter 6), technical progress depends not only on learning from past failures but also on learning to anticipate new opportunities. It is not only a matter of determined effort in the face of considerable uncertainty regarding the outcome, but also a matter of readily abandoning a chosen R&D project once there is a clear indication of its deficiencies. It involves both a constant striving to attain the possibilities that are intrinsic to the system and occasional assimilation of new ideas from outside.[6] Thus there is seldom a sole source of technical progress within any given field. Rather innovations depend on the interplay of many different factors.

This should not be taken to imply that technical progress is immune from regulation and control. It is true that our results point to a law of chance in the distribution of technical capabilities. Equally significant, however, is that this law is shown to arise from a deterministic mechanism. The conclusion to be drawn is that potential for technological innovation in a given area may well be a matter of chance. However, whether or not this potential can in fact be exploited depends on deliberate policy. Very generally, technical progress is neither a wholly laissez-faire affair nor a transaction in a closed shop.

The applications of the proposed relationships to various cases of technological innovation indicate that their fit to the data is generally very good. As is, of course, characteristic of simple laws (i.e., laws containing relatively few parameters), the proposed relationships do not hold over the entire range of data. They do, however, remain invariant over long periods of time. Specifically, the results of this study point to the existence of certain relationships that are found to remain stable over various periods of time, ranging from a minimum of about 20 years to a maximum of about 50 years. The corresponding range of the relevant variables, such as system size, is frequently even larger by several orders of magnitude. Further, the deviations from these relationships at the extreme range of the data tend to be generalizable. In summary, certain invariant patterns of innovative activity are, in fact, found to exist.

One cardinal feature of these patterns is that they are system specific. That is, they are primarily of a technical nature. Moreover, they remain unchanged amid a constant flux of socioeconomic conditions surrounding them.

---

[6]For a similar viewpoint see Steindl (1980).

This is not to imply that variables of a socioeconomic nature are unimportant in an understanding of technological innovation. Indeed, while system size, for example, is a technical variable, it is also representative of various institutional, sociological, and economic factors. The essential point, however, is that the proximate causes of technological innovation are always of a physical nature, although ultimately their origin may be traced to socioeconomic factors. From a theoretical point of view, the implication is that the process of innovation is endogenous to the technological system. From a policy point of view, the implication of our results is that technological planning cannot be expected to provide quick solutions to relevant problems inasmuch as patterns of innovation change very slowly over time. Instead, technological planning is properly regarded in the context of a long-term strategy.

Within the proposed framework of lawlike relationships, it has been shown that the innovation potential of various techniques can be systematically assessed. There are two principal determinants of technological innovation potential. One is the redundancy of the system. The other is its scale. The role of scale in determining technological innovation potential can be further specified in terms of its bearing on the form and physical complexity of the system. Earlier it was inferred that the observed interindustry differences in the growth of productivity are attributable to differences in the nature of the technology employed in different fields (Chapter 6). The results of the present investigation confirm this inference in an independent manner. Further, it is indicated that the "nature of technology" in turn has a concrete meaning in terms of its redundancy, form, and physical complexity. In light of these considerations, the hitherto elusive "productivity differential puzzle" becomes much more comprehensible. Specifically, the dichotomy implicit in the existence of high-growth versus stagnant sectors of the economy can be readily accounted for in terms of certain internal characteristics of alternative technologies.[7]

A number of interesting findings emerge from the results of the empirical analyses reported in this chapter. First, the development of any given technology generally takes place in well-defined stages characterized by systematic changes in its innovation potential over time. Put another way, technical progress occurs in a sequential manner. The finding seems to be one of great generality. It is certainly supported by the results of other unrelated studies. A good example of this is the progression of Mexican technical capability in heavy electrical equipment shown in Figure 9.15 (Vietorisz, 1972). The sequential character of technical progress in this case is clearly reflected in the introduction of electric motors of progressively larger size and progressively greater

---

[7]This possibility has often been suspected but it has never been adequately conceptualized or proved. For a most thought-provoking discussion of this point see Nelson and Winter (1977).

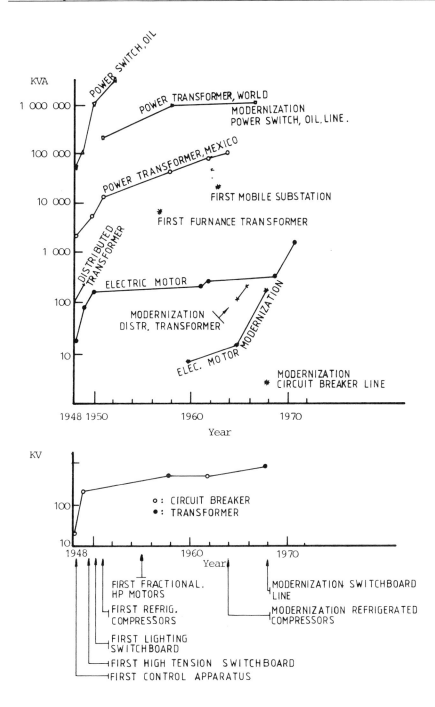

Fig. 9.15   Progression of Mexican technical capabilities in the heavy electrical equipment industry, 1948–1970 (after Vietorisz, 1972).

complexity over the course of time. According to the results of this chapter, such a pattern of industrialization is to be expected.

Second, at any given point in time, there exist significant differences in the innovation potential of comparable types of technology (e.g., boiling-water and pressurized-water reactors). These differences are attributable to differences in physical complexity due to change in the scale of the technology. Moreover, any given stage of technological development is characterized by marked differences in the capability of alternative designs. (Indeed, to a certain extent this is implicit in the Pareto distribution of technological capabilities.) That is, in the development of a technology, a few vital types of design account for the bulk of observed advances. This confirms the previously proposed principle of technological guideposts from an analytic point of view. Indeed, it is to be expected that innovative activity tends to utilize a relatively few, specific patterns of design rather than to try all the available alternatives inasmuch as there is a fundamental asymmetry in the capabilities of different technological layouts.

Third, the extent of asymmetry in the distribution of technical capabilities differs from field to field (i.e., the Pareto coefficients of different technical distributions differ from each other). That is, the importance of technological guideposts varies across different areas.

Fourth, learning determines the extent of scaling in technical progress. This result at the microlevel of analysis is essentially similar to earlier findings at the macrolevel of investigation (see Chapters 6, 8). The possibilities of learning are in turn influenced by the process of scaling.

Finally, there exist lower and upper bounds to the scale of any given technology below or above which its performance is generally expected to deteriorate with change in its size. These bounds in effect delineate an optimal range of variations in the scale of every technology.

There are profound implications of these findings for national technology strategy. In particular, they indicate that a country can leapfrog earlier stages of economic development by means of a judicious choice of technologies, that is, of techniques with the highest innovation potential. This strategy of development has, of course, been noted before (see, e.g., Pelc, 1980). The results of the present investigation provide a number of new insights into it. First, as is apparent, the exploration of technological frontiers is inherently full of risks. There is always the danger of having to begin again from square one for want of a backup system. Our results, however, indicate that the differences in the innovation potential of different technical systems are real, all-pervasive and highly significant. Unless a strategy of economic development is reconciled to a permanent state of relative backwardness, it can ill afford to ignore the advantages of leapfrogging. In short, the game is worth the candle.

Second, the relevant issue in technological planning is one not only of promoting specific sectors of the economy but also of choosing specific techniques to be utilized within any given industry. The choice of technology in turn must be based on explicit consideration of the learning and scaling processes involved. Moreover, in view of the sequential character of these processes, there is a strong case for protection of indigenous technology during the infant stage of its development.

Third, a technology strategy has to go beyond the choice of a set of techniques associated with the highest innovation potentials and appropriate scales. Additionally, a decision must be made as to how the resources are stretched toward the development of each of the chosen techniques. Thus, as an example, results of the present analysis indicate that the appropriate policy for R&D in the field of computer technology is one that aims at intensive development of a few carefully selected models. This strategy may be likened to that of putting all one's eggs in one basket. In contrast, nuclear technology calls for R&D into several alternative reactor types. In summary, the formulation of national technology strategy is more than just a matter of economic cost–benefit analysis. Technological planning must be based on multiple criteria, including both economic and engineering considerations rather than one at the exclusion of the other.

Chapter 10

PRINCIPAL THEORETICAL AND POLICY IMPLICATIONS

1. TWO PROPOSITIONS IN A THEORY OF TECHNOLOGICAL INNOVATION

This book has presented a general theory of technological innovation. In its essence, there are two key determinants of technical progress. One is the process of learning or the acquisition of production skills. The other is the process of scaling or patterning of the system to perform certain desired tasks. The role of these two processes in technical progress turns out to be of pivotal importance in a very interesting and somewhat surprising way.

It is generally recognized that accumulation of experience is a key factor in the improvement of productivity. This is exemplified by the concept of the manufacturing progress function—that the cost of production systematically declines with increase in cumulated output. The observed phenomenon is attributable to several factors, for example, improvement in plant layout, changes in the tolerances to which parts are machined, vocational training of the work force, improved communication, and better methods of management. Very generally speaking, the productivity of a system is gradually improved as it becomes possible to eliminate various bottlenecks involved in its operation through the accumulation of relevant experience. Thus the efficiency with which a *given* technology (i.e., an existing system of plant and equipment) can be utilized is a matter of learning.

A close examination of the phenomenon under consideration reveals that success in the development of a *new* technology is also a matter of learning. There are indeed very few innovations without a history of lost labor. What eventually makes the development of new techniques possible is the object lesson learned from the past failures. If there is one guiding dictum in innovative activity, it is to profit by example.

Devendra Sahal, Patterns of Technological Innovation   ISBN 0-201-06630-0

To begin with, the design of new techniques is almost always founded upon trial runs which are themselves systematically undertaken by means of wind tunnels, pilot plants, and the like. Even so, the performance of the new device invariably leaves much room for improvement. What was expected to be an impeccable design, in fact, turns out to be severely flawed. However, with the accumulation of further experience in the design and production process, it becomes possible to identify and remedy the weak links in the chain. The solutions thus reached in turn make it possible to perceive and exploit new technical opportunities. In this way, technical progress occurs in a cumulative manner.

In sum, according to the conventional notion of the progress function, the scope for *utilization* of a technique is determined by a process of learning. According to the considerations advanced here, learning is a central factor in the *evolution* of a technique in the first place. This type of learning occurs both in the design process (at the level of a single unit of physical equipment) and in the production activity (at the level of a plant). It is illustrative of what may be called meta progress functions underlying advances in technology. There are important differences between the dynamical forms of the progress functions at different levels. Nevertheless all of them have the common implication that learning is at the heart of growth in productivity.

The process of learning in the design of a technique is mediated through changes in its scale. It is commonly held that the scale of a technology can be more or less changed at will, provided it is justified by economic considerations. This belief apparently stems from many instances of indiscriminate increase in the size of technology, such as the phenomenal growth in the deadweight tonnage of tankers during the last two decades. While the conventional viewpoint is obviously true, it does not give a complete picture of innovative activity.

According to the results of the investigations reported herein, considerations of scale in innovation processes are hardly as peripheral as is commonly believed. Rather, determining the appropriate scale of a technique for the performance of a given task is the very crux of research and development activity at its grass roots level. This does not of course mean that changes of scale are always successful. Very often they prove unwarranted in the light of experience acquired in the actual utilization of technology. The important point is that the process of scaling is inherently an uncertain activity. This is because beyond a certain point, quantitative changes in the scale of an object are invariably transformed into certain qualitative changes with profound implications for its morphological, functional, and structural properties. Thus a great deal of learning in research and development activity has to do with solutions to the problems posed by contemplated changes in the scale of technology.

Contrary to popular opinion, changes in scale do not arise in the aftermath of successful innovation. Rather, innovations have their origin in the process of scaling.

As in the case of learning, the process of scaling plays an important role in innovative activity not only at the level of a single unit of physical equipment, as discussed above, but also at the plant level. It is commonly thought that innovations make it possible to change the scale of the larger system of their use. Conversely, the biggest obstacle in pushing forward the scale frontier is regarded to be lack of cost-effective technological means. Indeed, this viewpoint is so firmly entrenched in economic theory that it is regarded as gospel. Thus, as an example, it is often claimed that increased mechanization has made it possible to increase the plant size of certain industries over the course of time. According to the considerations advanced here, although this viewpoint is justified by commonplace evidence, it leaves out many essential aspects of the reality.

Specifically, there is considerable evidence to indicate that changes in plant size constitute an autonomous process governed by a multitude of factors, the effect of each being very small. Thus innovations tend to be more susceptible to change in the overall scale of their utilization than the other way around. This is especially true of advances in technology past the initial stage of its evolution. The statement that the mechanization of a system determines its scale does not take us very far. Once technological evolution is considered in its entirety, a much more interesting phenomenon emerges: The extent to which a system can be mechanized depends on the scope for utilization of its scale.

The roles of learning and scaling in technological innovation tend to be closely related. They are mutually interdependent processes. One without the other is powerless; together, the two are the driving force behind observed advances in technology. The *extent* of technical progress is largely determined by the learning capability of the system. The *content* of technical progress is, on the other hand, mainly determined by the possibilities of scaling. The process of learning tends to be of a *differential* nature: Its role in innovative activity tends to differ from one technique to the other. In comparison to this, changes of scale take the form of a comparatively *invariant* factor. This is not to say that scaling affects different techniques in the same way. Rather, learning is a process of variety generation, scaling a process of variety stabilization in the system of technical progress.

The results of our investigation then indicate that the process of scaling generally tends to precede learning in the microrealm of innovative activity (i.e., in the design process). In contrast, learning generally comes first, scaling afterward, in the macrorealm of innovative activity (i.e., in the production process). Further, it is evident that there is practically no limit to learning in innovative activity. In contrast, as we would expect on commonsensical grounds, the process of scale changes cannot continue indefinitely. Additionally, however, both processes involve certain pitfalls. This is exemplified by the loss of relevant experience (and possibly even the acquisition of misleading and retrogressive information) and "overshooting" the optimum scale of technology in certain cases of innovative activity.

Finally, it is apparent that the role of learning in technical progress is analogous to that of time, the role of scaling to that of space. This point, although obvious, deserves some attention. Hitherto, most studies of socioeconomic phenomena have been mainly concerned with the role of time in the process under consideration. In their preoccupation with the dynamic aspects of the process, they generally ignore the role played by space. If the results of our investigations are any guide, however, space plays as important a role in the process of development as does time. According to the findings reported here, a wide variety of technological innovation processes are demonstrably governed by the twin phenomena of learning and scaling. Moreover, the significance of these two factors turns out to be comparable. Thus both are of vital importance. Taken together, their implication is simply that technical progress takes place in time as well as over space.

2.   GENERAL PRINCIPLES OF TECHNOLOGICAL INNOVATION

A number of important principles of technological innovation emerge from the considerations advanced here. The first of these is the principle of technological guideposts. In its essence, the process of innovation invariably leads to a certain pattern of machine design. The pattern in turn sets the boundaries of further innovations. Consequently, a great deal of technical progress has its origin in bit-by-bit modifications of a given form of design rather than in the development of alternative techniques. Very simply, innovations are generally made possible by the gradual refinement of an essentially *invariant* pattern of design. This is a good example of the old proverb that slow but steady wins the race. Thus the gist of technical progress consists of an evolutionary system.

Examples of technological guideposts are ubiquitous, including innovations in such diverse cases as farm tractors, tunneling machines, aircraft, and electric motors. The evidence further indicates that there are two identifying characteristics of technological guideposts. First, their emergence tends to depend on the *synthesis* of a great many proven concepts from the past. Second, the greater the adaptability of a technique to its task environment, the more likely it is to become a vehicle for further advances. Indeed, technological evolution in a wide variety of cases, such as tractors, locomotives, and computers, is characterized by a "general-purpose design effect": Initially, the focal point of innovative activity tends to be the adaptation of technology to its task environment; subsequently, it is the adaptation of the task environment to the technology. This is to be expected: While an adaptive system often tends to be evolutionary, an evolutionary system does not necessarily tend to be adaptive.

In a nutshell, the proposed principle of technological guideposts is illustrative of a very general phenomenon. It is that evolutionary processes tend to be both self-generating and self-constraining. The implication of this for the management of R&D activity is evident: It is possible to assess the technological innovation potential of a new technique in advance.

The second general principle of technological innovation is the principle of creative symbiosis. It is a commonplace observation that there exist limits to every type of growth. The evolution of technology is, of course, no exception to this. The development of new techniques seldom continues unabated. Rather, it involves periods of both innovation and stagnation. A close examination of the phenomenon reveals that there are two main limiting factors in the evolution of a technology: One is the fixity of its form. The other is the complexity of its structure. The evidence further indicates that in certain instances, two or more technologies join forces such that the form of the overall system is greatly simplified. This makes it possible to overcome existing obstacles to further evolution. In this manner each member of the symbiotic system of technology can continue to evolve. We find therefore that very often what would seem to be the terminal points of evolution turn out, in reality, to be mere interludes. To be sure, the process of creative symbiosis does not always materialize, but when it does, the consequences are nothing short of radical in an otherwise evolutionary system of technical progress. In this respect, the nature of long-term technical progress is admirably illustrated by the saying that "united we stand, divided we fall." Indeed, we would expect this principle to be very generally applicable to evolutionary processes.

The third general principle of technological innovation is the putty–clay principle of technical progress. In a nutshell, technical know-how lacks the monolithic nature of pure scientific knowledge. It is often acquired in bits and pieces and it is compartmentalized to a far greater extent than is commonly believed. Specifically, innovations depend on object-specific learning and scaling processes. In consequence, even within closely related fields, the development of one technique leaves the state of many other techniques relatively unaffected. This is evidenced by the fact that different techniques vary greatly in the extent of advances that result from learning and scaling processes. Such differences would not of course exist if the technical know-how were all-embracing. In reality, however, it seems that each type of innovative activity occupies a distinct niche of its own. Each field of technical endeavor has its own traditions and rules of thumb derived from its unique experience. It is therefore to be expected that technical advances do not always occur in extenso. Often they draw a wedge in the state of art.

In sum, while we can choose any technique as the focal point of R&D activity, a substantial part of the resulting know-how is specific to one particular product or process. In this respect, *ex ante* technical know-how has the pliancy of putty, *ex post* the fixity of clay. As discussed below, there are a number of important policy implications of this principle. For the present it should be noted that, contrary to the popular view, the relevant form of technical know-how is seldom at one's beck and call. The reason is that, far from being a maid for all work, technical know-how tends to be, in no small measure, tied to the system within which it is generated.

The fourth general principle of technological innovation is the principle of technological insularity. It is closely related to the putty–clay principle of technical progress but it is sufficiently distinct to deserve an independent status. In its essence, the know-how derived from the development of one technique cannot be wholly retrieved for the development of another. It is generally recognized that the fruits of innovative activity often spill over from one field to the other. Indeed, one central feature of technical progress that distinguishes it from biological evolution is the possibility of combining otherwise disparate techniques in a single device. Invariably there tend to be far greater possibilities of forming new patterns of evolution in the realm of engineering than in the organic world. Thus most innovations tend to be a composite of various techniques. We could even say that the development and transfer of technology are two sides of the same coin. This viewpoint is nevertheless erroneous. It is wrong in the Pickwickian sense of being less than right.

For one thing, while in principle it is possible to keep changing the patterns of technological evolution, this is very rarely the case in practice. Indeed, this much is implied by the very existence of technological guideposts. What is more, the process of technology transfer is inherently prone to errors. True, success in interindustry transmission of technical know-how has been the very basis of such important and glamorous innovations as electron microscopes and numerically controlled machine tools. However, failures in technology transfer are just as common, although they are often overlooked. Witness the fact that the attempts to use microelectronic circuits to reduce fuel consumption and exhaust emissions in automobiles have made comparatively little headway despite their successful applications in other areas. Likewise, the use of photovoltaic solar systems for such a relatively mundane task as home heating remains impractical despite their successful application in aerospace technology. This list of examples can be almost indefinitely multiplied. The lesson to be drawn is that technology transfer is seldom an omnicompetent factor in the process of innovation.

This is not to minimize the significance of technology transfer. The interindustry transmission of relevant know-how is evidently one of the most fruitful sources of technological innovation. The important point, however, is that the time constants of technology transfer processes tend to be several orders of magnitude higher than is commonly believed. To take one among many case studies presented earlier, the use of diesel engines and electric transmission in rail transportation could not have been made possible were it not for the technology originally developed for marine transportation and street railways, respectively. Equally important, however, is that it took some three decades to develop the diesel locomotive from already existing technology in other fields. As another example, it is well known that gas turbine and steam turbine combined-cycle technology holds considerable promise in that it can achieve higher overall efficiency than either alone. However, only recently has it been possible to develop practicable combined-cycle plants with efficiencies at par with conventional steam plants despite nearly two decades of R&D effort in this area.[1] The moral of these and many other similar examples is that technological spinoffs seldom materialize at once. Rather, they involve a great deal of painstaking R&D activity of their own. Interindustry transmission of technical know-how is almost always a long haul.

In summary, it is widely believed that the essence of innovative activities lies in the transmission of technical know-how from one field to the other. This is surely a justified viewpoint insofar as it implies that technology transfer is a process of transfer by deed. It is, however, tautological and even misleading if it is taken to mean that technological spinoffs arise ad lib. Indeed, if the results of our investigations are any guide, one central feature of technical know-how that sharply distinguishes it from pure scientific knowledge is its taciturnity. Its transmission from one field to the other is inherently slow.

It is apparent that the proposed principles of technological innovation are closely related. One common implication of the principle of technological guideposts and the putty–clay principle of technical progress is that a large part of technical progress is specific to the system of its origin. In turn, these two principles are complementary to the principle of technological insularity. Thus the observed lack of interindustry transfusion of technical know-how and the system-specific character of technical progress are mutually reinforcing processes.

---

[1]The development of combined-cycle plants has been hampered by two main problems. One has to do with differences in the availability of the steam generator, gas turbine, and turbogenerator. The other has to do with the limit to the maximum scale of the gas turbine. These problems are further illustrative of the role played by learning and scaling processes in technological innovation.

Two general conclusions emerge from the considerations advanced here. First, the intrinsic nature of an individual technique plays a pivotal role in determining the effectiveness of the research undertaken toward its development. Indeed, one recurrent finding of the investigations reported in this work is that the role of learning and scaling processes in technological innovation is significantly conditioned by certain physical properties of the system such as its form and structural complexity. It may be, as Bernard Shaw observed, that in heaven an angel is nobody in particular. It seems otherwise with earthly techniques.

Second, it is apparent that there is a large element of latency in the evolution of technical know-how. In economic theory it is axiomatic to regard technical knowledge as a public good: Once generated, it permeates all firms. Not only is it available to everyone at a nominal cost, but it can also be obtained in a cut and dried form that is ready for use. According to the evidence presented here, nothing could be further from the truth. In reality, one has to go through a costly search process to obtain the requisite technical know-how. Moreover, once obtained, it is seldom possible to successfully transfer it from one field to the other without a great deal of additional groundwork.

We shall presently see that these findings have profound policy implications.

3. INTERNATIONAL TRANSFER OF TECHNOLOGY

It is a commonplace observation that technology transfer from the industrialized countries to the developing economies has made little visible progress in achieving its objectives. According to the considerations advanced here, this is to be expected because technical progress is never a matter of transferring ready-made techniques from one system to the other. Seeds of technical progress cannot just be sown and left. Rather, they have to be cultivated. As discussed earlier, the development and transfer of technology are two closely related processes. One cannot be considered in isolation from the other. In this respect existing mechanisms for technology transfer leave much to be desired. Their main deficiency is that they do not provide for R&D specifically directed toward the adaptation of existing techniques to meet the requirements of different task environments. Moreover, as discussed earlier, technology transfer is hardly as infallible as is commonly believed. It is undoubtedly one of the most potent factors in the development of new techniques but it is inherently slow and tedious. Contrary to some of the opinions voiced in the North–South dialogue, the reasons for the observed lack of technical and economic progress in the developing countries do not necessarily lie in their passivity. The process of technology transfer has yet to run its course. We have every reason to be optimistic in this regard.

A much more fundamental question is whether developing countries should opt for the import of technology from industrialized countries, or develop their own R&D facilities. To begin with, it is evident that both the industrialized and the developing countries have much to gain from joint effort in the development of new techniques. There are many examples of this: development of microbial techniques based on the use of sugar instead of petrochemicals (e.g., in the production of surfactants used in detergents), the use of jute for newsprint, research into thermoplastic natural rubbers for use in transportation systems, etc. However, while the importance of North–South cooperation cannot be overemphasized, it seems that there is a strong case for developing countries to undertake independent research efforts of their own. According to the results of our investigations, the role of learning in technical progress tends to be object specific. Thus the advanced state of technical know-how has comparatively little bearing on the development of alternative techniques suited to the needs of the third world. In conclusion, it is imperative that developing countries become masters of their own technological destiny.

This brings us to the next point. It is not just that the technology produced in the industrial world is inappropriate to the stipulations of certain environments (e.g., because of its labor-saving bias). What is more, its efficacy is often limited to a set of prespecified tasks. If the considerations advanced here are any guide, there is much less to ultramodern technical know-how than meets the eye. Advanced technology works supremely well in a world of familiar events. This is, however, no guarantee that it can also function adequately in the face of unexpected events. It is evidently capable of avoiding a breakdown of the system. It is hardly equipped to cope with a crisis that has in fact occurred. Ours is a "fail-proof," not a "proof-fail," sociotechnical system.

4.   NATIONAL TECHNOLOGY STRATEGY

The foregoing considerations raise the question as to what constitutes an appropriate national technology strategy. We have already noted that technical progress is first and foremost a matter of learning or the accumulation of practical experience in design and production processes. The results of our investigation further indicate that the crucial factor in the phenomenon under consideration is learning in the capital-producing rather than the capital-using sector. This conclusion from quantitative analyses is also supported by the qualitative evidence indicating that industries in the former sector (e.g., machine tools and electrical equipment) tend to be far more interconnected than industries in the latter sector (e.g., paper products and textiles). Learning in the capital-producing sector is especially important because of its synergistic effect on what is demonstrably a highly integrated system of industries. Thus one consistent result that emerges from a wide variety of case studies presented here is that (cumulated) gross investment is a key determinant of technical progress. That is, all investment has *ipso facto* the character of investment in R&D activity.

There are several policy implications of this finding. First, technical progress is not just a matter of conscious R&D activity. It is also crucially dependent on the growth of the capital-producing sector of the economy. It is customary to draw a line between innovation and routine engineering and production activity. If the results of our investigations are any guide, both are equally important endeavors toward a common goal. Technical innovation is not just a prerogative of industrial research laboratories and productivity centers. It is also conditional upon day-to-day chores performed by small-scale workshops and the like.

Second, it will not suffice to pay lip service to the role of learning in technical progress. It is essential to protect the capital-producing sector during the infant stage of its development by all available means, such as tariffs and investment subsidies. In contrast, there is very little if any justification for protection of the capital-using sector. Rather, it should be adequate to establish certain pilot plants for vocational training of the work force.

Third, and finally, some caution is necessary in the protection of the capital-producing sector so as to ensure that it does not subsidize the process of "reinventing the wheel." The act of investment is analogous to R&D activity only insofar as the fruits of learning are equally available to all. This suggests that attempts to facilitate the industry-wide transmission of relevant know-how have a valuable role to play in the process of technological development. Indeed, mobility of technical personnel is one of the mainsprings of innovation.

Another point deserves attention. Popular opinion sometimes has it that advances in pure scientific research are synonymous with advances in technology. According to the considerations advanced here, this viewpoint is both inaccurate and misleading. A close examination of the evidence reveals that pure scientific knowledge is only one among a host of other factors in technical progress. In general, the role of pure science in innovation processes tends to be relatively limited. Exceptions to this—such as advances in the chemical, electronic, and pharmaceutical industries—merely prove the rule. There are several reasons for this.

The objective of scientific activity is to know *why* a phenomenon is what it is. In contrast, the objective of technical activity is to determine *how* a phenomenon can be tamed toward a desired end. Science relies mainly on analysis, technology mainly on synthesis. The value of science lies in providing an indication of what is unfeasible. The essence of cybernetics is that under certain conditions the capacity of a system as a regulator cannot exceed its capacity as a channel of communication. However, it leaves largely unresolved the more down-to-earth problem of actually designing a viable control system. Science is the knowledge of the impossible. In contrast, technology is the art of the possible. Innovations seldom originate in scientific discoveries except in the case of a few industries. Rather, as discussed earlier, they generally depend on the gradual modification of existing techniques.

The implications of this for national education policy are obvious. The roots of technical progress lie in the activities of polytechnics and technical universities rather than in studies undertaken at institutes of fundamental research and centers of higher scientific education. This is not to minimize the importance of basic research as a backup system to industrial R&D activity. Indeed, a national technology strategy must, above all, foster a climate of close cooperation among universities, R&D facilities, science centers, engineering colleges, and manufacturing enterprises. However, basic scientific research effort can never be an adequate substitute for actual engineering explorations. The Mecca of economic and technological development lies in vocational training.

From what has been said thus far, it is apparent that the capital-producing sector is the mainspring of technical progress. The importance of its growth cannot be overemphasized. However, a national technology strategy must go beyond the promotion of a selected set of industries toward this end. Additionally, it must provide a mechanism for the choice of techniques employed within any given industries. A mere consideration of industrial growth is necessarily inadequate. Rather, the essence of a national R&D strategy lies in the choice of the technology itself. According to the findings of our analyses, a country can successfully leapfrog many of the earlier stages of economic development by a judicious choice of technology. What is more, the converse is equally true. There is a severe penalty for the indiscriminate choice of technology in the case of both developing and industrialized countries. The initial choice of a technique is of paramount importance because it will continue to influence the course of its subsequent development. Thus, it is generally recognized that there are certain inevitable consequences of the choice between, say, biogas plants, synthetic-fuel processing systems, and nuclear reactors. What is often overlooked and needs to be emphasized in the light of the evidence presented here is that the consequences of an initial choice of a technique cannot be easily repealed. Moreover, the choice has to be both flexible and distinctive. For example, if biogas is chosen to be one of the main sources of energy for a country, it is mandatory to discriminate further between various relevant techniques, such as pyrolysis (involving baking of the fuel), gasification, digestion by bacteria, or the fermentation of plants. The point is that the initial choice of technology is like giving a hostage to fortune.

In essence, the objective of technological planning must be not only to develop a *systemic* set of *selected* industries in the capital-producing sector but also to promote an *appropriate* class of techniques. The central task of all those involved in the process of technological development is to build what may be called an "integrated technological niche" as an essential framework for innovative activity.

5.   TOWARD A SCIENCE OF TECHNOLOGY POLICY

It is sometimes argued that the most modern techniques, having survived the process of selection, are also the most efficient ones. This is the viewpoint of technological determinism: The scope for the choice of alternative techniques tends to be inherently limited because there can be only one optimal trajectory of innovations. The proponents of this thesis thus claim that there is no viable alternative to the advanced capital-intensive techniques developed in the industrialized world.

If the results of our investigations are any guide, however, the viewpoint of technological determinism must be rejected as patently untenable. To reiterate, there are two main criteria of success in technological innovation. The first criterion has to do with the adaptability of technology to its task environment. This is essentially a consideration of whether the scale of the chosen technique is appropriate to its task. The second criterion has to do with the adaptability of technology to the existing production possibilities. This is essentially a consideration of whether the technique is amenable to modifications and upgrading through learning. Contrary to the generally held view in the literature on the subject, all other criteria, such as the labor and capital intensity of a technique, are of secondary importance in determining its technological innovation potential.

Thus it seems apparent that there are virtually no restrictions on the range of technological choice. The essence of technological progress does not lie in the singularity of the initial choice. Rather, it is the irreversibility of the consequences. Artifacts do not evolve of themselves. Human beings do.

This is not to say that innovations occur in a totally haphazard manner. Indeed, our investigations reveal that there are a number of significant regularities in the process of technological innovation. However, they also indicate that these regularities are themselves outcomes of what are essentially stochastic processes. Nature does seem to abhor a vacuum, as is commonly recognized. However, it is not preprogrammed to do so. The potential for technical progress is always there. Whether it can in fact be tapped in a matter of conscious policy.

This brings us to the conclusion. Just as war is too important to be left to generals, technological innovation is too significant a process to be left to economists and engineers. What is needed is an independent science of technology. My attempt in this book has been to provide the essentials of this emerging science. Further work in this area should prove to be fruitful.

Appendix

THE DATA AND THEIR SOURCES

This appendix provides the data underlying the empirical analyses reported in this book. It also presents a brief explanation of the variables employed and references to the sources utilized in the compilation of the relevant data.

Virtually all the data underlying the investigations of this book can be found here. The only exceptions are a few cases in which the data are equally accessible elsewhere in the published literature. The description of every statistical table is given in the corresponding text (e.g., see note A1-3.1 for a description of Table A1-3.1). The data concern the United States unless otherwise indicated.

Often, the nature of the data had to be clarified by scanning volumes of trade journals, company records, laboratory reports, technical manuals, and the like. Detailed information on the various case studies of technological innovation is, for the most part, provided in the text and need not be reiterated. However, those interested in further information will find additional details in the references cited in the individual chapters and those listed here.

The following short forms and abbreviations are used.

Historical Statistics (1957): *Historical Statistics of the United States, Colonial Times to 1957,* Washington, D.C., 1960.

Historical Statistics (1970): *Historical Statistics of the United States, Colonial Times to 1970,* Washington, D.C., 1975.

IAEA   International Atomic Energy Agency

NIER   National Institute Economic Review

NTTR   Nebraska Tractor Test Reports

USDA   United States Department of Agriculture

USDC   United States Department of Commerce

USFPC United States Federal Power Commission

SAE   Society of Automotive Engineers

ACKNOWLEDGMENTS

I am grateful to a number of individuals for their help in compiling the data reported here. Mr. Olavi Koponen of Valmet Corporation, Mr. Paul E. Strickler of the U.S. Department of Agriculture, Mr. L.F. Larsen of the tractor testing laboratory of the University of Nebraska–Lincoln, Mr. R.C. Bechler of the Ford Motor Corporation, and Mr. Geo. H. Seferovich, editor of the journal *Implement & Tractor,* were most generous in providing relevant information on technological change in farm machinery along with numerous helpful suggestions in the clarification of data. R.R. Powers of the General Electric Company gave some very useful historical information on the development of locomotive technology. Mr. John Rabbitt of the Sun Oil Company made available to me many of their records on changes in tank ship technology over the course of time. Allen B. Hamilton of the Electro-motive Division of General Motors and David R. Winans of the International Naval Research Organization were kind enough to share with me their knowledge of developments in locomotive and tank ship technology, respectively. The data on technological change in turbogenerators were compiled by Mr. Brian Tucker and made available to me by the permission of Mr. Frank Doyle, both of the Ministry of Industry, Trade and Commerce, Canada. Brown Boveri Company provided much further useful information on changes in the design of traction motors. They each have my sincere gratitude but I am alone responsible for any errors I may have committed herein.

THE EMPIRICAL MATERIAL

A1-1.1 *Major Innovations in Tractor Technology, 1800–1971* The data are obtained by means of an extensive literature survey and via correspondence. They are presented here in Table A1-1.1.

Sources
1. R.B. Gray, *Development of the Agricultural Tractor in the United States,* Vol. I, 1954, and Vol. II, 1958. USDA, Washington, D.C.
2. R.B. Gray and E.M. Diffenbach, "Fifty Years of Tractor Development in the U.S.A.," *Agricultural Engineering* June 1957, pp. 388–397.
3. R.H. Casterton and O.A. Smith, "Historical and Current Development in the Utilization of Tractor Power," Paper 710684, SAE, New York, 1971.
4. W.H. Worthington, "50 Years of Agricultural Tractor Development," Paper 660584, SAE, New York, 1966.
5. R.M. Wick, *Steam Power on the American Farm,* University of Pennsylvania, Philadelphia, 1953.
6. D. Sahal, "Evolution of Technology: A Case Study of the Farm Tractor," International Institute of Management, dp 78-99, Berlin, 1978.

A1-1.2 *Innovations in Farm Equipment Industry, 1902–1949* The data are obtained from ref. 1 and reproduced here in Table A1-1.2. While they could have been further refined (see, e.g., ref. 2), the author has deferred to the judgment underlying the original collection, rather than use his own criteria.

Sources
1. W.A. Cromarty, "The Demand for Farm Machinery and Tractors," Bulletin 275, Michigan Agricultural Experiment Station, 1959.
2. G.O. Seferovich, "Symbols of Modern Mechanization," *Implement & Tractor* February 1, 1961 (Diamond Jubilee Number), pp. 30–31.

A1-1.3 *Major Inventions in Farm Equipment, Railroad, Paper Making, and Petroleum Refining Industries, 1800–1957* The data are obtained verbatim from ref. 1 where the reader will find further details. Suffice here to note that the inventions included in this collection are those deemed to be either economically important in themselves or the basis for later inventions that have been economically important [1, p. 198].

Source

1. Jacob Schmookler, *Invention and Economic Growth,* Harvard University Press, Cambridge, Mass., 1966 (Table A-1, pp. 218–222; pp. 269–328).

A1-1.4 *Annual Number of Patented Inventions in Farm Equipment and Railroad Industries, 1837–1957* The data are obtained from ref. 1.

Source

1. Jacob Schmookler, *Inventions and Economic Growth,* Harvard University Press, Cambridge, Mass., 1966 (Table A-2, pp. 223–226).

A1-1.5 *Annual Number of Patented Inventions in the Construction Industry, 1855–1952* The data are obtained from ref. 1.

Source

1. Jacob Schmookler, *Patents, Inventions and Economic Change (Data and Selected Essays)* (Zvi Griliches and Leonid Hurwicz, eds.), Harvard University Press, Cambridge, Mass., 1972 (Series 122, pp. 138–139).

A1-2.1 *Diffusion of the Farm Tractor in the United States, 1920–1970* The data are laid out in Table A1-2.1. The sources of data for the variables employed are as follows.

Variables and Sources

V_1 Number of tractors on farms in thousands (at the end of the year)

1. *Implement & Tractor* November 21, 1972, p. 45.

V_2 Average belt horsepower of tractor purchased

1. A. Fox, "Demand for Farm Tractors in the U.S.," Economic Research Service, USDA, Report 103, November 1966, p. 33 [for the data from 1920–1939].
2. P.E. Strickler, File #18400, Farm and Industrial Equipment Institute, 1975 [for the data from 1940–1970].

V_3 Stock of tractor horsepower on farms in millions (at the end of the year)

1. A. Fox, *loc. cit.* [for the years 1920–1940 and 1946–1960].
2. P.E. Strickler, *loc. cit.* [for the years 1941–1945 and 1961–1970].

V_4 Number of horses on farms in thousands

1. *Historical Statistics,* 1975 (Series K-570, p. 519).

V_5 Gross investment in farm tractors in hundreds

1.   W.M. Hurst, "Power and Machinery in Agriculture," Miscellaneous Publication No. 157, USDA, April 1933, p. 33 [for the data from 1921–1930].

2.   W.A. Cromarty, *loc. cit.* [for the data from 1931–1956].

3.   P.E. Strickler, *loc. cit.* [for the data from 1957–1970].

A1-2.2 *The Growth of Hydro- and Thermoelectric power in Canada, 1917–1972* The variables employed are total installed capacity and maximum capacity of turboelectric generators in megawatts. The data are presented here in Table A1-2.2.

Source

1.   The data on all variables come from the records of the Canadian Ministry of Industry, Trade and Commerce. Additional information is provided by *The 1970 National Power Survey,* Part IV, USFPC, Washington, D.C., 1972.

A1-2.3 *Diffusion of Combines with Cornheads and of Mechanical Cornpickers in the United States, 1956–1971.* The variables employed are stocks and annual sales of combines and cornpickers, respectively. The data are presented here in Table A1-2.3. The following sources of data are employed.

Variables and Sources

V_1, V_2 Stock of combines and cornpickers, respectively

1.   *Implement & Tractor* November 21, 1972, p. 45.

V_3, V_4 Annual sales of combines and cornpickers, respectively

1.   "Farm Machines and Equipment," Bulletin No. 419, USDA, March 1968.

2.   "Current Industrial Reports. Farm Machines and Equipment," MA-35 A(71)-1, USDC, September 1972.

A1-2.4 *Diffusion of the Diesel-Powered Tractor in the United States 1955–1971* The variables employed are annual production of diesel-powered tractors and total tractor production. The data are presented in Table A1-2.4.

Source

1.   *Implement & Tractor* November 21, 1972, p. 56.

A1-2.5 *Diffusion of the Oxygen Steel Process in Austria, 1952–1967* The data concern the share of oxygen steel process in total steel output and are presented in Table A1-2.5.

Source

1.   G.F. Ray, "The Diffusion of New Technology. A Study of Ten Processes in Nine Countries," NIER, London, *48*, May 1969, Table 1.

A1-2.6   *Diffusion of the Continuous Process of Casting Steel in Austria, 1952–1968* The data concern the share of continuously cast steel in total steel output and are presented here in Table A1-2.6.

Source

1.   G.F. Ray, *op. cit.*, Table 9.

A1-2.7   *Diffusion of Shuttleless Looms in the Weaving of Cotton and Man-Made Fibers in Germany, 1958–1968* The data concern the share of shuttleless looms to all looms and are presented here in Table A1-2.7.

Source

1.   G.F. Ray, *op. cit.*, Table 27.

A1-2.8   *Diffusion of Tunnel Kilns in Brickmaking in Sweden, 1950–1968* The data concern the share of tunnel kilns in total output of bricks and are presented in Table A1-2.8.

Source

1.   G.F. Ray, *op. cit.*, Table 32.

A1-2.9   *Diffusion of New Methods of Steel-Plate Marking and Cutting in Shipbuilding in Sweden, 1950–1968* The data concern the share of steel cut by new methods in the total and are presented in Table A1-2.9.

Source

1.   G.F. Ray, *op. cit.*, Table 38.

A1-2.10   *Diffusion of Automatic Transfer Lines for Car Engines in Sweden, 1955–1968* The data concern the share of total output produced by ATL and are presented in Table A1-2.10.

Source

1.   G.F. Ray, *op. cit.*, Table 42.

A1-3.1   *The Development of Farm Tractor Technology, 1920–1968*

$V_1 - V_5$ These variables represent the *average* values of these measures: fuel-consumption efficiency (defined in terms of drawbar horsepower-hours per gallon of fuel consumed), mechanical efficiency (ratio of drawbar horsepower to belt or power takeoff horsepower), ballasted weight in pounds, belt or power takeoff horsepower per 1000 lb of ballasted tractor weight, and belt or power takeoff horsepower per 1000 lb of unballasted tractor weight, respectively. The data on all five variables have been obtained from nearly 1000 reports of tractor tests conducted by the Department of Agricultural Engineering at the University of Nebraska [1]. These tests, which first began in 1920 and have been conducted ever since (except from 1942 to 1947, because of the war), provide an exceptionally rich body of data on technological innovation in this area. In particular, the test results represent the entire universe of tractor models produced in North American and most West European countries from an early period in the development of technology. Furthermore, they are based on clearly defined experimental procedures, in marked contrast with the excessive and sometimes differing claims found in the advertising literature and trade directories [2–5]. The test reports have been leafed through issue by issue with the resulting time series data shown in Table A1-3.1. Specifically, the data on fuel-consumption efficiency have been derived from the results of a 10-hr. test based on a fixed percentage of the drawbar pull at maximum power of each tractor under ballasted conditions (test H). The data on mechanical efficiency have been derived from the results of a 1-hour test of maximum power and fuel consumption of the tractor (test C, and test H as described above). The data on horsepower-to-weight ratio have been derived from the results of test C and a test of tractor weight under unballasted conditions (test J). The horsepower variable in the chosen measures (V_2, V_4, V_5) refers to belt horsepower during the period 1920–1958, and to PTO horsepower during the period 1959–1968. This is simply a reflection of the fact that modern tractors no longer have integral belt pulley power takeoffs. Note also that tractors were first designed to be ballasted in 1941. All together, the data on the chosen measures of technology provide precise estimates of tractor performance under both ballasted and unballasted conditions.

Variables and Sources
1.   NTTR 1-1000 (1920–1968), Lincoln, Nebraska.
2.   O.W. Sjogren, "Tractor Testing in Nebraska," *Agricultural Engineering* (February 1921), 2:34–37.

3.   E.L. Brackett, "The Nebraska Tractor Tests," Agricultural Engineering (June 1931), 12:205–206.

4.   H.L. Wallace, "Recent Changes in Tractors as noted from the Nebraska Tractor Tests," *Agricultural Engineering* (March 1928), 9:85–91.

5.   L.W. Hurburt, L.F. Larsen, G.W. Steinbruegge and J.J. Sulek, "The Nebraska Tractor Tests," Agricultural Engineering (April 1960), pp. 229–231.

V_6 Cumulated tractor production quantities in hundreds

1.   W.M. Hurst, *op. cit.*, p. 31 [for the data from 1909, the earliest year from which the production figures are cumulated, up to 1930].

2.   W.A. Cromarty, *loc. cit.*, pp. 70–72 [for the data from 1931–1942].

3.   *Implement & Tractor* November 21, 1971, p. 48 [for the data from 1943–1968].

V_7 Average acreage per farm

1.   *Historical Statistics* 1975 (Series K-7, p. 457).

V_8 Relative price ratio—an index of tractor price in relation to the price of labor. It is based on these series: (1) prices received by farmers for all products (1910–1914 = 100) obtained from refs. 1 and 2. It was changed to the base 1957–1958 = 100 (U_1). (2) Tractor prices in relation to the prices received by the farmers (1957–1958 = 100) obtained from ref. 3 (U_2). (3) Composite farm wage rate obtained from ref. 4 (U_3). The relative price ratio index is obtained as $V_8 = U_1 \times U_2/U_3$. Some inaccuracy is inherent in the use of differing series for the present purpose. However, the procedure employed here will be defended on the grounds of empirical necessity.

1.   *Historical Statistics,* 1957, Series K-129, p. 283 [for the data from 1921–1950].

2.   "Annual Price Summary," Crop Reporting Board, SRS, USDA, p. 6 [for the data from 1951–1971].

3.   A. Fox, *loc. cit.,* p. 33 [for the data from 1921–1961].

4.   *Historical Statistics,* 1970, Series K-177, p. 468 [for the data from 1921–1970].

A1-3.2   *The Development of Locomotive Technology, 1904–1967.*
Variables and Sources

V_1 This variable represents the average tractive effort of locomotives *in service* and is measured in terms of pounds.

1.   *Historical Statistics,* 1970, Series Q-300, p. 728.

V_2 This variable represents the total track-operated railroad mileage.

1.   *Historical Statistics,* 1970, Series Q-288, p. 728.

V_3 This variable represents the total number of all types of locomotives in service, including steam, electric, and diesel.

1.   *Historial Statistics,* 1970, Series Q-295, p. 728.

V_4 This variable represents cumulated locomotive production. The production figures are cumulated from 1899, the earliest year for which data are available.

1.   *Historical Statistics,* 1970, Series P-293, p. 696.

A1-3.3   *The Development of Tank Ship Technology, 1899–1973*
Variables and Sources

$V_1–V_3$ These variables represent the average deadweight tonnage, average service speed in knots, and total number of tank ships in service, respectively. All three variables concern the world fleet of ocean-going vessels of 2000 gross tons and over.

1.   The data on all three variables were obtained from the historical records of the Sun Oil Company, St. Davids, Pennsylvania.

V_4 This variable represents cumulated tank ship production in the United States. The data on worldwide production of tank ships are not available. The production figures are cumulated from 1914, the earliest year for which the data are available.

1.   *Historical Statistics,* 1970, Series Q-446, p. 752.

A1-3.4   *The Development of Aircraft Technology, 1932–1965* The data on all the following variables pertain to change in the characteristics of scheduled domestic air carriers only.

Variables and Sources

V_1, V_2, V_4 These variables represent the average seating capacity, airspeed in miles per hour, and total route miles, respectively. The data on speed refer to the "wheels-off wheels-on" speed; that is, they are based on time from takeoff to landing.

1.   A. Phillips, "Air Transportation in the United States," in *Technological Change in Regulated Industries,* (W.M. Capron, ed.), p. 130. The Brookings Institution, Washington, D.C., 1976.

V_3, V_6 These two variables represent the number of aircraft in service and cumulated addition of new aircraft, respectively.

1.   A. Phillips, *Technology and Market Structure,* pp. 70 and 158. D.C. Heath, Lexington, Mass., 1971.

V_5 This variable represents the total number of airports and landing fields.

1.   *Historical Statistics,* 1970, Series Q-604, p. 772.

A1-3.5 *The Development of Computer Technology, 1944–1967*
Variables and Sources

V_1-V_3 These variables represent speed characteristics of computers developed for scientific and commercial use and cumulated number of computers produced. The speed characteristic (V_1, V_2) is based on Knight's general definition of homogeneous computing power. It includes the measurement of and interaction between three main variables: (1) internal arithmetic capabilities; (2) memory capacity; and (3) input–output capabilities. For further details see [1, Table 1]. The sources of data come from refs. 1 and 2.

1.   K.E. Knight, "Changes in Computer Performance," *Datamation* (September 1966), 12(9):40–54.

2.   K.E. Knight, "Evolving Computer Performance 1963–67," *Datamation* (January 1968), 14(1):31–35.

A1-3.6 *Technological Change in Steam-Powered Electricity-Generating Plants, 1920–1970*
Variables and Sources

V_1 Average fuel-consumption efficiency in kilowatt-hours per pound of coal

1.   *Historical Statistics,* 1970, Series S-104, p. 826.

V_2 This variable is a measure of average scale of plant utilization. It represents the ratio of net production of steam-powered electrical energy in millions of kilowatt-hours to number of steam-powered plants.

1.   *Historical Statistics,* 1970, Series S-38, p. 820 and Series S-55, p. 822.

V_3 Cumulated net production of steam-powered electrical energy

1.   *Historical Statistics,* 1970, Series S-38, p. 820.

V_4 Production per kilowatt of capacity in kilowatt-hours

1.   *Historical Statistics,* 1970, Series S-57, p. 822.

A1-3.7 *Technological Change in Internal-Combustion Type Electric Power Plants, 1920–1970*
Variables and Sources

V_1, V_2 These variables represent fuel-consumption efficiencies in kilowatt-hours per gallon of oil and per cubic foot of gas, respectively.

1.   *Historical Statistics,* 1970, Series S-105 and S-106, p. 826.

V_3 This variable is a measure of average scale of plant utilization. It represents the ratio of net production of electrical energy by internal-combustion type plants in millions of kilowatt-hours to total number of these plants.

1.   *Historical Statistics,* 1970, Series S-39, p. 820, and Series S-56, p. 822.

V_4 Cumulated net production of electrical energy by internal-combustion type power plants

1.   *Historical Statistics,* 1970, Series S-39, p. 820.

A1-4.1   *Speed Characteristics of Computers Introduced in 1961* In all, 25 computer models developed for both scientific and commercial use are included. The speed characteristic is based on homogeneous computing power and is measured in terms of the number of operations per second.

Source

1.   K.E. Knight, *op. cit.,* 1966.

A1-4.2   *Technical Characteristics of Passenger Ships, 1840–1961* This is a highly selected set of ships built during the period 1840–1961. The characteristics included are certain dimensions in length, tonnage, and horsepower.

Source

1.   E.P. Harnack, *All About Ships and Shipping,* Faber and Faber, London, 11th ed., 1964, pp. 5–7.

A1-4.3   *Technical and Performance Characteristics of Nuclear Reactors* Two principal types of reactors are considered: boiling-water and the pressurized-water reactors. In all, data on 15 morphological and functional characteristics of 48 reactors of both types have been obtained. The principal source of data comes from ref. 1. Other sources come from refs. 2–4.

Sources

1.   *Nuclear Engineering* 18:327–362 (1973).

2.   "Small Nuclear Plants. Design, Construction and Operating Experience," U.S. Atomic Energy Commission Report COO-284, October 1966.

3.   M. Ristic and M. Khan, "Operating Experience with Nuclear Stations in Member States," IAEA, 127, Vienna, 1970.

4.   D. Chase, J. Iljas, and M. Khan, "Operating Experience with Nuclear Power Stations in Member States until end of 1970," IAEA, 130, Vienna, 1971.

A1-4.4 *Efficiencies of Energy Use* In all, 19 major sources of energy are considered in terms of efficiency of their use. Principal source of data comes from [1].

Source

1.   *World Energy Requirements in 1975 and 2000.* Proceedings of International Conference on Peaceful Uses of Atomic Energy, 8–20 August, 1955, Vol. I, United Nations, 1956, New York, P/902, Table XXIIC.

Table A1-1.1

Major Innovations in Tractor Technology, 1800–1971

	Innovation	Date
1	Stationary steam engine for farm use	1808
2	Portable steam engine for farm use	1849
3	Self-propelled steam engine for farm use	1870
4	Gas traction engine for farm use	1903
5	Track-laying steam engine for farm use	1904
6	Gasoline track-type tractor	1908
7	Water under control of governor to control combustion	1911
8	Frameless or unit design in farm tractors	1916
9	Farm tractor equipped with power takeoff	1918
10	Cultivating or tricycle-type general tractor	1924
11	Farm tractor equipped with mechanical power lift	1929
12	Diesel-powered track-type tractor	1931
13	Low-pressure rubber tires for tractors	1934
14	Hydraulic lift equipment on tractor	1934
15	High-compression tractors to burn leaded gasoline	1935
16	Three-point hitch for tractor	1938
17	Continuous running or independent power takeoff for tractor	1947
18	Hydraulic remote control of drawn implements	1947
19	Power steering for tractor	1953
20	On-the-go power shifting for tractor	1954
21	Dual rear wheels	1965
22	Auxiliary front-wheel drive and four-wheel drive	1967
23	Electric remote control of implements	1970

Table A1-1.2

Major Innovations in the Farm Equipment Industry, 1902–1952

Innovation	Date
Portable grain elevator	1902
Cylinder corn sheller	1902
Gas traction engine	1903
Track-type traction engine	1904
Steel thresher	1904
Cornpicker	1909
Rod weeder for fallow-land farming	1910
Pressure regulator and air chamber for power sprayer	1911
Rotary hoe	1912
Frameless, or unit, design in farm tractors	1913
Reel-type side-delivery rake	1914
Power-operated milking machine	1914
Motor-cultivator-type tractor	1915
Garden tractor	1916
Two-row potato planter	1917
Farm tractor equipped with power takeoff	1919
Hammermill	1920
Starter and lights for tractor	1920
Tool-bar idea for mounting tillage tools on tractors	1921
Combination tool-carrier for heavy implements	1923
Mounted-type tractor implements	1924
Electric ventilating system for dairy barns	1924
Cultivating or tricycle-type tractor	1924
Two-row, tractor-drawn rolling stalk cutter	1925
Cotton stripper for the high plains jet-type pump	1926
One-way or wheatland disk plows popularized	1927
Two-row, power takeoff cornpicker, and one-row mounted picker	1928
Land leveler for irrigation farming	1928
Crankcase oil filters	1928
Two-row mounted cornpicker	1929
Attachments for placing fertilizer in bands	1929
Portable sprinkler irrigation	1930
Farm tractor equipped with power lift	1930

Table A1-1.2 (continued)

Innovation	Date
First diesel-powered, track-type tractor	1931
Field pickup baler	1932
Low-pressure rubber tires for farm tractors	1932
Lift-type mounted plow	1935
One-man, power takeoff combine	1935
Factory-built high-compression tractor to burn leaded gasoline	1935
Hydraulic lift equipment on tractors	1935
Forage harvester	1936
Automatic barn cleaner	1937
Self-propelled combine	1938
Tractor loader	1938
Automatic self-tying pickup baler	1940
Potato harvester	1940
Side-delivery rake driven by power takeoff	1940
Successful cotton picker built	1941
Hydraulic remote control of drawn implements	1941
Precision planting of vegetable seed	1941
Factory-built LP gas tractors	1941
First appreciable use of commercial sugar-beet harvesters	1943
Tractor-mounted cotton stripper	1943
Duster with attachment to inject liquids	1944
Low-pressure, low-volume sprayer	1944
Power takeoff and drawbar dimensions standardized	1944
Mow-hay finisher	1945
Self-propelled cornpicker	1946
Silo unloader	1946
Transmission clutch for live power takeoff	1946
Rear-engine tractor	1946
Equipment to apply fertilizer in vapor form (anhydrous ammonia)	1947
Hydraulic remote-control cylinder dimensions standardized	1949

Table A1-2.1

Diffusion of the Farm Tractor in the United States, 1920–1970[a]

T	V_1	V_2	V_3	V_4	V_5
1920	343		7.00	19767	
1921	372	20.60	7.74	19369	650
1922	428	21.00	9.03	18764	964
1923	496	21.30	10.61	18125	1107
1924	549	21.60	11.97	17378	922
1925	621	22.00	13.66	16651	1142
1926	693	22.30	15.45	16083	1169
1927	782	22.60	17.75	15388	1471
1928	827	23.00	19.27	14792	869
1929	920	23.60	21.80	14234	1363
1930	997	24.00	23.93	13742	1132
1931	1022	24.30	24.83	13195	540
1932	1019	24.60	25.07	12664	220
1933	1016	24.90	25.30	12291	220
1934	1048	25.20	26.41	12052	600
1935	1125	25.50	28.69	11861	1187
1936	1230	25.80	31.73	11598	1584
1937	1370	26.10	35.62	11342	2162
1938	1445	26.30	38.00	10995	1437
1939	1545	26.60	41.68	10629	1575
1940	1675	26.30	44.79	10444	2157
1941	1885	26.60	50.00	10193	2869
1942	2100	26.80	53.00	9873	1617
1943	2215	26.60	58.00	9605	770
1944	2354	28.60	62.00	9192	2059
1945	2480	27.00	65.00	8715	1712
1946	2613	25.60	70.79	8081	1960
1947	2821	25.80	76.17	7340	3393
1948	3123	27.10	84.57	6704	4197
1949	3399	28.30	92.59	6096	4362
1950	3678	29.00	101.07	5548	4176
1951	3907	29.00	108.26	7036	4601
1952	4100	31.20	114.92	6150	3478
1953	4243	34.60	121.18	5403	3282
1954	4345	38.70	126.44	4791	2059
1955	4480	40.60	130.00	4309	2847
1956	4570	39.60	134.00	3958	1710
1957	4620	43.20	139.00	3632	1860
1958	4673	45.70	143.00	3415	1940
1959	4688	46.20	146.00	3189	2150
1960	4743	50.50	150.00	3089	1240
1961	4763	53.80	154.00		1380
1962	4778	56.00	160.00		1530
1963	4786	58.00	166.00		1550
1964	4787	59.30	171.00		1570
1965	4783	63.10	178.00		1620
1966	4786	65.90	186.00		1850
1967	4766	68.20	190.00		1770
1968	4712	69.50	196.00		1580
1969	4619	72.80	201.00		1440
1970	4562	74.80	204.00		1340

[a]Definition of variables: T, years; V_1, number of tractors on farms in thousands (at the end of the year); V_2, average belt horsepower of tractors purchased; V_3, stock of tractor horsepower on farms in millions (at the end of the year); V_4, number of horses on farms in thousands; V_5, gross investment in farm tractors in hundreds.

Table A1-2.1.2

Maximum and Minimum Belt Horsepower of Farm Tractor, 1920–1968[a]

T	V_1	V_2
1920	75.60	2.37
1921	66.13	16.76
1922	63.04	19.61
1923	91.42	21.84
1924	72.51	26.97
1925	50.57	20.50
1926	50.05	22.28
1927	69.76	1.08
1928	63.00	21.61
1929	75.88	18.10
1930	90.23	23.24
1931	66.53	22.50
1932	91.93	20.39
1933	92.85	14.59
1934	58.89	15.28
1935	71.81	2.31
1936	118.29	18.98
1937	91.56	21.69
1938	109.64	10.42
1939	108.84	15.18
1940	145.39	18.82
1941	78.03	14.34
1947	44.07	4.30
1948	129.08	10.07
1949	76.90	16.98
1950	97.83	1.47
1951	117.68	1.56
1952	57.89	1.95
1953	63.81	2.01
1954	65.19	17.88
1955	133.83	18.34
1956	121.70	10.39
1957	101.97	21.38
1958	85.37	21.16
1959	81.39	1.77
1960	89.35	25.96
1961	67.09	22.55
1962	108.67	31.11
1963	103.06	23.79
1964	119.90	18.51
1965	127.75	88.10
1966	117.35	23.79
1967	105.24	26.64
1968	97.81	29.78

[a]Definition of variables: T, years; V_1, maximum belt horsepower of tractors produced; V_2, minimum belt horsepower of tractors produced.

Table A1-2.2

The Growth of Hydro- and Thermoelectric Power in Canada, 1917–1972

Year	Installed capacity (MW)			Thermal total	Largest machine size (MW)	
	Hydro	Thermal	Total		Hydro	Thermal
1917	1,232	144	1,376	10.47	18.0	0.8
1918	1,255	119	1,374	8.66	18.0	2.0
1919	1,295	216	1,511	14.30	18.0	5.0
1920	1,309	208	1,517	13.71	18.0	5.0
1921	1,362	212	1,574	13.47	18.0	5.0
1922	1,575	221	1,796	12.31	36.0	5.0
1923	1,703	215	1,918	11.21	36.0	5.0
1924	2,021	230	2,251	10.22	44.0	5.0
1925	2,548	244	2,792	8.74	46.75	5.0
1926	2,691	251	2,942	8.53	46.75	5.0
1927	2,965	256	3,221	7.95	46.75	5.0
1928	3,317	257	3,574	7.19	46.75	5.0
1929	3,521	283	3,804	7.44	46.75	10.0
1930	3,837	320	4,157	7.70	46.75	10.0
1931	—	—	4,394	—	46.75	10.0
1932	—	—	4,870	—	46.75	10.0
1933	4,704	375	5,079	7.38	46.75	10.0
1934	4,894	374	5,268	7.10	46.75	10.0
1935	5,079	375	5,454	6.88	46.75	10.0
1936	5,080	379	5,459	6.94	46.75	10.0
1937	5,239	384	5,623	7.30	46.75	10.0
1938	5,337	385	5,722	6.73	46.75	15.0
1939	5,402	418	5,820	7.18	46.75	15.0
1940	5,646	419	6,065	6.91	46.75	25.0
1941	5,807	424	6,231	6.80	46.75	25.0
1942	6,143	430	6,573	6.54	60.0	25.0
1943	6,867	443	7,310	6.06	60.0	25.0
1944	6,914	471	7,385	6.38	60.0	25.0
1945	6,876	465	7,341	6.33	60.0	25.0
1946	6,997	465	7,462	6.23	60.0	25.0
1947	6,812	490	7,302	6.71	60.0	25.0
1948	7,064	559	7,623	7.33	60.0	25.0
1949	7,440	681	8,121	8.39	60.0	30.0
1950	8,228	706	8,934	7.90	60.0	30.0
1951	8,793	927	9,720	9.54	60.0	100.0
1952	9,365	1,245	10,610	11.73	60.0	100.0
1953	10,014	1,670	11,684	14.29	60.0	100.0
1954	10,788	1,686	12,474	13.52	97.60	100.0
1955	11,591	1,825	13,416	13.60	97.60	100.0
1956	13,426	2,424	15,850	15.29	114.0	100.0
1957	14,518	2,650	17,168	15.44	114.0	100.0
1958	15,684	2,874	18,558	15.49	114.0	100.0
1959	17,537	3,572	21,109	16.92	114.0	200.0
1960	18,657	4,391	23,048	19.05	148.5	200.0
1961	19,019	5,072	24,091	21.05	148.5	300.0
1962	19,339	5,629	24,968	22.54	148.5	300.0
1963	20,101	6,199	26,300	23.57	148.5	300.0
1964	20,315	6,715	27,030	24.84	148.5	300.0
1965	21,772	7,575	29,347	25.81	148.5	300.0
1966	22,438	8,326	30.764	27.06	148.5	300.0
1967	23,354	9,613	32,967	29.16	161.5	300.0
1968	24,960	10,951	35,911	30.49	227.0	300.0
1969	27,033	12,560	39,593	31.72	227.0	500.0
1970	28,299	14,527	42,826	33.92	227.0	500.0
1971	30,601	16,074	46,675	34.44	475.0	540.0
1972	32,345	16,987	49,332	34.43	475.0	540.0

Table A1-2.3

Diffusion of Combines with Cornheads and of Mechanical Cornpickers

Years	Stock [a] of combines (000)	Stock [a] of cornpickers (000)	Annual sales of combines	Annual sales of cornpickers
T	V_1	V_2	V_3	V_4
1920	4	10		
1930	61	50		
1940	190	110		
1941	225	120		
1942	275	130		
1943	320	138		
1944	345	146		
1945	275	168		
1946	420	203		
1947	465	236		
1948	535	299		
1949	620	372		
1950	714	456		
1951	810	522		
1952	887	588		
1953	930	630		
1954	965	660		
1955	980	688		
1956	1005	715	4015	34170
1957	1015	740	5264	40933
1958	1030	755	6094	35711
1959	1045	775	5767	35242
1960	1042	792	10373	25554
1961	980	740	7751	22014
1962	960	730	8360	22565
1963	940	720	12944	21255
1964	920	705	17346	17447
1965	910	690	20653	13355
1966	888	686	25576	13232
1967	867	680	22829	13560
1968	847	673	18616	13797
1969	820	657	16828	8976
1970	790	635	15855	6813
1971	760	613	20230	7579
1972	725	593		

[a]On January 1.

Table A1-2.4

Diffusion of the Diesel-Powered Tractors in the United States

Year	Annual Production of Diesel-Powered Tractors	Total Tractor Production	$F/1-F^a$
1955	41,506	330,141	0.144
1956	26,762	214,654	0.142
1957	37,352	229,050	0.195
1958	55,864	241,269	0.301
1959	79,548	259,916	0.441
1960	62,033	152,187	0.688
1961	80,920	171,417	0.894
1962	80,293	188,101	0.745
1963	98,609	203,449	0.941
1964	115,439	213,221	1.181
1965	138,247	244,050	1.307
1966	157,352	270,687	1.389
1967	158,741	242,215	1.902
1968	147,744	213,199	2.257
1969	135,118	195,704	2.230
1970	122,622	171,603	2.503
1971	126,857	167,484	3.122

$^a F$ is the ratio of diesel-powered tractor production to total tractor production.

Table A1-2.5

Diffusion of the Oxygen Steel Process in Austria, 1952–1967

Year	Share of oxygen process in total steel output (%) = 100F	$F/1-F$
1952	0.6	0.006
1954	35.9	0.560
1956	40.9	0.692
1958	48.7	0.949
1960	56.1	1.278
1962	61.6	1.604
1964	61.5	1.597
1966	61.3	1.584
1967	67.0	2.030

Table A1-2.6

Diffusion of the Continuous Process of Casting Steel in Austria, 1952–1968

Year	Share of continuous cast steel in total crude steel output (%) = 100F	F/1 − F
1952	0.05	0.000500
1954	0.25	0.002506
1956	0.27	0.002707
1958	0.51	0.005126
1960	0.79	0.007963
1962	1.10	0.011122
1964	1.27	0.012863
1966	1.24	0.012556
1968	2.80	0.028807

Table A1-2.7

Diffusion of the Shuttleless Looms in the Weaving of
Cotton and Man-Made Fibers in Germany, 1958–1968

Year	Shuttleless looms as percentage of all looms	F/1 − F
1958	0.5	0.00503
1960	0.8	0.00806
1962	1.3	0.01317
1964	1.9	0.01937
1966	2.4	0.02459
1968	3.1	0.03199

Table A1-2.8

Diffusion of Tunnel Kilns in Brickmaking in Sweden, 1950–1968

Year	Share of tunnel kilns in total output (%) = 100F	F/1 − F
1950	6	0.06383
1954	12	0.13636
1958	25	0.33333
1960	31	0.44928
1962	48	0.92308
1964	85	5.66667
1966	92	11.5
1968	94	15.66667

Table A1-2.9

Diffusion of New Methods of Steel-Plate marking
and Cutting in Shipbuilding in Sweden, 1950–1968

Year	Share of steel cut by new methods in total (%) = 100F	F/1 – F
1950	2	0.02041
1955	17	0.20482
1960	52	1.08333
1962	62	1.63158
1964	71	2.44828
1966	80	4
1968	80	4

Table A1-2.10

Diffusion of Automatic Transfer Lines (ATL) for Car
Engines in Sweden, 1955–1968

Year	Share of total output produced by ATL (%) = 100F	F/1 – F
1955	15	0.17647
1960	55	1.22222
1962	65	1.85714
1964	68	2.12500
1966	91	10.11111
1968	97	32.33333

Table A1-3.1

Historical Data on the Development of Farm Tractor Technology, (1920–1968)[a]

T	V_1	V_2	V_3	V_4	V_5	V_6	V_7	V_8
1920	4.49	52.17	7998.7	4.72	4.72	6604.00	149	—
1921	4.84	50.95	7767.9	5.31	5.31	7284.00	146	249.97
1922	5.54	54.19	7182.0	6.06	6.06	8230.00	145	222.08
1923	5.53	52.25	8437.1	5.51	5.51	9499.00	144	197.95
1924	6.04	53.99	10257.9	5.35	5.35	10621.00	144	210.57
1925	5.83	53.09	7895.7	4.65	4.65	12201.00	145	192.99
1926	5.77	48.03	6743.4	5.72	5.72	13904.00	145	195.68
1927	5.82	48.62	6507.2	5.63	5.63	15750.00	147	198.05
1928	5.78	54.95	6801.9	5.65	5.65	17272.00	149	194.19
1929	6.01	56.10	6181.9	6.13	6.13	19231.00	150	194.19
1930	6.31	57.99	7487.7	6.21	6.21	20992.00	157	203.71
1931	5.98	60.64	6933.6	5.22	5.22	21661.00	151	262.30
1932	7.57	68.49	13646.4	3.84	3.84	21832.00	152	330.53
1933	8.25	65.58	13114.1	4.51	4.51	22053.00	152	351.57
1934	7.36	63.99	6885.0	4.85	4.85	22654.00	154	343.27
1935	8.42	63.94	9546.1	4.92	4.92	24035.00	155	328.41
1936	8.16	64.09	11323.3	5.23	5.23	25974.00	157	319.00
1937	7.67	62.19	9214.5	5.64	5.64	28352.00	159	286.14
1938	8.95	67.01	6654.7	5.68	5.68	30076.00	162	322.67
1939	9.28	68.61	9203.5	4.65	4.65	31932.00	165	297.69
1940	10.51	69.35	10230.0	4.82	4.82	34426.00	175	277.68
1941	10.03	70.79	6031.7	6.73	6.73	37560.00	171	234.12
1947	8.77	70.25	5207.1	5.00	7.53	52185.17	196	125.19
1948	9.19	71.45	7966.8	5.21	7.37	57481.04	199	136.21
1949	9.93	70.42	7227.1	5.28	6.94	63036.27	202	155.65
1950	10.05	68.95	8838.1	5.23	7.65	68023.95	216	156.46
1951	10.45	69.56	9393.4	4.67	6.45	73698.41	222	150.71
1952	9.47	72.54	7870.7	5.01	7.55	77844.01	232	148.38
1953	9.99	72.12	6616.0	5.22	7.45	81747.86	242	139.69
1954	10.99	69.57	7938.0	5.48	8.35	84205.41	242	141.96
1955	10.90	71.77	9550.8	5.21	7.73	87506.82	258	141.20
1956	10.22	72.54	10348.6	4.90	6.38	89653.36	265	140.72
1957	10.62	74.22	8360.9	5.23	8.05	91943.86	273	145.35
1958	10.59	74.08	8613.8	5.38	8.69	94356.55	280	143.44
1959	9.54	73.12	7240.4	5.26	7.98	96955.71	303	144.23
1960	10.62	74.55	9680.8	5.50	8.04	98477.58	297	139.76
1961	10.49	79.55	7855.8	5.07	7.37	100191.75	306	136.64
1962	10.95	75.41	9765.0	5.45	8.41	102072.76	314	
1963	10.15	76.03	11840.9	5.82	8.66	104107.25	322	
1964	11.15	82.26	12824.6	5.72	8.23	106239.46	352	
1965	11.03	83.09	9326.2	5.65	8.50	108679.96	340	
1966	10.22	75.34	10207.2	6.02	8.22	111386.83	348	
1967	11.01	66.05	10090.0	6.02	8.56	113808.98	355	
1968	11.52	73.97	11657.1	6.15	9.15	115940.97	363	

[a]Definition of variables: T, years; V_1, fuel-consumption efficiency in horsepower-hours per gallon; V_2, mechanical efficiency (ratio of drawbar horsepower to belt or PTO horsepower); V_3, ballasted weight in pounds; V_4, belt or PTO horsepower per 1000 lb of ballasted tractor weight; V_5, belt or PTO horsepower per 1000 lb of unballasted tractor weight; V_6, cumulated tractor production quantities (in hundreds); V_7, acreage per farm (acres); V_8, an index of tractor price in relation to price of labor.

Table A1-3.2

Historical Data on the Development of Locomotive Technology (1904–1967)[a]

T	V_1	V_2	V_3	V_4
1904	22804	297073	46743	8593
1905	23666	306797	48357	14084
1906	24741	317083	51672	21036
1907	25781	327975	55388	28398
1908	26356	333646	57698	30740
1909	26601	342351	58219	33627
1910	27282	351767	60019	38382
1911	28291	362824	62463	41912
1912	29049	371238	63463	46827
1913	30258	379508	65597	52159
1914	31006	387208	67012	54394
1915	31501	391142	66502	56479
1916	32380	394944	65314	60554
1917	33932	400353	66070	66000
1918	34995	402343	67936	72475
1919	35789	403891	68977	75747
1920	36365	406580	68942	79419
1921	36935	407531	69122	81242
1922	37441	409359	68518	82776
1923	39177	412993	69414	86561
1924	39891	415028	69486	88597
1925	40666	417954	68098	89882
1926	41886	421341	66847	91652
1927	42798	424737	65348	92828
1928	43838	427750	63311	93575
1929	44801	429054	61257	94736
1930	45225	429883	60189	95870
1931	45764	429823	58652	96092
1932	46299	428402	56732	96215
1933	46916	425664	54228	96278
1934	47712	422401	51423	96388
1935	48367	419228	49541	96593
1936	48972	416381	48009	96795
1937	49412	414572	47555	97410
1938	49803	411324	46544	97756
1939	50395	408350	45172	98111
1940	50905	405975	44333	98671
1941	51217	403625	44375	99778
1942	51811	399627	44671	100796
1943	52451	398730	45406	101960
1944	52822	398437	46305	103398
1945	53217	398054	46253	106611
1946	53735	398037	45511	
1947	54506	397355	44344	
1948	55170	397203	44474	
1949	56333	397232	43272	
1950	57075	396380	42951	
1951	58476	395831	42473	

Table A1-3.2 (continued)

Historical Data on the Development of Locomotive Technology (1904–1967)[a]

T	V_1	V_2	V_3	V_4
1952	59966	394631	39697	
1953	61339	393736	37251	
1954	63152	392580	35033	
1955	65005	390965	33533	
1956	68745	389668	32593	
1957	61515	386978	32391	
1958	61312	385264	31616	
1959	61408	383912	31539	
1960	61314	381745	31178	
1961	61969	379415	30889	
1962	61415	376290	30701	
1963	61533	374522	30506	
1964	62311	372300	30296	
1965	63096	370636	30061	
1966	70900	370104	30124	
1967	65267	368030	29874	

[a]Definition of variables: T, years; V_1, tractive effort in pounds; V_2, total railroad mileage; V_3, number of locomotives in service; V_4, cumulated locomotive production.

Table A1-3.3

Historical Data on the Development of Tank Ship Technology (1899–1973)[a]

T	V_1	V_2	V_3	V_4
1899	4869.04	9.03	109	
1900	5015.57	9.13	116	
1901	5181.22	9.21	126	
1902	5322.92	9.26	135	
1903	5596.33	9.33	153	
1904	5566.94	9.35	159	
1905	5562.52	9.35	162	
1906	5572.57	9.35	167	
1907	5623.30	9.33	171	
1908	5914.43	9.35	194	
1909	6000.85	9.38	208	
1910	6035.10	9.40	213	
1911	5992.22	9.40	215	
1912	6103.93	9.48	245	
1913	6489.82	9.56	300	
1914	6881.99	9.63	347	8
1915	7071.29	9.52	347	12
1916	7288.77	9.63	372	36
1917	7659.59	9.75	417	68
1918	7883.19	9.96	467	102
1919	8179.70	10.01	540	144
1920	8445.48	10.06	645	224
1921	8924.35	10.13	791	328
1922	9039.11	10.14	856	334
1923	9082.68	10.16	888	336
1924	9098.72	10.17	896	337
1925	9104.19	10.16	933	337
1926	9126.76	10.17	973	338
1927	9082.29	10.19	1044	341
1928	9172.89	10.23	1147	345
1929	9286.98	10.29	1191	346
1930	9519.09	10.38	1273	357
1931	9709.06	10.46	1366	362
1932	9775.22	10.49	1373	362
1933	9867.22	10.52	1363	362
1934	9923.89	10.55	1357	362
1935	10041.90	10.61	1381	364
1936	10208.47	10.73	1421	372
1937	10419.51	10.87	1486	387
1938	10566.39	11.01	1571	405
1939	10739.92	11.18	1637	416
1940	10179.99	11.28	1689	432
1941	10944.43	11.40	1550	460
1942	11165.79	11.75	1388	521
1943	11633.56	12.20	1556	752
1944	12255.45	12.67	1768	992
1945	12255.45	12.67	1768	1180
1946	12255.45	12.67	1768	1188

Table A1-3.3 (continued)

Historical Data on the Development of Tank Ship Technology (1899–1973)[a]

T	V_1	V_2	V_3	V_4
1947	12255.45	12.67	1768	1191
1948	12722.11	12.99	1872	1197
1949	12722.11	12.99	1872	1230
1950	12722.11	12.99	1872	1253
1951	12722.11	12.99	1872	1257
1952	12722.11	12.99	1872	1265
1953	14281.37	13.63	2502	1287
1954	15041.12	13.93	2602	1314
1955	15525.18	14.00	2681	1316
1956	16156.59	14.15	2778	1322
1957	17070.07	14.40	2954	1338
1958	18004.13	14.60	3146	1359
1959	19156.89	14.82	3276	1385
1960	20153.19	15.06	3264	1396
1961	21187.38	15.20	3250	1403
1962	22091.44	15.34	3259	1406
1963	23419.03	15.32	3279	1412
1964	25342.66	15.58	3359	1416
1965	27116.41	15.61	3436	1418
1966	29202.33	15.69	3524	1419
1967	31100.47	15.74	3613	1419
1968	33941.19	15.77	3775	1422
1969	37510.66	15.75	3893	1430
1970	41964.02	15.81	4002	1437
1971	46087.71	15.77	4207	
1972	50945.25	15.76	4342	
1973	56260.33	15.77	4563	

[a]Definition of variables: T, years; V_1, average deadweight tonnage; V_2, average service speed in knots; V_3, number of tank ships in service; V_4, cumulated tank ship production.

Table A1-3.4

Historical Data on the Development of Aircraft Technology 1932–1965[a]

T	V_1	V_2	V_3	V_4	V_5	V_6
1932	6.6	109	374	28956	2117	
1933	7.6	116	336	28283	2188	
1934	8.9	127	334	28609	2297	
1935	10.3	142	321	29190	2368	
1936	10.7	149	257	29797	2342	
1937	12.5	153	235	32006	2299	
1938	13.9	153	231	34879	2374	
1939	14.7	153	261	36654	2280	
1940	16.5	155	338	42757	2331	
1941	17.5	160	342	45163	2484	
1942	17.9	159	164	41596	2809	
1943	18.4	154	191	42537	2769	
1944	19.1	156	271	47384	3427	
1945	19.7	153	397	48516	4026	
1946	25.3	169	634	53981	4490	
1947	29.9	170	723	62215	5759	105
1948	32.4	176	762	68702	6414	220
1949	34.7	178	782	72667	6484	278
1950	37.1	180	792	77440	6403	329
1951	39.1	183	821	78913	6237	392
1952	42.2	189	914	77894	6042	540
1953	45.6	196	950	78384	6760	667
1954	49.6	204	964	78294	6977	742
1955	51.5	208	1010	78992	6839	797
1956	52.1	210	1098	84189	7028	922
1957	53.7	214	1211	87550	6412	1101
1958	55.5	219	1244	89569	6018	1183
1959	58.7	223	1258	95063	6426	1349
1960	65.4	235	1226	98008	6881	1446
1961	72.9	252	1196	102309	7715	1571
1962	79.4	274	1124	104673	8084	1637
1963	83.4	286	1064	105003	8814	1660
1964	86.1	296	1105	105059	9490	1779
1965	89.2	314	1122	104870	9566	1914

[a]Definition of variables: T, years; V_1, seating capacity; V_2, airspeed in miles per hour; V_3, number of aircraft in service; V_4, total route miles; V_5, number of airports and landing fields; V_6, cumulated addition of new aircraft.

Table A1-3.5

Historical Data on the Development of Computer Technology (1944–1967)[a]

T	V_1	V_2	V_3
1944	0.0379	0.035	1
1945	0.0068	0.035	2
1946	7.448	44.65	3
1947	0.0674	0.0296	4
1948	0.1712	0.774	5
1949	7.98	8.79	8
1950	202.03	137.35	13
1951	139.08	111.09	19
1952	141.32	84.345	28
1953	205.43	147.75	47
1954	145.22	113.86	66
1955	96.86	164.16	83
1956	1523.65	866.52	94
1957	2901.23	1907.08	111
1958	12452.1	9089.49	122
1959	7281.67	4977.39	137
1960	19819.99	10211.23	172
1961	24871.10	33049.21	197
1962	37737.24	16825.77	218
1963	69250.11	26975.56	237
1964	420995.84	268443.24	258
1965	296239.87	191866.88	283
1966	714200.35	557332.77	302
1967	570031.00	511924.75	310

[a]Definition of variables: T, years; V_1, capability for scientific computations in operations per second; V_2, capability for commercial computations in operations per second; V_3, cumulated production quantities.

Table A1-3.6

Technological Change in the Steam-Powered Electricity-Generating Plants (1920–1970)[a]

T	V_1	V_2	V_3	V_4
1920	.328	9.70	23489	3101
1921	.368	9.60	45800	2839
1922	.397	11.68	72379	3145
1923	.418	14.43	104472	3434
1924	.450	16.12	139427	3276
1925	.493	19.64	178794	3138
1926	.526	22.11	222216	3094
1927	.549	24.94	268831	3111
1928	.578	28.75	318201	3127
1929	.602	34.83	377166	3197
1930	.625	36.47	436459	2926
1931	.658	36.58	494144	2646
1932	.671	29.57	540066	2337
1933	.685	31.51	587775	2374
1934	.690	37.10	641714	2540
1935	.694	39.43	697858	2777
1936	.694	51.88	767217	3145
1937	.694	57.59	841108	3364
1938	.714	54.65	909531	3110
1939	.725	69.27	992314	3346
1940	.746	80.66	1085316	3601
1941	.746	100.64	1197635	4003
1942	.769	109.53	1318114	4257
1943	.769	129.32	1460495	4687
1944	.775	140.78	1612823	4699
1945	.769	132.86	1753258	4487
1946	.775	136.15	1895670	4441
1947	.763	166.99	2070170	4984
1948	.769	188.45	2267098	5191
1949	.806	187.74	2464976	4862
1950	.840	218.40	2694519	4984
1951	.877	255.01	2961771	5124
1952	.909	281.93	3252156	5051
1953	.943	320.40	3585697	5098
1954	1.010	345.30	3946531	4862
1955	1.053	411.60	4376650	5037
1956	1.064	457.62	4851202	5108
1957	1.075	476.71	5348414	5056
1958	1.111	476.46	5849178	4748
1959	1.124	536.62	6418533	—
1960	1.136	564.39	7023564	4635
1961	1.163	600.22	7661000	4540
1962	1.163	637.96	8342340	4583
1963	1.163	694.59	9088332	4354
1964	1.163	748.05	9890239	4427
1965	1.163	801.79	10746551	4469
1966	1.149	870.44	11690981	4617
1967	1.149	859.87	12678972	4510
1968	1.149	913.57	13780739	4568
1969	1.136	932.71	14967149	4602
1970	1.099	960.97	16245240	4490

[a]Definition of variables: T, years; V_1, average fuel-consumption efficiency in kilowatt-hours per pound of coal; V_2, average scale of plant utilization (i.e., the ratio of net production of steam-powered electrical energy in millions of kilowatt-hours to number of steam-powered plants); V_3, cumulated net production of steam-powered electrical energy in millions of kilowatt-hours; V_4, production per kilowatt of capacity in kilowatt-hours.

Table A1-3.7

Technological Change in Internal-Combustion Type Electric Power Plants (1920–1970)[a]

T	V_1	V_2	V_3	V_4
1920	3.94	0.027	0.549	156
1921	4.55	0.032	0.589	322
1922	4.78	0.032	0.586	500
1923	5.13	0.034	0.555	696
1924	5.49	0.038	0.555	914
1925	6.06	0.042	0.591	1200
1926	6.37	0.044	0.668	1528
1927	6.54	0.047	0.610	1857
1928	6.99	0.048	0.740	2407
1929	7.30	0.051	0.750	2974
1930	7.58	0.053	0.648	3603
1931	7.81	0.056	0.638	4240
1932	8.20	0.057	0.585	4833
1933	8.20	0.058	0.565	5407
1934	8.33	0.058	0.591	6042
1935	8.47	0.059	0.687	6813
1936	8.47	0.058	0.826	7712
1937	8.40	0.058	0.868	8721
1938	8.85	0.058	0.947	9831
1939	10.00	0.061	1.031	11126
1940	8.93	0.061	1.173	12640
1941	8.93	0.059	1.242	14246
1942	8.70	0.060	1.244	15875
1943	9.01	0.059	1.292	17621
1944	9.17	0.060	1.429	19637
1945	9.17	0.061	1.572	21618
1946	9.26	0.061	1.788	23978
1947	8.93	0.062	2.098	26791
1948	9.35	0.063	2.414	30091
1949	10.20	0.067	2.537	33564
1950	10.64	0.071	2.695	37224
1951	10.64	0.074	2.760	40895
1952	10.53	0.075	2.975	44632
1953	11.11	0.077	3.140	48522
1954	11.24	0.081	3.166	52305
1955	11.76	0.083	3.397	56249
1956	11.76	0.084	3.610	60336
1957	12.05	0.085	3.646	64398
1958	12.35	0.087	3.802	68470
1959	12.66	0.090	4.068	72908
1960	12.82	0.092	4.202	77455
1961	13.16	0.093	4.329	82135
1962	12.99	0.093	4.576	87013
1963	13.33	0.094	4.833	92059
1964	13.33	0.094	4.859	97069
1965	13.33	0.095	5.135	102158
1966	13.33	0.096	5.227	107322
1967	13.16	0.096	4.770	112178
1968	13.16	0.097	5.103	117363
1969	13.16	0.095	5.514	122943
1970	12.99	0.095	6.026	129005

[a]Definition of variables: T, years; V_1, average fuel-consumption efficiency in kilowatt-hours per gallon of oil; V_2, average fuel-consumption efficiency in kilowatt-hours per cubic foot of gas; V_3, average scale of plant utilization (i.e., the ratio of net production of electrical energy by internal-combustion type plants in millions of kilowatt-hours to total number of these plants); V_4, cumulated net production of electrical energy produced by internal-combustion type plants in millions of kilowatt-hours.

Table A1-4.1

Speed Characteristics of Computers Introduced in 1961

Number	Computer name	Scientific computation speed[a]	Commercial computation speed[a]
1	AN/TYK 7V	4713.000	9077.000
2	Bendix G 20 and 21	37260.000	17060.000
3	BRLESC	47240.000	28550.000
4	CCC-DDP 19 (card)	5159.000	3027.000
5	CCC-DDP 19 (magnetic tape)	7908.000	8073.000
6	CDC 160A	1015.000	1780.000
7	GE 225	6566.000	7131.000
8	General Mills Apsac Jan.	16.220	7.084
9	George II	298.000	675.100
10	Honeywell 290	354.300	182.800
11	Honeywell 400	1354.000	2752.000
12	IBM 1410	1673.000	4638.000
13	IBM Stretch (7030)	371700.000	631200.000
14	IBM 7074	41990.000	31650.000
15	IBM 7080	27090.000	30860.000
16	ITT Bank Loan Process	492.600	1916.000
17	NCR 390	2.034	10.430
18	RCA 301	323.000	1055.000
19	Recomp III	48.280	35.760
20	Rice University	7295.000	2378.000
21	RW 530	13460.000	5086.000
22	Univac Solid State 80/90 II	3199.000	3044.000
23	Univac 490	17770.000	15050.000
24	Univac 1000 and 1020	3861.000	3292.000
25	Univac 1206	20990.000	17700.000

[a]Speed is expressed as operations per second.

Table A1-4.2

Technical Characteristics of Passenger Ships 1840–1961

built	Vessel	Length (feet)	Beam (feet)	Tonnage	Horsepower
1840	*Acadia*	206	34	1,136	425
1850	*Atlantic*	276	45	2,860	850
1855	*Persia*	376	45	3,300	900
1858	*Great Eastern*	680	80	18,914	7,675
1861	*Scotia*	379	47	3,871	1,000
1881	*City of Rome*	560	52	8,415	17,500
1884	*Umbria*	502	57	8,128	13,000
1889	*Teutonic*	565	57	9,984	17,000
1893	*Campania*	601	65	12,884	30,000
1897	*Kaiser Wilhelm der Grosse*	627	66	14,349	31,000
1899	*Oceanic*	686	68	17,274	29,000
1900	*Deutschland*	661	67	16,502	35,000
1901	*Celtic*	681	75	21,179	14,000
1902	*Kaiser Wilhelm II*	684	72	19,361	40,000
1904	*Carmania*	650	72	19,566	18,900
1905	*Kaiserin Auguste Victoria*	678	77	25,160	18,000
1906	*Adriatic*	709	76	24,563	17,000
1906	*Kronprinzessin Cecile*	685	72	19,503	45,000
1907	*Mauretania*	762	88	30,696	68,000
1911	*Olympic*	853	93	46,439	55,000
1912	*Imperator*[a]	884	98	52,101	62,000
1914	*Aquitania*	869	97	45,647	60,000
1914	*Britannic*	852	94	48,158	50,000
1914	*Vaterland*[b]	908	100	48,943	72,000
1921	*Bismarck*[c]	916	100	56,621	80,000
1929	*Bremen*	899	102	51,656	95,000[d]
1930	*Europa*	890	102	49,746	95,000[d]
1931	*Empress of Britain*	733	98	42,348	60,000
1932	*Rex*	817	97	51,062	100,000[d]
1932	*Conte Di Savoia*	786	96	48,502	100,000[d]
1935	*Normandie*	1029	118	83,423	160,000
1936	*Queen Mary*	1018	119	81,235	200,000
1940	*Queen Elizabeth*	1031	119	83,673	200,000
1961	*France*	1035	111	66,348	160,000

[a]Later Berengaria.
[b]Later Leviathan.
[c]Later Majestic.
[d]Designed.

Table A1-4.3

Technical and Performance Characteristics of Nuclear Reactors[a]

Index[b]	V_1	V_2	V_3	V_4	V_5	V_6	V_7	V_8	V_9	V_{10}	V_{11}	V_{12}	V_{13}	V_{14}	V_{15}
1	13.8	12.5	1.43	1.27	66.7	8.53	2.08	3.66	2.6	272	277	2	16	1	3000
2	29.6	250	3.3	2.75	121	16.8	3.7	3.55	2.22	264	286	3	89	1	1500
3	31.8	460	3.66	3.44	160	19	5	4.35	2.30	177	285	2	97	1	1500
4	32.5	682	3.68	3.96	146	19.6	5.69	5.4	2.07	274	285	3	145	1	1800
5	33.6	801	3.66	4.03	140	18.8	15.6	5.38	2.15	186	285	2	137	1	1800
6	31.9	784	3.66	4.03	137	22	5.5	3.92	2.20	196	286	2	137	1	1500
7	33.7	849	3.66	4.07	170	21.1	5.56	4.97	2.17	277	286	2	137	1	1800
8	33.5	762	3.70	4.40	150	6.0	20	5.2	2.60	180	286	6	157	2	3000
9	32.8	550	3.66	3.3	117	20.2	4.65	3.94	1.9	216	283	2	89	1	1800
10	33.6	805	4.5	3.97	200	20.7	5.58	4.4	2.26	215	285	8	129	1	1500
11	33.7	847	3.66	4.07	145	21.13	5.54	5.4	2.25	215	285	2	137	1	1800
12	33.6	900	3.66	4.5	143	20.7	5.85	4.4	2.63	215	285	9	145	1	1500
13	31.9	784	3.66	4.03	140	22	5.5	3.92	2.20	196	286	2	137	1	1500
14	31.9	540	3.66	3.28	120	21	4.7	4.92	2.20	276	286	2	89	1	1800
15	31.9	784	3.66	4.03	137	22	5.5	3.92	2.20	196	286	2	137	1	1500
16	31.9	840	4.1	3.7	140	22	5.6	4.92	2.2	276	286	2	137	1	1800
17	32.9	524	3.66	3.3	120	21	4.7	4.9	2.2	276	286	2	89	1	1500
18	34	600	3.65	3.58	129	20	5.2	5.1	2.5	180	286	4	109	1	3000
19	31.9	784	3.66	4.03	137	22	5.5	3.92	2.2	196	286	2	137	1	1500
20	33.7	847	3.66	4.07	140.5	21.13	5.54	5.4	2.25	215	285	2	137	1	1800
21	33	727	3.66	3.79	134	20.96	5.4	3.35	2.68	215	286.4	6	113	1	3000
22	33.8	907	3.66	4.18	146	20.7	5.85	4.4	2.6	215	286	8	145	1	1500
23	33.6	636	3.66	3.47	140	21.3	5.1	5.2	2.2	215	295	2	97	1	1800
24	34.2	1316	3.66	4.99	166	21.85	6.7	5.1	2.6	215	286.4	10	205	1	1500
25	33.6	900	3.66	4.5	143	20.7	5.85	4.25	1.9	215	285	9	145	1	1500
26	33.5	866	3.66	4.07	175	5.5	21.1	3.37	2	218	286	2	137	1	1800
27	32.4	1118	3.6	4.8	159	6.4	22.1	4.97	2.17	216	282	2	185	1	1800
28	32.4	1118	3.6	4.8	159	6.4	22.1	4.97	2.17	216	282	2	185	1	1800

Table A1-4.3 (Continued)

Technical and Performance Characteristics of Nuclear Reactors[a]

Index[b]	V_1	V_2	V_3	V_4	V_5	V_6	V_7	V_8	V_9	V_{10}	V_{11}	V_{12}	V_{13}	V_{14}	V_{15}
29	31.9	450	3.07	2.87	248	11.3	4.05	4.29	3.4	291	314.5	3	45	1	1800
30	32.2	822	3.66	3.04	203	12.3	3.99	4.2	2.48	284	318	3	53	1	1800
31	33.2	622	2.99	3.05	199	12.5	4.4	3.85	3.18	288	316	4	49	1	1500
32	31.6	902	3.66	3.37	218	13.4	4.4	3.62	2.8	284	313	4	61	1	1800
33	32	822	3.66	3.04	203	12.3	3.99	4.2	2.5	284	318	3	53	1	1800
34	32.5	583	3.7	2.5	165	11.5	3.35	3.44	3.4	284	319	2	33	1	1800
35	31.4	1132	3.7	3.4	215	12.9	4.4	3.3	2.7	285	320	4	61	1	1800
36	32.9	477	2.65	2.676	185	7.5	3.81	4.9	3.1	295.4	319.4	2	28	1	3000
37	31	861	3.66	3.02	203	13	4.4	3.8	3.35	285	322	3	53	1	1800
38	33.3	526	3.66	3.28	214	11.4	4.77	5.7	2.45	290	318	4	57	1	1800
39	34.4	1218	3.7	3.4	279	13.3	4.4	1.05	2.8	292	322	4	61	1	1800
40	33.8	855	3.66	3.28	214	11.4	4.77	5.7	2.45	290	318	4	57	1	1800
41	34.8	1295	3.66	3.52	235	12.23	5.1	5	3.27	296	329	2	76	1	1500
42	32.5	1158	3.7	3.4	215	12.9	4.4	4.3	2.7	285	320	4	61	1	1800
43	33	1171	3.7	3.4	215	12.9	4.4	0.84	2.7	285	321	4	61	1	1800
44	31	861	3.66	3.02	203	13	4.4	3.8	3.35	285	322	3	53	1	1800
45	32.4	947	3.66	3.04	203	13.6	3.99	4.2	2.7	284	322	3	53	1	1800
46	32.5	805	2.99	3.24	210	10.4	4.8	4.8	2.5	295	323	3	45	1	3000
47	34.1	1218	3.7	3.4	279	13.3	5.2	1.05	2.8	292	322	4	61	1	1800
48	33.2	657	3.66	2.47	165	11.2	3.4	2.6	2.5	287.5	327	2	37	1	1800

[a] V_1 thermal efficiency (%)
V_2 plant size (MWe)
V_3 core length (m)
V_4 core size: height × diameter (m²)

V_5 vessel-wall thickness (mm)
V_6 vessel height (m)
V_7 vessel diameter
V_8 containment pressure (kg/cm²)

V_9 enrichment (%)
V_{10} coolant inlet temperature (°C)
V_{11} coolant outlet temperature (°C)
V_{12} no of primary pumps

V_{13} no. of control rods
V_{14} no. of turbines
V_{15} turbine speed (rpm)

[b] Index: Numbers 1–28 represent boiling-water reactors; numbers 29–48, pressurized-water reactors.

Appendix to Table A1-4.3

Reactors Included

Index	Reactor name	Country	Years[a]
1	JPDR II	Japan	8.63
2	Gundermmingen, Krb	West Germany	12.66
3	Fukushima No. 1	Japan	3.71
4	Millstone No. 1	United States	12.70
5	Copper	United States	11.73
6	Fukushima No. 2	Japan	5.73
7	Fitzpatrick	United States	10.73
8	Ringhals No. 1	Sweden	6.73
9	Arnold	United States	1.74
10	Brunsbuttel Kkb	West Germany	.74
11	Brunswick No. 1	United States	.76
12	Kkp No. 1 Philippsburg	West Germany	.74
13	Fukushima No. 3	Japan	12.74
14	Hamoaka No. 1		
	Chuba No. 1	Japan	11.74
15	Fukushima No. 4	Japan	8.76
16	Hamoaka No. 2	Japan	6.77
17	Onagawa No. 1	Japan	12.75
18	Barseback No. 1	Sweden	7.75
19	Fukushima No. 5	Japan	12.75
20	Brunswick No. 2	United States	.75
21	GKT, Tullnerfeld	Austria	.76
22	Isar, Kki	West Germany	.76
23	Chinshan No. 2	Taiwan	12.76
24	Geesthacht, Krummel	West Germany	.77
25	Kkp No. 2 Philippsburg	West Germany	.77
26	Shoreham	United States	—
27	Newbold Island No. 1	United States	5.79
28	Newbold Island No. 2	United States	10.80
29	San Onofre No. 1	United States	1.68
30	Surry No. 1	United States	12.72
31	Kks No. 1 Stade	West Germany	5.72
32	Indian Point No. 2	United States	9.73
33	Surry No. 2	United States	.73
34	Kewaunee Dukepower 2	United States	9.73
	Ocanee		
35	Burlington No. 1 Salem	United States	3.75
36	Borssele	Netherlands	7.73
37	Farley No. 2	United States	4.77
38	Midland No. 1	United States	2.80
39	Watts Bardam 2	United States	.78
40	Midland No. 2	United States	2.79
41	Mucheim-Karlich	West Germany	.78
42	Burlington No. 2 Salem	United States	3.76
43	Sequoyah No. 2	United States	12.75
44	Farley No. 1	United States	4.75
45	North Anna No. 2	United States	.76
46	Neckar, Gkn	West Germany	2.76
47	Watts Bardam No. 1	United States	5.77
48	Angra Dos Reio	Brazil	3.77

[a]The number of years since the reactor was planned/built is shown in each case.

Table A1-4.4

Efficiencies of Energy Use

Source of energy	Efficiency of use (%)
Electricity	100.0
Fuel wood	20.4
Lumber mill waste	40.0
Bagasse	39.9
Other vegetal fuels	20.0
Coal and briquettes	39.4
Lignite	48.6
Lignite briquettes	44.6
Peat	29.6
Coke	52.4
Liquefied petroleum gas	57.1
Aviation gasoline and jet fuel	20.0
Motor spirit	19.9
Kerosene	45.8
Distillate fuel oil	47.3
Residual fuel oil	41.7
Refinery gas	80.0
Natural gas	71.6
Oven gas	74.2

References

CHAPTER 1

Abramovitz, M.
1956.   "Resource and Output Trends in the U.S. since 1870," *American Economic Review* Papers and Proceedings, **46**:5–23.

Boulding, K. E.
1977.   "The Universe as a General System," *Behavioral Science* **22**:299–306.

Bruton, H. J.   "Contemporary Theorizing on Economic Growth," in *Theories of Economic*
1956.   *Growth"* (B. F. Hoselitz et al., eds.), New York: Free Press. pp. 239–298.

Christensen, L. R., and Jorgenson, D. W.
1970.   "U.S. Real Product and Real Factor Input, 1929–1967," *Review of Income and Wealth* **16**:19–50.

Davies, J. T.   "The Simple Laws of Science and History," in *The Critical Approach to Science*
1964.   *and Philosophy* (*Essays in Honour of Karl R. Popper*) (M. Bunge, ed.), New York: Free Press. pp. 255–265.

Ehrenberg, A. S. C.
1975.   *Data Reduction*. New York: Wiley.

Gilfillan, S. C.   *The Sociology of Invention* (reprinted 1963). Chicago: Follett; MIT paperback
1935.   edition, March 1970.

Gilfillan, S. C.   "Prediction of Technical Change," *Review of Economics and Statistics* **34**:
1952.   368–385.

Griliches, Z.   "Hybrid Corn: An Exploration in the Economics of Technological Change,"
1957.   *Econometrica* **25**:501–522.

Griliches, Z.   "The Sources of Measured Productivity Growth: U.S. Agriculture 1940–1960,"
1963.   *Journal of Political Economy* **71**:331–346.

Ijiri, Y., and Simon, H. A.
1977.   *Skew Distributions and the Sizes of Business Firms*. Amsterdam: North Holland.

Israd, W.   *Location and Space Economy*. London: Chapman & Hall.
1956.

Jorgenson, D. W., and Griliches, Z.
1967.   "The Explanation of Productivity Change," *Review of Economic Studies* **34**: 249–283.

Kendall, M. G.   "Natural Laws in Social Sciences," *Journal of the Royal Statistical Society*
1961.   **A124**:1–19.

Mansfield, E. "Technical Change and the Rate of Imitation," *Econometrica* **29**:741–765.
1961.

Nelson, R. R., and Winter, S. G.
1974.   "Neoclassical versus Evolutionary Theories of Economic Growth: Critique and Prospectus," *Economic Journal* **84**:886–905.

Ogburn, W. F. *Social Change.* New York: Viking Press.
1922.

Richardson, L. F.
1960.   *Statistics of Deadly Quarrels.* Pittsburgh: Boxwood Press.

Rogers, E. M., and Shoemaker, F.
1971.   *Communication of Innovation: A Cross-Cultural Approach.* New York: Free Press.

Sahal, D.   "Evolving Parameter Models of Technology Assessment," *Journal of the*
1975.   *International Society for Technology Assessment* **1**:11–20.

Sahal, D.   "The Multidimensional Diffusion of Technology," *Journal of Technological*
1977.   *Forecasting and Social Change* **10**:277–298; errata in **11**:391–392 (1978).

Schmookler, J. *Invention and Economic Growth.* Cambridge: Harvard University Press.
1966

Schumpeter, J. *The Theory of Economic Development.* Cambridge, Mass.: Harvard University
1934.   Press.

Schumpeter, J. *Business Cycles: A Theoretical, Historical and Statistical Analysis of Capitalist*
1939.   *Process.* New York: McGraw-Hill.

Simon, H. A.   "On Judging the Plausibility of Theories," in *Logic, Methodology and*
1968.   *Philosophy of Sciences* Vol. 3, pp. 25–45. (Van Roostelaar and Staal, eds.), Amsterdam: North Holland. Reprinted in H. A. Simon's *Models of Discovery*, 1977. Dordrecht: D. Reidel.

Smith A.   *The Wealth of Nations.* New York: Random House.
1937.

Solow, R.   "Technical Change and the Aggregate Production Function," *Review of Eco-*
1957.   *nomics and Statistics* **39**:312–320.

Star, S.   "Accounting for Growth of Output," *American Economic Review* **64**:123–135.
1974.

Steindl, J.   *Random Processes and the Growth of Firms.* London: Griffin.
1965.

Zipf, G. K.   *Human Behavior and the Principle of Least Effort.* Reading, Mass.: Addison-
1949.   Wesley.

CHAPTER 2

Atkinson, A. B., and Stiglitz, J. E.
1969.   "A New View of Technological Change," *Economic Journal* **79**:573–578.

Baker, C. L.   "Trends to Bigger Models," *Power Farming* November, p. 15.
1970.

Blaug, M.   "A Survey of the Theory of Process-Innovations," *Economica* **30**:13–32.
1963.

Blaug, M.   *The Cambridge Revolution: Success or Failure?* London: The Institute of
1975.   Economic Affairs.

Bloom, G. F.   "Union Wage Pressure and Technological Discovery," *American Economic*
1951.   *Review* **41**:602–617.

Brozen, Y.   "Trends in Industrial Research and Development," *Journal of Business*
1960.   **33**:204–217.

Carlsson, B.   "Economies of Scale and Technological Change: An International Comparison
1978.   of Blast Furnace Technology," in *Welfare Aspects of Industrial Markets* (A. P.
                 Jacquemin and H. W. de Jong, eds.), pp. 303–325. Reprinted in *Publications of*
                 the Industrial Institute for Economic and Social Research, Stockholm.

Chenery, H. B.
1949.   "Engineering Production Functions," *Quarterly Journal of Economics*
                 63:507–531.

Clausen, H.
1961.   "The Borderland Between Science and Engineering," *The Engineer* June 2, pp.
                 896–898.

Cramer, J. S.   *Empirical Econometrics.* Amsterdam: North Holland.
1969.

Eckaus, R. S.   "Notes on Invention and Innovation in Less Developed Countries," *American*
1963.   *Economic Review* **56**:98–109.

Enos, J. L.   "A Measure of the Rate of Technological Progress in the Petroleum Refining
1957–1958.   Industry," *Journal of Industrial Economics* **6**:180–197.

Fellner, W.   "Two Propositions in a Theory of Induced Innovations," *Economic Journal*
1961.   **71**:305–308.

Francis, A. J.   "Engineering Education and Engineering Practice," *The Engineer* May 26,
1961.   pp. 848–852.

Freeman, C.   "Measurement of Output of Research and Experimental Development,"
1969.   Unesco, COM/CONF. 22/8, Paris.

Garg, M. K.   "The Upgrading of Traditional Technologies in India," in *Appropriate Tech-*
1976.   *nology* (N. Jéquier, ed.) pp. 156–170. Paris: OECD.

Gazis, D. C.   "Influence of Technology on Science: A Comment on Some Experiences at
1979.   IBM Research," *Research Policy* **8**:244–259.

Griliches, Z.   *Price Indexes and Quality Change,* 2nd ed. Cambridge, Mass.: Harvard Univer-
1971.   sity Press.

Gustafson, W. E.
1962.   "Research and Development, New Products and Productivity Change," *Amer-*
                 ican Economic Review **52**:177–189.

Harrod, R. F.   *Towards a Dynamic Economics.* London: Macmillan.
1948.

Hicks, J. R.   *The Theory of Wages* (revised 1963). London: Macmillan.
1932.

Hollander, S.   *The Sources of Increased Efficiency.* Cambridge, Mass.: MIT Press.
1965.

Hughes, W. R.   "Scale Frontiers in Electric Power," in *Technological Change in Regulated In-*
1971.   *dustries* (W. M. Capron, ed.), pp. 44–85. Washington, D.C.: The Brookings In-
                stitution.

Hunter, L. C.   *Steamboats on the Western Rivers, An Economic and Technological History.*
1949.   Cambridge, Mass.: Harvard University Press.

International Harvester Spokesman.
1972.   "Fifty Years of the Farmall," *Implement & Tractor* May 21, pp. 9–10.

Kaldor, N.   "A Model of Economic Growth," *Economic Journal.* **67**, December.
1957.

Kaldor, N.   "Capital Accumulation and Economic Growth," in *The Theory of Capital*
1961.   (F. A. Lutz and D. C. Hague, eds.), pp. 177–220. New York: Macmillan.

Kuznets, S.   "Innovative Activity: Problems of Definition and Measurement," in *The Rate*
1962.   *and Direction of Inventive Activity* (R. R. Nelson, ed.), pp. 19–51. Princeton,
                N.J.: Princeton University Press.

Laithwaite, R. E.
1977.   "Biological Analogues in Engineering Practice," *Interdisciplinary Science
                Reviews* **2**:100–108.

Lenz, R., Jr.   *Technological Forecasting,* ASD-TDR-62-414. Dayton, Ohio: Aero-
1962.   nautical Systems Division, Wright-Patterson Air Force Base.

Mak, J., and Walton, G. M.
1972.   "Steamboats and the Great Productivity Surge in River Transportation," *Jour-
                nal of Economic History* **32**:619–640.

Martino, J. P.   *Technological Forecasting for Decision Making.* New York: American Elsevier.
1972.

May, K. O.   "Quantitative Growth of the Mathematical Literature," *Science* **154**:1672–
1966.   1673.

Merton, R.   "Fluctuations in the Rate of Industrial Invention," *Quarterly Journal of Eco-
1935.   nomics* **49**:454–474.

Miller, R., and Sawers, D.
1968.   *The Technical Development of Modern Aviation.* London: Routledge & Kegan
                Paul.

Moravcsik, M. J.
1973.   *"Measures of Scientific Growth,"* Research Policy **2**:266–275.

Nelson, R. R.   "Less Developed Countries—Technology Transfer and Adaptation: The Role
1974.   of the Indigenous Science Community," *Economic Development and Cultural
                Change* **23**:61–77.

Pearl, D. J., and Enos, J. L.
1975.   "Engineering Production Functions and Technological Progress," *Journal of
                Industrial Economics* **24**:55–72.

Peck, M. J.   "Inventions in Postwar Aluminum Industry," in *The Rate and Direction of In-
1962.   ventive Activity* (R. R. Nelson, ed.), pp. 279–298. Princeton, N.J.: Princeton
                University Press.

Phillips, A.   *Technology and Market Structure.* Lexington, Mass.: D. C. Heath.
1971.

Piaget, J.   *Biology and Knowledge* (translated by B. Walsh). Chicago: University of
1974.   Chicago Press.

Price, De Solla, D. J.
1965.   "Is Technology Historically Independent of Science? A Study in Statistical
Historiography," *Technology and Culture* **6**:553–568.

Price, De Solla, D. J.
1980.   "A Theoretical Basis for Input–Output Analysis of National R&D Policy," in
Research, Development and Technological Change (D. Sahal, ed.), pp. 251–260.
Lexington, Mass.: D. C. Heath.

Reece, A. R.   "The Shape of the Farm Tractor," *Proceedings of the Institution of Mechani-*
1969–1970.   *ical Engineers* **184** (Pt. 3Q):125–131.

Reekie, W. D.   "Patent Data as a Guide to Industrial Activity," *Research Policy* **2**:246–264.
1973.

Robinson, J.   "The Production Function and the Theory of the Capital," *Review of Eco-*
1954.   *nomic Studies* **21**:81–106.

Robinson, J.   *The Accumulation of Capital* (reprinted 1965). London: Macmillan.
1956.

Sahal, D.   "A Theory of Measurement of Technology," *International Journal of Systems*
1977.   *Science* **8**:671–682.

Sahal, D.   "Structure and Self-Organization," International Institute of Management,
1978.   dp/78–87, Berlin.

Sahal, D.   *Research, Development and Technological Innovation: Recent Perspectives on*
1980.   *Management*. Lexington, Mass.: D. C. Heath.

Salter, W. E. G.
1969.   *Productivity and Technical Change,* 2nd ed. New York and London: Cambridge
University Press.

Sanders, B. S.   "Some Difficulties in Measuring Inventive Activity," in *The Rate and Direction*
1962.   *of Inventive Activity* (R. Nelson, ed.), pp. 53–83. Princeton, N.J.: Princeton
University Press.

Sandström, G. E.
1963.   *A History of Tunnelling*. London: Barrie and Rockliff.

Schmookler, J.   *Invention and Economic Growth*. Cambridge, Mass.: Harvard University Press.
1966.

Schumpeter, J.   *Business Cycles: A Theoretical, Historical and Statistical Analysis of Capitalist*
1939.   *Process*. New York: McGraw-Hill.

Schumpeter, J.   *History of Economic Analysis*. New York: Oxford University Press.
1961.

Stewart, F.   *Technology and Underdevelopment*. London: Macmillan.
1977.

Wagner, L. U.   "Problems in Estimating Research and Development and Stock," *Proceedings*
1968.   *of the American Statistical Association,* Business and Economics Section.

Zink, W. L.   "The Agricultural Power Take-Off," *Agricultural Engineering* **12**:209–210.
1931.

CHAPTER 3

Barnard, G. A.
1953.   "Time Intervals Between Accidents", *Biometrika* **40**:212–213.

Beall, G., and Rescia, R.
1953.   "A Generalization of Neyman's Contagious Distribution," *Biometrics* **9**: 354–386.

Bliss, C. I., and Fisher, R. A.
1953.   "Fitting the Negative Binomial Distribution to Biological Data," *Biometrics* **9**:176–196.

Blum, L. M., and Mintz, A.
1951.   "Correlation Versus Curve Fitting in Research on Accident Proneness," *Psychological Bulletin* **48**:413–418.

Catling, H.   "Transfer of Technology in Textile Industry," in *Transfer Processes in Tech-*
1978.   *nical Change* (F. Bradbury, P. Jervis, R. Johnston, and A. Pearson, eds.). Alphen aan den Rijn; Sijthoff and Noordhoff.

Cooper, C., Kaplinsky, R., Bell, R., and Satyarakwit, N.
1975.   "Choice of Techniques for Can Making in Kenya, Tanzania, and Thailand," in *Technology and Employment in Industry* (A. S. Bhalla, ed.), pp. 85–121. Geneva: International Labor Office.

Cox, D. R.   "Some Statistical Methods Connected with Series of Events," *Journal of the*
1955.   *Royal Statistical Society* **B17**:129–157.

Cox, D. R., and Lewis, P. A. W.
1966.   *The Statistical Analysis of Series of Events.* London: Methuen.

Cramer, H.   *Mathematical Methods of Statistics.* Princeton, N.J.: Princeton University
1946.   Press.

Cramer, J. S.   *Empirical Econometrics.* Amsterdam: North Holland.
1969.

Ehrenberg, A. S. C.
1959.   "The Pattern of Consumer Purchases," *Applied Statistics* **8**:26–41.

Feller, W.   "On a General Class of Contagious Distributions," *Annals of Mathematical*
1943.   *Statistics* **14**:389–400.

Fisher, R. A.   *Contributions to Mathematical Statistics.* New York: Wiley.
1950.

Gilfillan, S. C.   *Inventing the Ship.* Chicago: Follett.
1935a.

Gilfillan, S. C.   *The Sociology of Invention* (reprinted 1963). Chicago: Follett; MIT paperback
1935b.   edition, March 1970.

Greenwood, M., and Yule, G. U.
1920.   "An Inquiry into the Nature of Frequency Distributions Representatives of Multiple Happenings, with Particular Reference to Occurrence of Multiple Attacks of Disease or of Repeated Accidents," *Journal of the Royal Statistical Society* **83**:255–279.

Jeffreys, H.   *Theory of Probability*, 3rd ed. New York: Oxford University Press.
1961.

Kendall, M. G.   "Natural Laws in Social Sciences," *Journal of the Royal Statistical Society*
1961.   **A124**:1–19.

Kendall, M. G., and Stuart, A.
1947.   *The Advanced Theory of Statistics*, Vol. I, 3rd ed.; Vol. II, 2nd ed. New York:
                 Hafner.

Kuznets, S.   "Innovations and Adjustments in Economic Growth," *Swedish Journal of*
1972.   *Economics* **74**:431–451.

Magee, S.   "Multinational Corporations, the Industry Technology Cycle and Develop-
1977.   ment," *Journal of World Trade Law* 11.

Maguire, B. A., Pearson, E. S., and Wynn, A. H. A.
1952.   "The Time Intervals Between Industrial Accidents," *Biometrika* **39**:168–180.

Myers, S., and Marquis, D. G.
1969.   "Successful Commercial Innovations," National Science Foundation Report
                 69–71, NSF, Washington, D.C.

Ogburn, W. F., and Thomas, D. S.
1922.   "Are Inventions Inevitable?" *Political Science Quarterly* Appendix 37:83–98.

Phillips, A.   *Technology and Market Structure*. Lexington, Mass.: D. C. Heath.
1971.

Quenouille, M. H.
1949.   "A Relationship Between the Logarithmic, Poisson, and Negative Binomial
                 Series," *Biometrics* **5**:162–164.

Richardson, L. F.
1944.   "The Distribution of Wars in Time," *Journal of the Royal Statistical Society*
                 107:242–250.

Sahal, D.   "Generalized Poisson and Related Models of Technological Innovation," *Jour-*
1974.   *nal of Technological Forecasting and Social Change* **6**:403–436.

Scherer, F. M.   "Firm Size, Market Structure, Opportunity, and the Output of Patented Inven-
1965.   tions," *American Economic Review* **55**:1097–1125.

Schmookler, J.   *Invention and Economic Growth*. Cambridge, Mass.: Harvard University
1966.   Press.

Schumpeter, J.   *The Theory of Economic Development*. Cambridge, Mass.: Harvard University
1934.   Press.

Schumpeter, J.   *Business Cycles,* Vols. I and II. New York: McGraw-Hill.
1939.

Stafford, A. B. "Is the Rate of Invention Declining?" *American Journal of Sociology* **57**:
1952.   539–545.

Student,   "An Explanation of Deviations from Poisson's Law in Practice," *Biometrika*
1919.   **12**:211–215.

Taton, R.   *Reason and Chance in Scientific Discovery* (translated by A. J. Pomerans).
1957.   New York: Philosophical Library.

Teece, D. J.   "Technology Transfer by Multinational Firms: The Resource Cost of Trans-
1977.   ferring Technological Know-How," *Economic Journal* **87**:242–261.

Welles, J. G. 1974.   "Contributions to Technology and Their Transfer: The NASA Experience," in *Technology Transfer* (H. F. Davidson, M. J. Cetron, and J. D. Goldhar, eds.). Leiden: Noordhoff.

Yule, G. U. 1910.   "On the Distribution of Deaths with Age When the Causes of Death Act Cumulatively," *Journal of the Royal Statistical Society* **73**:26-35.

CHAPTER 4

Alexander, C. 1964.   *Notes on the Synthesis of Form.* Cambridge Mass.: Harvard University Press.

Arrow, K. J., and Hahn, F. H. 1971.   *General Competitive Analysis.* Edinburgh: Oliver and Boyd.

Baker, C. L. 1970.   "Trends in Tractor Development," *Farm Engineering Industry* November, p. 391.

Bonner, J. T. 1952.   *Morphogenesis,* Princeton N.J.: Princeton University Press.

Bruni, L. 1963-1964.   "Internal Economies of Scale with a given Technique," *Journal of Industrial Economics* **12**:175-190.

David, P. A. 1975.   *Technical Choice, Innovation and Economic Growth.* New York and London: Cambridge University Press.

Denyer, S. 1978.   *African Traditional Architecture.* London: Heinemann.

Dobzhansky, T. 1950.   "Heredity, Environment, and Evolution," *Science* **111**:161-166.

Dunn, E. S., Jr. 1971.   *Economic and Social Development.* Baltimore, Md.: John Hopkins Press.

Galileo, G. 1914.   *Dialogues Concerning Two New Sciences* (translated by H. Crew and A. De Salvio). New York: Macmillan.

Gould, S. J. 1966.   "Allometry and Size in Ontogeny and Phylogeny," *Biological Review* **41**:587-640.

Hirsch, S. 1965.   "The United States Electronics Industry in International Trade," *National Institute Economic Review* **34**:92-97.

Huxley, J. S. 1932.   *Problems of Relative Growth.* London: Methuen. Reprinted 1972; New York: Dover.

Nelson, R. R., and Winter, S. G. 1973.   "Towards an Evolutionary Theory of Economic Capabilities," *American Economic Review* **63**:440-449.

Nordbeck, S. 1965.   "The Law of Allometric Growth," Discussion Paper No. 7, Michigan Inter-University Community of Mathematical Geographers, Ann Arbor.

Rapoport, A. 1974.   *Conflict in Man-Made Environment.* Harmondsworth: Penguin.

Sahal, D.   "Structure and Self-Organization," International Institute of Management,
1978.   dp/78-87, Berlin.

Thompson, D'Arcy W.
1917.   *On Growth and Form*. Cambridge: Cambridge University Press. (Abridged edi-
                   tion, J. T. Bonner, ed., Cambridge University Press, New York and London,
                   1961.)

Vernon, R.   "International Investment and International Trade in Product Cycle," *Quar-
1966.   terly Journal of Economics* **80**:190-207.

Went, F. W.   "The Size of Man," *American Scientist* **56**:400-413.
1968.

CHAPTER 5

Bain, A. D.   "The Growth of Demand for New Commodities," *Journal of the Royal Sta-
1963.   tistical Society* **A126**:285-299.

Chow, G. C.   "Technological Change and the Demand for Computers," *American Economic
1967.   Review* **57**:1117-1130.

David, P.   "A Contribution to the Theory of Diffusion," Memorandum No. 71 (June). Re-
1969.   search Center in Economic Growth, Stanford University.

Davies, S.   "Diffusion, Innovation and Market Structure," in *Research, Development and
1980.   Technological Innovation* (D. Sahal, ed.). Lexington, Mass.: D. C. Heath.

Fisher, J. C., and Pry, R. H.
1971.   "A Simple Substitution Model of Technological Change," *Technological
                   Forecasting and Social Change* **3**:75-88.

Freeman, C.   *The Economics of Industrial Innovation*. Harmondsworth: Penguin.
1974.

Gilfillan, S. C.   *Inventing the Ship*. Chicago: Follett.
1935.

Graham, C. S.   "The Ascendancy of the Sailing Ship, 1850-1885," *Economic History Review*
1956.   (Series 2), **9**:74-88.

Griliches, Z.   "Hybrid Corn: An Exploration in the Economics of Technological Change,"
1957.   *Econometrica* **25**:501-522.

Griliches, Z.   "Hybrid Corn and the Economics of Innovation," *Science* July, pp. 275-280.
1960.

Mansfield, E.   *Industrial Research and Technological Innovation*. New York: Norton.
1968a.

Mansfield, E.   *The Economics of Technological Change*. New York: Norton.
1968b.

Nabseth, L., and Ray, G., eds.
1974.   *The Diffusion of New Industrial Processes*. New York and London: Cambridge
                   University Press.

Nelson, R. R., Peck, M. J., and Kalacheck, E. D.
1967.   *Technology, Economic Growth and Public Policy*. Washington, D.C.: The
                   Brookings Institution.

Ray, G. F.
1969.
"The Diffusion of New Technology: A Study of Ten Processes in Nine Indus-tries," *National Institute Economic Review* **48**:40–83.

Reeve, E. C. R., and Huxley, J. S.
1945.
"Some Problems in the Study of Allometric Growth," in *Essays on Growth and Form* (W. E. Le Gros Clark and P. B. Medawar, eds.). Oxford: Clarendon Press.

Rogers, E. M.
1978.
"Re-invention during the innovation process," in *The Diffusion of Innovations: An Assessment* (M. Radnor, ed.). Document PB-287687/AS, NSF, Washington, D.C.

Rosenberg, N.
1976.
"On Technological Expectations," *Economic Journal* **86**:523–535.

Sahal, D.
1977a.
"Substitution of Mechanical Corn Pickers by Field Shelling Technology—An Econometric Analysis," *Journal of Technological Forecasting and Social Change* **10**:53–60.

Sahal, D.
1977b.
"The Multidimensional Diffusion of Technology," *Journal of Technological Forecasting and Social Change* **10**:277–298; errata in **11**:391–392 (1978).

Salter, W. E. G.
1969.
Productivity and Technical Change, 2nd ed. New York and London: Cambridge University Press.

Schumpeter, J.
1939.
Business Cycles, Vols. I and II. New York: McGraw-Hill.

Walker, H. B.
1929.
"Engineering Applied to Agriculture," *Agricultural Engineering* **10**:341–349.

Ward, W. H.
1967.
"The Sailing Ship Effect," *Bulletin of the Institute of Physics and the Physical Society* **18**, 169.

Weiss, L.
1971.
"Quantitative Studies of Industrial Organization" in *Frontiers of Quantitative Economics* (M. Intriligator, ed.) Amsterdam: North Holland.

CHAPTER 6

Abernathy, W. J., and Wayne, K.
1974.
"Limits to the Learning Curve," *Harvard Business Review* **52**:109–119.

Addie, A. N.
1977.
"The History of the Diesel-Locomotive in the United States," paper presented to the American Society of Mechanical Engineers, General Motors, Lagrange, Illinois.

Albu, A.
1976.
"Causes of the Decline in British Merchant Ship-Building and Marine Engineer-ing," *Omega* **4**:513–525.

Alchian, A.
1963.
"Reliability of Progress Curves in Airframe Production" *Econometrica* **31**:679–693.

Allen, T. J.
1964.
"The Use of Information Channels in R&D Proposal Preparation," MIT Sloan School of Management, Working Paper.

Allen, T. J., Tushman, M. L., and Lee, D. M. S.
1979.
"Technology Transfer as a Function of Position in the Spectrum from Research through Development to Technical Services," *Academy of Management Journal* **22**:694–708.

Al-Timimi, W.   "Innovation-Led Expansion: The Shipbuilding Case," *Research Policy* 4:160–
1975.   171.

Andress, F. J.   "The Learning Curve as a Production Tool," *Harvard Business Review*
1954.   32:87–97.

Archer, R. D.   "Evolution of the Supersonic Shape," *Space/Aeronautics* 48:89–104.
1967.

Arrow, K. J.   "The Economic Implications of Learning by Doing," *Review of Economic*
1962.   *Studies* 29:155–173.

Arrow, K. J.   "Classificatory Notes on the Production and Transmission of Technological
1969.   Knowledge," *American Economic Review* Papers and Proceedings, 59:29–35.

Atkinson, A. B., and Stiglitz, J. E.
1969.   "A New View of Technological Change," *Economic Journal* 79:573–578.

Baker, C. L.   In "Tractor Design: Looking Back and Thinking Forward," *Farm Machine*
1970a.   *Design Engineering* December, p. 32.

Baker, C. L.   In "Tractor Power and Speed Are not Decisive but 4-Wheel-Drive and Quality
1970b.   of Work will Influence Design," *Farm Engineering Industry* November, pp.
                 391–392.

Baker, E. J.   "A Quarter Century of Tractor Development," *Agricultural Engineering*
1931.   12:206–207.

Baker, N., Siegmann, J., and Rubenstein, A.
1967.   "Effects of Perceived Means and Needs on the Generation of Ideas in R&D
                 Projects," *IEEE Transactions on Engineering Management* 14:156–163.

Baldamus, W.   "Mechanization, Utilization and Size of Plant," *Economic Journal* 63:50–69.
1953.

Barger, H.   *The Transportation Industries 1889–1946* (A Study of Output, Employment, and
1951.   Productivity). New York: National Bureau of Economic Research.

Bruce A. W.   *The Steam Locomotive in America*. New York: Norton.
1952.

Cantley, M. F., and Glagolev, V. N.
1978.   "Problems of Scale"—The Case of IIASA Research, RM-78-47. International
                 Institute for Applied Systems Analysis, Laxenburg, Austria.

David, P. A.   *Technical Choice, Innovation and Economic Growth*. New York and London:
1975.   Cambridge University Press.

Dommen, A. J.   "The Bamboo Tube Well: A Note on an Example of Indigenous Technology,"
1975.   *Economic Development and Cultural Change* 23:483–487.

Dullemeijer, P. *Concepts and Approaches in Animal Morphology. Assen: van Gorcum.*
1974.

Eckaus, R. S.   "The Factor Proportions Problem in Underdeveloped Areas," *American Eco-*
1960.   *nomic Review* 50:642–648.

Farris, G.   "The Effect of Individual Roles on Performance in Innovative Groups," *R&D*
1972.   *Management* 3:23–28.

Fellner, W.   "Two Propositions in the Theory of Induced Innovations," *Economic Journal*
1961.   71:305–308.

Fellner, W.
1969.
"Specific Interpretations of Learning by Doing," *Journal of Economic Theory* 1:119–140.

Fisher, J. C.
1974.
Energy Crisis in Perspective. New York: Wiley.

Fishlow, A.
1966.
"Productivity and Technological Change in the Railroad Sector, 1840–1910," in *Output, Employment, and Productivity in the U.S. after 1800*, National Bureau of Economic Research Studies in Income and Wealth, No. 30, pp. 583–646. New York: National Bureau of Economic Research.

Fox, A.
1966.
"Demand for Farm Tractors in the U.S.," Agriculture Economic Report 103. Washington, D.C.: U.S. Department of Agriculture.

Frankel, M.
1955.
"Obsolescence and Technological Change in a Maturing Economy," *American Economic Review* **45**:296–319.

Frear, H. P.
1945.
"History of Tankers," *Historical Transactions 1893–1943, The Society of Naval Architectures and Marine Engineers* pp. 135–144.

Fusfeld, A. R.
1970.
"Technological Progress Function: A New Technique for Forecasting," *Technological Forecasting and Social Change* 1:301–312.

Gellman, A. J.
1971.
"Surface Freight Transportation," in *Technological Change in Regulated Industries* (W. M. Capron, ed.), pp. 166–196. Washington, D.C.: The Brookings Institution.

Girardier, J. P., and Vergnet, M.
1976.
"The Solar Pump and the Problems of Integrated Rural Development," in *Appropriate Technology* (N. Jéquier, ed.), pp. 253–259. Paris: OECD.

Gohlke
1942.
"Thermal-Air Jet Propulsion," *Aircraft Engineering* **14**:32–39.

Goldbeck, G.
1970.
"Marine Internal Combustion Engines: From Petrol and Gas to Diesel Oil," *The Motorship* **51**:71–73.

Goldstine, H. H.
1972.
The Computer from Pascal to von Neumann. Princeton, N.J.: Princeton University Press.

Graham, C. S.
1956.
"The Ascendancy of the Sailing Ship, 1850–1885," *Economic History Review* (Series 2), **9**:74–88.

Gray, R. B.
1954, 1958.
Development of Agricultural Tractor in the United States, Vols. I and II. Washington, D.C.: U.S. Department of Agriculture.

Gray, R. B., and Dieffenbach, E. M.
1957.
"Fifty Years of Tractor Development in the U.S.A.," *Agricultural Engineering* pp. 388–397.

Hagstrom, W.
1965.
The Scientific Community. New York: Basic Books.

Hall, G. R., and Johnson, R. E.
1970.
"Transfer of United States Aerospace Technology to Japan," in *The Technology Factor in International Trade* (R. Vernon, ed.), pp. 305–358. New York: National Bureau of Economic Research.

Hartman, E. P.
1970.
Adventures in Research, A History of Ames Research Center 1940–1965. Washington, D.C.: National Aeronautics and Space Administration.

Hicks, J. R.   *The Theory of Wages* (revised 1963). London: Macmillan.
1932.

Hirschman, W. B.
1964.   "Profit from the Learning Curve," *Harvard Business Review* **42**:125–139.

Hughes, W. R.   "Scale Frontiers in Electric Power," in *Technological Change in Regulated In-*
1971.   *dustries* (W. M. Capron, ed.), pp. 44–85. Washington, D.C.: The Brookings In-
                 stitution.

Hunter, L. C.   *Steamboats on the Western Rivers, An Economic and Technological History.*
1949.   Cambridge, Mass.: Harvard University Press.

Ishikawa, T.   "Conceptualization of Learning by Doing: A Note on Paul David's 'Learning
1973.   by Doing and ... the Ante-Bellum United States Cotton Textile Industry'," *Jour-*
                 nal of Economic History **33**:851–861.

Israd, W.   *Location and Space Economy.* London: Chapman & Hall.
1956.

Jewkes, J., Sawers, D., and Stillerman, R.
1970.   *The Sources of Invention.* New York: St. Martin's Press.

Joseph, E. C.   "Future Computer Architectures—Polysystems," *COMPCOM-72 Digest of*
1972.   *Papers* IEEE Computer Society, pp. 149–154.

Keith, V. F.   "Analysis and Statistics of Large Tankers," Report No. 4, Department of Naval
1968.   Architecture and Marine Engineering, University of Michigan, Ann Arbor.

Kendrick, J.   *Productivity Trends in the United States.* Princeton, N.J.: Princeton University
1961.   Press.

Kendrick, J.   *Postwar Productivity Trends in the United States, 1948–1969.* Princeton, N.J.:
1973.   Princeton University Press.

Kennedy, C.   "Induced Bias in Innovation and the Theory of Distribution," *Economic Jour-*
1964.   *nal* **74**:541–548.

Khan, A. U.   "Mechanisation Technology for Tropical Agriculture," in *Appropriate Tech-*
1976.   *nology* (N. Jéquier, ed.), pp. 213–230. Paris: OECD.

Kimberly, J. R. "Organizational Size and the Structuralist Perspective: A Review, Critique and
1976.   Proposal," *Administrative Science Quarterly* **21**:571–597.

Kirchmayer, L., Mellor, A. G., O'Mara, J. F., and Stevenson, J. R.
1955.   "An Investigation of the Economic Size of Steam-Electric Generating Units,"
                 AIEEE Transactions Part 3, **74**:600–609.

Knight, K. E.   "Changes in Computer Performance," *Datamation* **12**:40–54.
1966.

Knight, K. E.   "A Descriptive Model of the Intra-firm Innovation Process," *Journal of Busi-*
1967.   *ness* **40**:478–496.

Koyck, L. M.   *Distributed Lags and Investment Analysis.* Amsterdam: North Holland.
1954.

Leonard, W.   "Research and Development in Industrial Growth," *Journal of Political Econ-*
1971.   *omy* **79**:232–256.

Leontief, W.   "Domestic Production and Foreign Trade; the American Capital Position Re-
1953.   Examined," *Proceedings of the American Philosophical Society.* Vol. 197.

Lisle, B. O.   *Tanker Technique, 1700–1936.* London: World Tankship Publications.
1936.

Lundberg, E.   *Produktivitet och Rantabilitet.* Stockholm: Norstedt and Söner.
1961.

Mansfield, E.   "Research and Technological Change," *Industrial Research* February.
1964.

Mansfield, E.   *Industrial Research and Technological Innovation,* New York: Norton.
1968.

Meek, J. E.   "Concentration in the Electric Power Industry: The Impact of the Anti-Trust
1972.   Policy," *Columbia Law Review* **72**:64–130.

Meurer, E. h. J. S.
1969.   "The Present and Future of Marine Propulsion Engineering," *The Motorship*
                  50:19–23.

Miller, R., and Sawers, D.
1968.   *The Technical Development of Modern Aviation.* London: Routledge & Kegan
                  Paul.

Mostert, N.   *Supership.* London: Penguin.
1975.

Mowery, D., and Rosenberg, N.
1979.   "The Influence of Market Demand Upon Innovation: A Critical Review of
                  Some Recent Empirical Studies," *Research Policy* **8**:102–153.

Mynt, H.   *Southeast Asia's Economy.* Harmondsworth: Penguin.
1972.

Naroll, R. S.   "An Index of Social Development," *American Anthropologist* **58**:687–715.
1956.

Nelson, R. R.   "Issues in the Study of Industrial Organization in a Regime of Rapid Technical
1972.   Change," in *A Roundtable on Policy Issues and Research Opportunities in In-
                  dustrial Organization* (V. Fuchs, ed.). New York: National Bureau of Economic
                  Research.

Nelson, R. R., and Winter, S. G.
1975.   "Growth Theory from an Evolutionary Perspective: The Differential Produc-
                  tivity Puzzle," *American Economic Review* Papers and Proceedings,
                  65:338–344.

Organization for Economic Cooperation and Development.
1969.   *Gaps in Technology—Electronic Computers.* Paris: OECD.

Phillips, A.   "Air Transportation in the U.S.," in *Technological Change in Regulated In-
1971a.   dustries* (W. M. Capron ed.), pp. 123–166. Washington, D.C.: The Brookings
                  Institution.

Phillips, A.   *Technology and Market Structure.* Lexington, Mass.: D. C. Heath.
1971b.

Pinkepank, J. A.
1966.   "Where Would Dieselization Be Without GE?," *Trains* November.

Qvarnström, R. G. E.
1970.   "From Polar to UDAB—A Swedish Enterprise in Diesel Engines," *The Motor-
                  ship* **51**:41–44.

Raj, K. N.   "Role of the 'Machine-Tools Sector' in Economic Growth," in *Socialism, Cap-*
1969.   *italism and Economic Growth* (C. H. Feinstein, ed.), pp. 217–226. New York
                  and London: Cambridge University Press.

Rapping, L.   "Learning and World War II Production Function," *Review of Economics and*
1965.   *Statistics* 47:81–86.

Reguero, M. A.   "An Economic Study of the Military Airframe Industry," Wright-Patterson Air
1957.   Force Base, Ohio, Department of the Air Force, October.

Reynolds, L. G.   "Discussion," *American Economic Review* Papers and Proceedings, **56**:112–
1966.   114.

Richards, A.   "Technical and Social Change in Egyptian Agriculture, 1890–1914," *Economic*
1978.   *Development and Cultural Change* **26**:725–745.

Robinson, H. F., Roeske, J. F., and Thaeler, A. S.
1948.   "Modern Tankers," *Transactions of the Society of Naval Architects and Marine
                  Engineers* **56**:422–471.

Rosen, S.   "Electronic Computers: A Historical Survey," *Computing Surveys ACM,* 1
1969.   (No. 1): 7–36.

Rosenberg, N.   "Capital Goods, Technology, and Economic Growth," *Oxford Economic Pa-*
1963.   *pers* November, 217–227.

Rosenberg, N.   *Technology and American Economic Growth.* New York: Harper and Row.
1972.

Sahal, D.   "A Reformulation of Technological Progress Function," *Jc 'rnal of Tech-*
1975.   *nological Forecasting and Social Change* **8**:75–90. Errata in Sahal (1980).

Sahal, D.   "Substitution of Mechanical Corn Pickers by Field Shelling Technology—An
1977.   Econometric Analysis," *Journal of Technological Forecasting and Social
                  Change* **10**:53–60.

Sahal, D.   "Evolution of Technology: A Case Study of Farm Tractor," dp/78-99. Berlin:
1978.   International Institute of Management.

Sahal, D.   "Technical Progress and Policy" in *Research, Development, and Technological
1980.   Innovation* (D. Sahal, ed.), pp. 171–198. Lexington, Mass: D. C. Heath.

Salter, W. E. G.
1969.   *Productivity and Technical Change.* 2nd ed. New York and London: Cambridge
                  University Press.

Sargen, N. P.   *"Tractorization in the United States and Its Relevance for the Developing Coun-*
1975.   *tries.* Ph.D. dissertation, Stanford University.

Schlaifer, R., and Heron, S. D.
1950.   *The Development of Aircraft Engines and Fuels.* Cambridge, Mass.: Harvard
                  Business School.

Schmookler, J.   "Economic Sources of Inventive Activity," *Journal of Economic History* **22**:
1962.   1–20.

Schmookler, J.   *Invention and Economic Growth.* Cambridge, Mass.: Harvard University Press.
1966.

Schmookler, J., and Brownlee, O.
1962.   "Determinants of Inventive Activity," *American Economic Review* Papers and
                  Proceedings, **52**:165–176.

Sen, A. K.
1962.
The Choice of Techniques. Oxford: Basil Blackwell.

Serrell, R., Astrahan, M. M., Patterson, G. W., and Pyne, I. B.
1962.
"The Evolution of Computing Machines and Systems," *Proc. IRE* **50** (No. 5): 1039-1058.

Shehinski, E.
1967.
"Test of the Learning by Doing Hypothesis," *Review of Economics and Statistics* **49**:568-578.

Snell, J. B.
1971.
Railways: Mechanical Engineering. London: Longman.

Solo, R.
1966.
"The Capacity to Assimilate an Advanced Technology," *American Economic Review* Papers and Proceedings, **56**:91-97.

Terleckyj, N.
1974.
The Effect of R&D on Productivity Growth in Industries. Washington, D.C.: National Planning Association.

Turn, R.
1974.
Computers in the 1980's. New York: Columbia University Press.

Usher, A. P.
1954.
A History of Mechanical Inventions. Cambridge, Mass.: Harvard University Press.

Vietorisz, T., ed.
1969.
Engineering Industry. Unido monograph no. 4 on industrial development. New York: United Nations.

Weinberg, G. W.
1975.
An Introduction to General Systems Thinking. New York: Wiley.

Williams, J. E. D.
1964.
The Operation of Airliners. London: Hutchinson.

Worthington, W. H.
1966.
"Fifty years of Agricultural Tractor Development," Paper 660584, Society of Automotive Engineers, New York.

Wright, T. P.
1936.
"Factors Affecting the Cost of Airplanes," *Journal of the Aeronautical Sciences* **3**:122-128.

Yelle, L. E.
1979.
"The Learning Curve: Historical Review and Comprehensive Survey," *Decision Sciences* **10**:302-328.

Zipf, G. K.
1949.
Human Behavior and the Principle of Least Effort. Reading, Mass.: Addison-Wesley.

CHAPTER 7

Adelman, I.
1956.
"Long Cycles—Fact or Artifact?," *American Economic Review* **55**:444-463.

Carey, W. D.
1980.
"Innovation: The Next Step," *Technology Review* **82**:12,86,87.

Fishman, G. S.
1969.
Spectral Methods in Econometrics. Cambridge, Mass.: Harvard University Press.

Frisch, R. "Propagation and Impulse Problem in Dynamic Economics," in *Economic*
1933. *Essays in Honour of Gustav Cassel*, pp. 171–205. London: George Allen & Unwin. Reprinted in *Readings in Business Cycles* (R. A. Gordon and L. R. Klein, eds.), pp. 155–185. Homewood, Ill.: Richard D. Irwin, 1965.

Gordon, R. A., and Klein, L. R., eds.
1965. *Readings in Business Cycles*. Homewood, Ill.: Richard D. Irwin.

Granger, C. W. J.
1966. "The Typical Spectral Shape of an Economic Variable," *Econometrica* **34**:150–161.

Granger, C. W. J., and Hatanaka, M.
1964. *Spectral Analysis of Economic Time Series*. Princeton, N.J.: Princeton University Press.

Gudmundsson, G.
1971. "Time Series Analysis of Imports, Exports, and Other Economic Variables," *Journal of the Royal Statistical Society* **134**:383–412.

Harkness, J. P. "A Spectral Analytic Test of Long-Swing Hypotheses in Canada," *Review of*
1968. *Economics and Statistics* **50**:429–436.

Heertje, A. *Economic and Technical Change*. London: Weidenfeld and Nicholson.
1973.

Jenkins, G. M. "A Survey of Spectral Analysis," *Applied Statistics* pp. 2–47.
1965.

Jenkins, G. M., and Watts, D.
1968. *Spectral Analysis and Its Application*. San Francisco: Holden-Day.

Kuznets, S. "Equilibrium Economics and Business-Cycle Theory," *Quarterly Journal of*
1954a. *Economics* **44**:381–415 (1930). Reprinted in S. Kuznets's *Economic Change* (*Selected Essays in Business Cycles, National Income, and Economic Growth*), pp. 3–31. London: Heinemann.

Kuznets, S. "Schumpeter's Business Cycles," *American Economic Review* **30**:250–371
1954b. (1940). Reprinted in S. Kuznets's *Economic Change* (*Selected Essays in Business Cycles, National Income, and Economic Growth*), pp. 105–124. London: Heinemann.

Morgenstern, O.
1961. "A New Look at Economic Time Series Analysis," in *Money, Growth and Other Essays in Honour of Johan Åkerman* (H. Hegeland, ed.), pp. 261–272. Lund: Berlingska Boktryckeriet.

Naylor, T. H., Wertz, K., and Wonnacott, T. H.
1969. "Spectral Analysis of Data Generated by Simulation Experiments with Econometric Models," *Econometrica* **37**:333–352.

Parzen, E. "Mathematical Considerations in the Estimation of Spectra," *Technometrics*
1961. **3**:167–190.

Ray, G. F. "Innovation as the Source of Long Term Economic Growth," *Long Range*
1980. *Planning* **13**:9–19.

Schmookler, J. *Invention and Economic Growth*. Cambridge, Mass.: Harvard University Press.
1966.

Schmookler, J. *Patents, Invention, and Economic Change.* Cambridge, Mass.: Harvard Uni-
1972.   versity Press.

Schumpeter, J. *The Theory of Economic Development.* Cambridge, Mass.: Harvard University
1934.   Press.

Schumpeter, J. *Business Cycles,* Vols. I and II. New York: McGraw-Hill.
1939.

Slutzky, E.   "The Summation of Random Causes as a Source of Cyclic Processes,"
1937.   *Econometrica* **5**:105–146.

Volcker, P. A. *The Rediscovery of the Business Cycle* (*The Charles C. Moskowitz Memorial*
1978.   *Lectures*). New York: Free Press.

CHAPTER 8

Abdulkarim, A. J., and Lucas, N. J. D.
1977.   "Economies of Scale in Electricity Generation in the United Kingdom," *Energy
                Research* **1**:223–231.

Ashby, W. Ross.
1952.   *Design for a Brain.* London: Chapman and Hall.

Beckwith, S.   "Super-charged Hydrogen Cooling of Generators," *AIEE Transactions* **71**:
1952.   168–175.

Fisher, J. C.   "Size-Dependent Performance of Subcritical Fossil Generating Units," Electric
1978a.   Power Research Institute, October.

Fisher, J. C.   "Economies of Scale in Electric Power Generation," paper presented at IIASA,
1978b.   Laxenburg, October 10.

Hughes, W. R.   "Scale Frontiers in Electric Power," in *Technological Change in Regulated In-
1971.   dustries* (W. M. Capron, ed.), pp. 44–85. Washington, D.C.: The Brookings In-
                stitution.

Krick, N.   "Turbogenerators Today," *Brown Boveri Review* **56**:368–379.
1969.

Lee, T.   "Optimization of Size in Power Generation," paper presented at IIASA, Laxen-
1978.   burg, December 10.

Ling, S.   *Economies of Scale in the Steam-Electric Power Generating Industry.* Amster-
1964.   dam: North Holland.

Martino, J. P., and Conver, S. K.
1972.   "The Step-Wise Growth of Electric Generator Size," *Technological Forecasting
                and Social Change* **3**:465–471.

Meek, J. E.   "Concentration in the Electric Power Industry: The Impact of the Antitrust
1972.   Policy," *Columbia Law Review* **72**:64–130.

Rosenberg, L. T.
1974.   "Evolution of the Turbogenerator," paper presented at the 1974 Joint IEEE/
                ASME Power Generation Technical Conference, Miami Beach, Fla., September
                15–19.

Salter, W. E. G.
1969.   *Productivity and Technical Change,* 2nd ed. New York and London: Cambridge
                         University Press.

Schroeder, T. W., and Wilson, G. P.
1958.   "Economic Selection of Generating Capacity Additions," *AIEE Transactions*
                         77:1133–1145.

Simmonds, W. H. C.
1969a.   "Step-wise expansion and profitability," *Chemistry in Canada* September, pp.
                         16–18.

Simmonds, W. H. C.
1969b.   "The Canada–U.S. Scale Problem," *Chemistry in Canada* October, pp. 39–41.

Simmonds, W. H. C.
1972.   "The Analysis of Industrial Behavior and Its Use in Forecasting," *Technological
                         Forecasting and Social Change* **3**:205–224.

Snowden, D. P. "Superconductors for Power Transmission," *Scientific American* April, pp.
1972.   84–92.

U.S. Federal Power Commission.
1970.   *The 1970 National Power Survey,* Part IV (Vols. 1–4: 1970–1972). Washington,
                         D.C.

Wiedemann, E. "The Design of Very Large Turbo-Alternators," *Brown Boveri Review* **45**:3–13.
1958.

CHAPTER 9

Arbib, M.   "Self Producing Automata—Some Implications for Theoretical Biology," in
1969.   *Towards A Theoretical Biology*, (C. H. Waddington, ed)., pp. 204–206.
                Chicago: Aldine.

Bridgman, P. W.
1931.   *Dimensional Analysis*, rev. ed. New Haven, Conn.: Yale University Press.

Buckingham, E. "On Physically Similar Systems: Illustrations of the Use of Dimensional Equa-
1914.   tions," *Physical Review* **4**:345–376.

Causey, R. L.   "Derived Measurement, Dimensions, and Dimensional Analysis," *Philosophy
1969.   of Science* **36**:252–270.

Conway, R. W., and Schultz, A.
1959.   "The Manufacturing Progress Functions," *Journal of Industrial Engineering*
                         10:39–54.

Freeman, C.   *The Economics of Industrial Innovation.* Harmondsworth: Penguin.
1974.

Gunther, B., and Leon de Labarra, B.
1966.   "A Unified Theory of Biological Similarities," *Journal of Theoretical Biology*
                         13:48–59.

Huettner, D. A. "Scale, Costs and Environmental Pressures," in *Technological Change: Eco-
1975.   nomics, Management, and Environment* (Bela Gold, ed.), pp. 65–105. Oxford:
                Pergamon.

Huxley, J. S.   *Problems of Relative Growth.* London: Methuen. Reprinted 1972; New York:
1932.   Dover.

Ijiri, Y., and Simon, H. A.
1977.   *Skew Distributions and the Sizes of Business Firms.* Amsterdam: North Hol-
                 land.

Johnston, J.   *Econometric Methods.* New York: McGraw-Hill.
1963.

Kendall, M. G.   "Natural Laws in Social Sciences," *Journal of the Royal Statistical Society*
1961.   **A124**:1-19.

Langhaar, H. D.
1951.   *Dimensional Analysis and Theory of Models.* New York: Wiley.

Macfadyen, A.   *Animal Ecology: Aims and Methods.* London: Pitman & Sons.
1963.

Miller, R., and Sawers, D.
1968.   *The Technical Development of Modern Aviation.* London: Routledge & Kegan
                 Paul.

Moran, M. J.   "A Generalization of Dimensional Analysis," *Journal of the Franklin Insti-
1971.   tute* **292**:423-432.

Nelson, R. R., and Winter, S. G.
1977.   "In Search of Useful Theory of Innovation," *Research Policy* **6**:36-76.

Pelc, K.   "Remarks on Formulation of Technology Strategy," in *Research, Development,
1980.   and Technological Innovation* (D. Sahal, ed.). Lexington, Mass.: D. C. Heath.

Sahal, D.   "On the conception and measurement of trade-off in engineering systems: A
1976a.   case study of aircraft design process," *Journal of Technological Forecasting and
                 Social Change* **8**:371-384.

Sahal, D.   "On the conception and measurement of technology: A case study of aircraft
1976b.   design process," *Journal of Technological Forecasting and Social Change*
                 8:385-400.

Sahal, D.   "The generalized distance measures of technology," *Journal of Technological
1976c.   Forecasting and Social Change* **9**:289-300.

Sahal, D.   "Structural Models of Technology Assessment," *IEEE Transactions on Sys-
1977.   tems, Man, and Cybernetics* **7**:582-589.

Sahal, D.   "A Formulation of the Pareto Distribution," *Environment and Planning*
1978a.   **A10**:1363-1376.

Sahal, D.   "Law-like Aspects of Technological Development." International Institute
1978b.   of Management, dp/78-85, Berlin.

Sahal, D.   "SELF": A Framework for Generation of Patterns," *Pattern Recognition* **10**:
1978c.   397-401.

Sahal, D.   "Systemic Similitude," *International Journal of Systems Science* **9**:1351-1357.
1978d.

Sahal, D.   "A Unified Theory of Self-Organization," *Journal of Cybernetics* **9**:127-142.
1979a.

Sahal, D.   "A Theory of Progress Functions," *Transactions of the American Institute of
1979b.   Industrial Engineers* **11**:23-29.

Simon, H. A.   *The Sciences of the Artificial.* Cambridge, Mass.: MIT Press.
1969.

Stahl, W. R.   "Similarity and Dimensional Methods in Biology," *Science* **137**:205–211.
1962.

Steindl, J.   *Random Processes and the Growth of Firms.* London: Griffin.
1965.

Steindl, J.   "Technical Progress and Evolution," in *Research, Development and Techno-*
1980.   *logical Innovation* (D. Sahal, ed.), pp. 131–142. Lexington, Mass.: D. C. Heath.

Vietorisz, T.   "Modelling the Engineering Sector: The Heavy Electrical Equipment Industry
1972.   in Mexico," paper submitted to the European Meeting of the Econometric So-
                ciety, Budapest, September.

White, G.   "Developments in Boiling-Water Reactors," *Proceedings of the Third Inter-*
1964.   *national Conference on Peaceful Uses of Atomic Energy* (Geneva), pp. 147–155.
                Published by the United Nations.

Wright, T. P.   "Factors Affecting the Cost of Airplanes," *Journal of the Aeronautical*
1936.   *Sciences* **3**:122–128.

Zipf, G. K.   *Human Behavior and the Principle of Least Effort.* Reading, Mass.: Addison-
1949.   Wesley.

Subject Index